Contemporary Catalysis

Contemporary Catalysis

Fundamentals and Current Applications

Julian R.H. Ross

University of Limerick, Limerick, Ireland

ELSEVIER

Elsevier
Radarweg 29, PO Box 211, 1000 AE Amsterdam, Netherlands
The Boulevard, Langford Lane, Kidlington, Oxford OX5 1GB, United Kingdom
50 Hampshire Street, 5th Floor, Cambridge, MA 02139, United States

British Library Cataloguing-in-Publication Data
A catalogue record for this book is available from the British Library

Library of Congress Cataloging-in-Publication Data
A catalog record for this book is available from the Library of Congress

ISBN: 978-0-444-63474-0

For Information on all Elsevier publications
visit our website at https://www.elsevier.com/books-and-journals

Publisher: Susan Dennis
Acquisition Editor: Kostas Marinakis
Editorial Project Manager: Carly Demetre
Production Project Manager: Bharatwaj Varatharajan
Cover Designer: Miles Hitchen

Typeset by MPS Limited, Chennai, India

Contents

Preface

For several centuries and even before Berzelius coined the term to describe the phenomenon in 1835 (see Chapter 1: An Introduction to Heterogeneous Catalysis and Its Development Through the Centuries–Chemistry in Two Dimensions), catalysis has been recognized as a rather special topic. In essence, a catalyst is a chemical substance (or combination of substances) that speeds up a chemical reaction without itself being used in the reaction. In other words, the chemical or chemicals making up the catalyst take part in the reaction but do not appear in the balanced equations for the overall process; they may, however, appear in the equations for the constituent steps. There are two general types of catalysis, homogeneous and heterogeneous catalysis, and many different types of chemicals can be encountered as catalysts: inorganic, organic, or biological (i.e., enzymes). This book is very largely concerned with heterogeneous catalysis, that is, catalysis occurring at the interface between two phases, with a focus on reactions at the gas–solid interface, and the catalysts considered are largely inorganic solids. However, many of the principles considered can be applied, sometimes with limitations, to reactions at other types of interface, for example, to a liquid–solid interface or to a three-phase interface (gas–liquid–solid) or even to enzyme catalysis. While the subject of catalysis is now much better understood than it was a century ago, it is one which is continuously evolving and in which there are many new developments.

This book is a revised and expanded version of the text by this author entitled "Heterogeneous Catalysis—Fundamentals and Applications" that was published in 2011. It has been produced as a response to very valuable comments made by the various users of that book. In the new volume, some of the topics previously handled have been expanded significantly and the order of their introduction has also been altered somewhat. More importantly, that part of the book dealing with specific catalysts and processes, previously summarized in a final chapter has been expanded into a sequence of separate chapters and the number of such systems discussed has been increased. Attempts have also been made to make fuller use of the possibilities offered by the web by providing more links to relevant material: for those working from a digital form of the text, the majority of the additional material is available by simple links provided in the text.[1]

[1] Further related material is being made available on a blog being written by the author (contemporarycatalysis.wordpress.com) and the reader is therefore encouraged to make use of this site.

Catalytic processes, being chemical reactions, are governed by all the considerations which can be applied to other chemical changes. They must not only conform with chemical thermodynamics but their kinetics are governed by parameters such as the concentrations of the reactants (including those of the catalyst in many cases) and the temperature. Further, their kinetic behavior may be affected by physical parameters such as mass transfer or heat transfer. On top of that, the solid catalyst has a whole chemistry of its own: e.g., special preparation methods and dedicated characterization techniques. And finally, unlike the situation with many homogeneous reactions, in which it is possible to isolate and measure intermediate species, it is very difficult or even impossible to isolate intermediate species on catalysts and so a whole range of methodologies have been developed to allow catalytic chemists to identify, or at least attempt to identify, reactive intermediates under reaction conditions. These topics are considered in the first half of this book, Chapters 2–8, these being similar in content to Chapters 2–7 of the earlier volume, "Heterogeneous Catalysis—Fundamentals and Applications." The second half of the book (Chapters 9–15) considers a number of examples of modern catalytic processes and the catalysts used for these processes, considering where appropriate some of the more basic research associated with these developments. The main difference between this text and that of the former volume is therefore the expansion of the various topics previously handled relatively briefly in the former Chapter 8 into separate chapters; several new topics not handled in the previous volume have also been added.

The aim of this text is to try to give a student embarking on a career in chemistry a good understanding of how modern catalysts are made, researched and used. The author believes that many aspects of chemistry come alive with, and can be better understood through, an understanding of catalysis. He also believes that further development of the subject will continue to have a central importance in all aspects of chemical technology, from fuel processing to pharmaceuticals, from environmental protection to fine chemicals, from biomass conversion to fuel cells. Further advances will only be made if a new generation of researchers is inspired to enter the field. It is hoped that this book will put across some of the excitement of the subject and that at least some of its readers will be inspired to embark on the challenging career of a researcher in catalysis.

The book can be used either for self-study or as the foundation for a variety of different types of courses on catalysis, from those that are very basic and general to those that are more specialized, the depth and breadth of study of each topic being selected to suit the approach and aims of the individual instructor. The approach adopted in the book is novel since it is expected that the reader will make as much use of related web-based material as of the printed book; indeed, it is likely that the reader will prefer to study the whole content using the electronic version of the book since this provides automatic links to much of the supplementary material. The literature studies suggested in the various tasks are formulated to make optimum use of the material available from citation sources such as Scopus or Web of Science while giving the student the chance to encounter some of the most up-to-date thinking on a range of important topics in catalysis. Each of the topics is introduced in outline in the main text and the reader is then given some guidance in the tasks as to how to approach that particular subject. In many cases, this involves the student being given several key references, often to well-cited reviews on the topic, and asked not only to read some of the key literature contained therein but also to follow through from the reviews to more up-to-date publications on the subject.[2] Further guidance on the use of the Scopus or similar databases is given in the Appendix and the reader intending to carry out some of the tasks in later chapters is advised to study that before proceeding.

In conclusion, it should be emphasized that the book is not a comprehensive guide to catalysis and is not intended to be used as a reference text by those who have had years of experience in the field. Instead, it is aimed more at the needs of new recruits to the field who will use it either in conjunction with a lecture course under the guidance of an instructor or for self-study. It is to be hoped that anyone who has worked carefully through it, carrying out many of the tasks, will have been encouraged to apply the methods used to explore the many other areas of the subject not covered in this text.

[2] How the student approaches these literature studies is very much up to the individual and will depend on personal preferences, aims, and objectives; for example, if the student is interested in catalyst preparation and characterization, he or she can concentrate on that aspect of the subject; if, on the other hand, he or she is interested in reactor design, that can be the focus of any literature search. When the book is used in conjunction with a lecture course, the instructor can give additional guidance regarding the direction to be followed in each literature study and can also assess the results on a continuing basis.

Acknowledgments

My understanding and knowledge of the field of heterogeneous catalysis have been garnered throughout my whole career by the efforts and the enthusiasm of the many undergraduate and graduate students and postdocs, too many to name individually, who have worked with me over that period. Through their work, I have been enabled to broaden my understanding of the subject and it has only been through their contributions that I have been able to undertake the writing of this book and its predecessor. I also wish to thank my colleagues from the various universities where I have worked (Bradford, Twente, and Limerick) for their help and encouragement during my time there. My acknowledgments would be incomplete if I did not also thank the many scientists and engineers with whom I have been fortunate enough to collaborate over the years and from whom I have been able to learn so much. Particular thanks are due to Donald Reinalda, Shell Chair Professor in Abu Dhabi, who has provided very useful feedback on the contents of Chapter 8, Mass and Heat Transfer Limitations and Other Aspects of the Use of Large Scale Catalytic Reactors, and to Michael Hayes, Research Professor at the University of Limerick, who commented very helpfully on the contents of Chapter 15, Catalysis in Biomass Conversion. I also want to thank various people in Elsevier who have been involved in the production of the book: Kostas Marinakis for suggesting that I write an expanded version of my earlier book rather than just bringing it up to date, Carliy Demetre (and her predecessor, Sarah Jane Watson) for continued encouragement and advice during the writing process, Sheela Bernadine Josy for her help in obtaining figure permissions, Bharatwaj Varatharajan for his help and guidance during the proof-reading process, and Miles Hitchen for his work on the cover design. Finally, and most importantly, I wish to thank my wife, Anne, for her continued support; if my first book would not have appeared without her help and understanding, this one even more certainly would never have materialized without her encouragement and patience.

PART I

Fundamentals of Heterogeneous Catalysis

CHAPTER 1

An Introduction to Heterogeneous Catalysis and Its Development Through the Centuries—Chemistry in Two Dimensions

Chapter Outline

1.1 Introduction

I assume that you are a chemist, or at least that you have enough understanding of chemistry to be able to understand its language and shorthand, and that you understand chemical equations such as:[1]

$$CO + H_2O = CO_2 + H_2 \tag{1.1}$$

or even abstract ones such as:

$$A + B = C + D \tag{1.2}$$

[1] This is the so-called water-gas shift reaction. This reaction is of importance in determining the equilibrium composition in many reactions such as steam reforming of methane and methanol synthesis. These reactions are discussed in more detail in Chapter 11, Production of Hydrogen and Syngas From Methane and Some Other Reactions of Methane, and Chapter 12, Catalytic Reactions Involving Syngas, Hydrogen, or Carbon Monoxide for the Production of Intermediates and Chemicals.

Contemporary Catalysis.
DOI: https://doi.org/10.1016/B978-0-444-63474-0.00001-1

You will recognize that the equals sign (=) means that the equation is balanced and that the reaction is (mostly) at equilibrium; in some cases, there is a reversible arrow ($\rightleftarrows$) indicating that both the forward and reverse reactions occur. You will probably also realize that the equations could have associated with them the enthalpies of reaction (for Eq. 1.1, $\Delta H^o = -40.6$ kJ mol^{-1}).[2] Further, you may recognize that should the equal sign be substituted by an arrow ($\rightarrow$), we are more likely to be dealing with a reaction controlled by kinetics than by thermodynamics. However, when we are taught organic chemistry or inorganic chemistry, we sometimes forget these niceties and just worry about what products are formed when we add two chemicals together. And when we read an equation such as:

$$C_2H_4 + H_2 \xrightarrow{Ni} C_2H_6 \tag{1.3}$$

we accept that the reaction is catalyzed by Ni—but often without asking why or how. Glance at your organic textbook (at least, if it is the type that I used when I was a student) and you will see many such "qualified" arrows, often without any explanation or rationale. In such descriptions, the catalyst is a "Black Box." My aim in writing this textbook, from which I hope you will be able to benefit significantly, with or without the help of a lecturer or instructor, is that you should, when you have finished studying it, be able to understand all the parameters associated with such equations and have a much deeper understanding of what a catalyst is, how it is made and applied, and how it works (or does not work); and you should also be in a position to delve into the literature in a critical way in order to gain more detailed information on any reaction about which you may have developed an interest.

Following a rather elementary description of how a catalyst works (using ethylene hydrogenation, Eq. (1.3), and some related reactions as examples), this chapter gives a description of some of the historical milestones in the development of the concepts and practical use of heterogeneous catalysis as well as a brief overview of some of the current industrial applications of heterogeneous catalysis.[3]

The approach used in this book, as discussed in the Preface (see also the Appendix), will be based largely on the use of literature accessible through the internet, this methodology

[2] The value given should be the standard enthalpy, ΔH^o_{298} (i.e., the heat of reaction for the reactants and products in their standard states at a reaction temperature of 298K) but it may also be the enthalpy at the temperature of reaction given. In the Russian literature, the enthalpy is often included within the equation; in that case, the enthalpy of reaction appears on the right-hand side of the equation as a positive quantity if the reaction gives out heat (i.e., it is exothermic) or with a negative sign if the reaction requires heat (i.e., it is endothermic).

[3] The importance of catalysis in the production of all the man-made commodities to which we have become so accustomed cannot be underestimated. It has been argued that there is at least one catalytic step (the majority of these involving heterogeneous catalysis) in the production of 80% of all such commodities. Catalysis is therefore one of the major cornerstones of the modern chemical industry. Because of this, every chemist and chemical engineer should have a thorough understanding of the basics and applications of catalysis. The aim of this book is to provide at least some of such an understanding.

being particularly important in those chapters dealing with modern catalytic research and processes. You, the reader, are encouraged throughout to read widely round the subject and, by carrying out various "tasks," to explore the most up-to-date literature on each subject considered. While in relation to these tasks you will sometimes read some of the original scientific articles, you will at this stage gain most benefit by concentrating on review articles, the so-called secondary literature.

1.2 What Is a Catalyst?

Before proceeding to give a brief introduction to the history and development of heterogeneous catalysis and its application in industry, it is important to provide a general understanding of what a catalyst is and how it works. Many definitions of catalysis have been given, some more precise and accurate than others. For the present purpose, a suitable definition is as follows:

> A catalyst is a material that brings about an increase in the rate of a chemical reaction without itself appearing in the allover stoichiometry of the reaction.

This does not mean that the catalyst does not take part in the reaction. Indeed, an essential aspect of catalysis is that, by forming intermediate species involving the catalyst surface and the reactants in question, the catalyst provides an alternative more energetically favorable route for the reaction in question. A commonly used analogy for this, as discussed in Box 1.1, is given by the flow of water down a hillside. The definition of catalysis given earlier does not exclude the possibility that there is a physical change of the catalyst during use; as will be seen from the example treated later in Box 1.7 and Task 1.5, a heterogeneous catalyst may change significantly in physical form during its time in the

BOX 1.1 An Analogy Between Catalysis and the Flow of Water Down a Hillside

A useful analogy to explain the action of a catalyst is to be found in the movement of water over the earth's surface, considering the water that falls as rain on a mountain slope and then moves downhill toward the sea. The water on the surface of the mountain possesses potential energy, this providing the driving force to transfer it, by way of streams and rivers, toward sea-level (where its potential energy is effectively zero); the potential energy of the water is lost as a result of either frictional forces or of a gain in its kinetic energy. The potential energy of the water on the mountainside can be harvested by creating dams with hydroelectric installations to make use of the potential energy of the stored water; alternatively, the water may be diverted to more arid regions for irrigation purposes. In a hydroelectric scheme, as shown in the accompanying diagram, the water passes to suitable turbines where electricity is generated before the water is passed back to the original streams and rivers: an alternative a more energetically favorable route to the same final destination (see figure).

(*Continued*)

BOX 1.1 An Analogy Between Catalysis and the Flow of Water Down a Hillside (Continued)

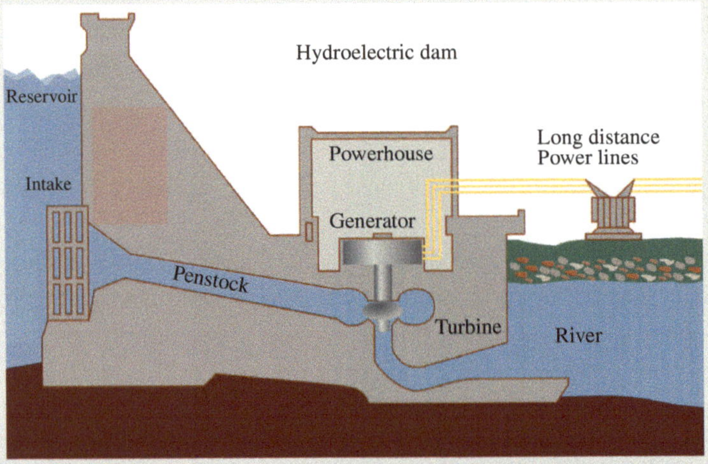

Electricity generation in a hydroelectric scheme. *Taken from https://commons.wikimedia.org/wiki/File:Hydroelectric_dam.svg*

We now consider the analogy between the flow of water and a chemical reaction. If there is no catalyst for a given reaction, the allover reaction rate of the (homogeneous) process is generally controlled by a single rate-determining step; in this, reaction only occurs if the reactant molecules acquire sufficient energy to overcome the activation energy of the reaction. (In the collision theory of gases, the activation energy required for reaction between two gas-phase molecules is explained by assuming that the reacting molecules gain this energy through molecular collisions.) The final state achieved in the reaction is generally the thermodynamically most stable one. This sequence is in many ways analogous to the water flowing over the dam: the water can only flow over the dam or through the spillway when it reaches the top (i.e., its potential energy is such that it is equivalent to that at the top of the dam) and the "head-pond" is full; a steady state will then be established in which the rate of flow over the dam will be equal to the rate at which water is supplied to the reservoir. This is equivalent in the chemical example to the rate of creation of activated molecules that then go on to react. Once the water passes over the dam, it passes down to the sea, this being the most thermodynamically stable level.

Now, what happens when there is a catalyst present? The catalyst provides an alternative route for the reaction that will then have associated with it an apparent activation energy lower than that of the homogeneous noncatalyzed reaction. In the water analogy, this is equivalent to the creation of the new channel for the water that allows it to pass out of the head-pond (e.g., to turbines or to an irrigation scheme) without reaching the top of the dam. The water traveling by the alternative spillway can end up in exactly the same place as did the water traveling over the top of the dam; this is analogous to a catalyzed reaction in which the

(*Continued*)

BOX 1.1 An Analogy Between Catalysis and the Flow of Water Down a Hillside (Continued)

products are the same as those in the uncatalyzed reaction. (In other words, the selectivity of the reaction is unchanged.) However, if the water ends up in a different location, its potential energy will be different to that of the water which has reached sea level; this is analogous to a chemical reaction in which a different product is formed than that which is thermodynamically most stable. (In other words, we are now talking about a selective catalytic process.) Many examples of selective catalytic processes exist and many of these will be discussed in some detail in later chapters. Suffice it to say here that the selective products are often less thermodynamically stable than those which are formed by noncatalytic routes and that this is rather similar to water from a reservoir ending up in a storage vessel well above sea level rather than passing to the sea. The various examples of the oxidation of methane, Eqs. (1.5)–(1.9), discussed in the main text give a good illustration of the principle of catalytic selectivity; total oxidation, Eq. (1.5), giving H_2O and CO_2 as the most thermodynamically stable products, can either occur homogeneously or be catalyzed; however, the selective reactions, Eqs. (1.6)–(1.9), giving thermodynamically less stable products, must all be carried out with a catalyst.

reactor. Furthermore, it may also sinter or be poisoned by the reactants or products or by impurities in the reactant stream. Hence, it can be concluded that the catalyst takes part in the reaction, forming intermediates that provide an alternative more energetically favorable route for the process being catalyzed, but that the catalyst is not itself consumed by the reaction.

A straightforward example to illustrate this definition of a catalyst will now be considered: the hydrogenation of ethylene using as a catalyst finely divided Ni, Eq. (1.3), a reaction studied first by Sabatier and Senderens. (See also Box 1.2).[4] In this reaction, the clean nickel surface adsorbs[5] both the hydrogen (as atoms—"dissociative adsorption") and the ethylene ("associative adsorption"); the surface C_2 surface species are then hydrogenated stepwise by the adsorbed hydrogen atoms and the product ethane is desorbed from the surface, freeing up surface Ni sites for further reaction.

$$H_{2(g)} \rightarrow 2H_{ads} \tag{1.3a}$$

$$C_2H_{4(g)} \rightarrow (CH_2-CH_2)_{ads} \tag{1.3b}$$

$$(CH_2-CH_2)_{ads} + H_{ads} \rightarrow (C_2H_5)_{ads} \tag{1.3c}$$

[4] A commonly used catalyst for hydrogenation is Raney Nickel, developed by Murray Raney in 1926 (Table 1.2); this is prepared from an Ni/Al alloy by dissolving out the majority of the Al using alkali.

[5] The topic of adsorption is covered in some detail in Chapter 2, Surfaces and Adsorption.

BOX 1.2 Sabatier and the Catalytic Hydrogenation of Double Bonds

The chemical process depicted in Eq. (1.3) is known as the Sabatier–Senderens reaction. It was developed in the laboratory of Paul Sabatier at the University of Toulouse (France) in the last decade of the 19th century. Sabatier was awarded a share of the Nobel Prize in Chemistry in December 1912 for his work on catalytic hydrogenation of unsaturated hydrocarbons and inorganic molecules such as CO (see, e.g., P. Sabatier and J.-B. Senderens, Comptes Rendus, Paris, 124 (1897) 1358), this work being the basis for many new industries including that of the hydrogenation of fats to give margarine.[a] His coawardee was Victor Grignard.[b] His close collaborator, J.-B. Senderens, did not share the glory of the prize; however, in his Nobel lecture, Sabatier stressed the important of the latter's contributions to the work. Sabatier also made important contributions to the theory of catalysis, stressing well before his time the importance of surface compounds as intermediates in catalytic reactions; see also Chapter 3 and also later chapters. Sabatier's life and scientific contributions have been described in some detail in a paper by M. Che [Nobel Prize in chemistry 1912 to Sabatier: Organic chemistry or catalysis, Catal. Today 218–219 (2013) 162–171]. The reader would benefit by reading this paper as it gives a very useful picture of the state of knowledge of catalysis at the end of the 19th century.

Paul Sabatier (1856–1941).

Abbé Jean-Baptiste Senderens (1856–1937).

(Continued)

BOX 1.2 Sabatier and the Catalytic Hydrogenation of Double Bonds (Continued)

The figure gives a sketch of the equipment used by Sabatier to examine the hydrogenation of olefins with a finely divided Ni catalyst and the insert shows a French postage stamp honoring Paul Sabatier, showing equipment and plants related to his prize. (From M. Che, reference mentioned earlier; reproduced with kind permission of Elsevier.)

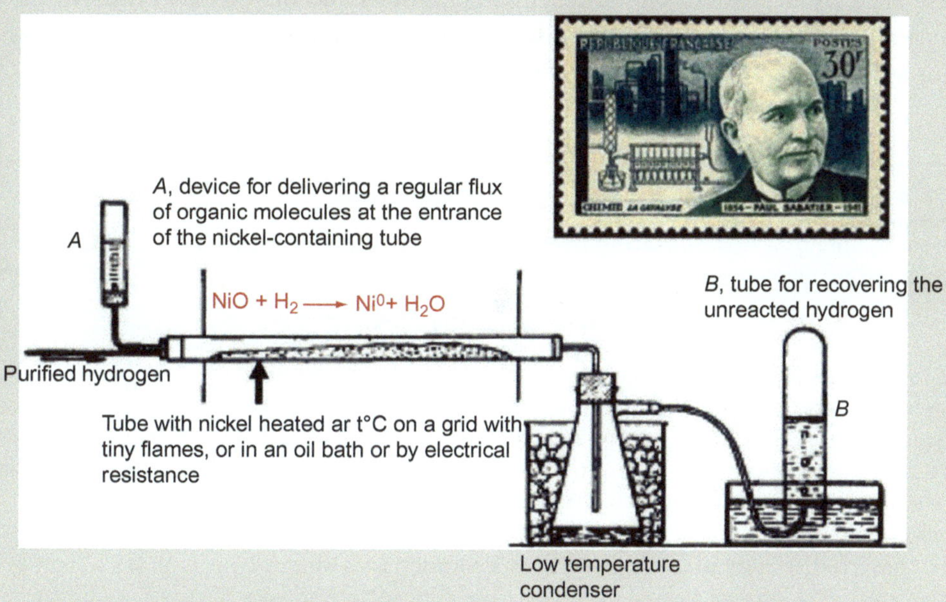

Schematic representation of the apparatus used by Sabatier and Sanderens and a stamp created to celebrate Sabatier's Nobel prize. *Reproduced from M. Che, Nobel Prize in chemistry 1912 to Sabatier: Organic chemistry or catalysis, Catal. Today 218–219 (2013) 162–171, with kind permission of Elsevier.*

[a]http://www.nobelprize.org/nobel_prizes/chemistry/laureates/1912/sabatier-bio.html

[b]Sabatier's coawardee was Victor Grignard who had developed so-called Grignard reagents (R-Mg-X; R = alkyl group, X = halide) for use in a wide variety of organic synthesis reactions. It is interesting to record that Grignard's Nobel citation included the word catalysis but that Sabatier's did not. However, the reactions involving Grignard reagents are not strictly catalytic processes since a Grignard reagent reacts stoichiometrically rather than catalytically.

http://www.nobelprize.org/nobel_prizes/chemistry/laureates/1912/grignard-bio.html

$$(C_2H_5)_{ads} + H_{ads} \rightarrow C_2H_{6(g)} \quad (1.3d)$$

We will return in Chapter 3 to a more detailed discussion of the various steps occurring in such a reaction and of the nature of adsorbed species involved. It is clear that the rate of the catalyzed reaction will depend on the rates of the constituent steps and we will find later that these are very dependent on the reaction conditions and the physical form of the Ni used to catalyze the reaction. We will also discover that many other metals (and even oxides or sulfides) can catalyze this and related reactions.

BOX 1.3 The Sabatier Reaction: CO_2 Hydrogenation

Another example of a hydrogenation reaction is the conversion of carbon dioxide to give methane and water:

$$CO_2 + 4H_2 = CH_4 + 2H_2O$$

At first sight, this reaction is unlikely to have any significant application since, as will be discussed in a later chapter (Chapter 11), the majority of the hydrogen in commercial use is produced by the reverse reaction, the steam reforming of methane. However, if the hydrogen can be provided economically by another route, for example by photochemical water splitting or by electrolysis using the electricity produced by a renewable route (e.g., wind power, hydroelectricity or photovoltaic generation), then the reaction provides a way in which CO_2 can be converted to methane. This can then be added to a suitable natural gas supply grid and hence no new infrastructure is needed. This concept is shown schematically in the following figure:

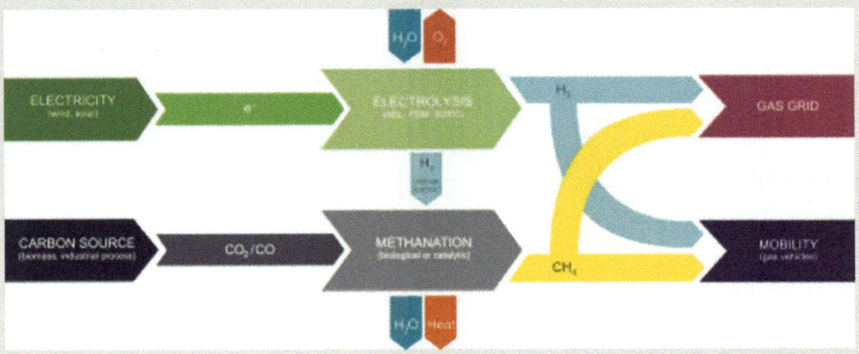

Schematic representation of the use of hydrogen generated by solar or hydroelectric sources to produce methane from CO_2. *Reproduced from M. Götz, J. Lefebvre, F. Mörs, A. McD. Koch, F. Graf, S. Bajohr, R. Reimert, T. Kolb, "Renewable power-to-gas: A technological and economic review, Renewable Energy 85 (2016) 1371–1390; with kind permission of Elsevier.*

(Continued)

BOX 1.3 The Sabatier Reaction: CO_2 Hydrogenation (Continued)

It is interesting to note that the Sabatier Reaction is also used in a reactor attached to the International Space Station; the reaction consumes CO_2 from the Space Station's atmosphere using hydrogen produced by the Station's Oxygen Generation System and produces drinking water to supplement water delivered by supply vehicles. Prior to the installation of this system, the hydrogen and CO_2 were vented from the Station. (Part of the reactor is shown in figure below.)

Doug Wheelock, Expedition 25, working with the Sabatier reactor on board the space station. *From http://www.nasa.gov/mission_pages/station/research/news/sabatier.html.*

Two other important aspects of catalysis also need to be stressed at this point. First, implicit in the definition given earlier is that the catalyst has no effect on the chemical composition of the species involved in the reaction. For example, the ethane formed in reaction (1.3) is the same ethane that would have been formed in the absence of a catalyst, albeit that if no catalyst were present the reaction would occur very slowly under the conditions required for the catalytic reaction. In other words, although the catalyst affects the kinetics of the reaction, it has no effect on the thermodynamic situation. Further, the potential yield of the product of reaction for a given reaction is unaffected by the presence of a catalyst. The other important aspect to remember, as discussed briefly earlier, is that the catalyst can cause a significant change in the products obtained in a reaction; in other words, it can bring about a significant change in the *selectivity* of the reaction. (For further discussion of selectivity in relation to the energetics of the reaction, see Chapter 3: How Does a Catalyst Work?) Again, the catalyst has no effect on the thermodynamics of the reaction in question but it can favor the formation of a product that is not necessarily the most stable one from the point of view of thermodynamics. Ethane is in fact a selective product in reaction (1.3), since the alternative possibility, the formation of methane by reaction (1.4), is more thermodynamically favored:

$$C_2H_4 + 2H_2 \rightarrow 2CH_4 \tag{1.4}$$

Reaction (1.4) (breaking or "cleavage" of C—C bonds, "hydrogenolysis" as opposed to "hydrogenation") can in fact also be catalyzed by nickel but the conditions will be different, hydrogenolysis generally requiring higher temperature.

As another example of selectivity in catalysis, let us consider the oxidation of methane. Without a catalyst (or with a total oxidation catalyst such as Pd supported on alumina), total oxidation to give carbon dioxide and water will occur according to reaction (1.5):

$$CH_4 + 2O_2 \rightarrow CO_2 + 2H_2O \tag{1.5}$$

However, with a catalyst, it is possible to obtain selectively a variety of other products by so-called partial oxidation reactions. The most important of these, to which we will return in Chapter 11, is the conversion of methane to syngas, a mixture of hydrogen and carbon monoxide:

$$CH_4 + 0.5O_2 \rightarrow CO + 2H_2 \tag{1.6}$$

Other possible partial oxidation reactions include the formation of formaldehyde:

$$CH_4 + O_2 \rightarrow HCHO + H_2O \tag{1.7}$$

or of a C_2 product (ethane or ethylene) by "methane coupling":

$$2CH_4 + ½O_2 \rightarrow C_2H_6 + H_2O \tag{1.8}$$

While reaction (1.6) is used commercially and there has also recently been a great deal of research on the topic, neither reaction (1.7) nor (1.8) has yet been carried out commercially since the yields attained are not sufficient to make them financially attractive.[6] Another possible product from methane is methanol (reaction 1.9); however, although this reaction is thermodynamically permissible, kinetic restraints mean that also in this case there have as yet been no reports of any heterogeneous catalyst that has sufficiently high selectivity to make the reaction viable.

$$CH_4 + ½O_2 \rightarrow CH_3OH \tag{1.9}$$

Which of these possible products is formed is determined not only by which catalyst is used but also quite strongly by the reaction conditions: important parameters are the partial pressures of both the methane and the oxygen, the reaction temperature, and even the

[6] For all these reactions, the economics may be changed by the very significant decrease in the price of methane brought about by the exploitation of shale gas reserves. As a result, the subject of methane coupling has recently attracted renewed interest.

reactor design. How a catalyst speeds up a reaction and how it can do that selectively for these and other reactions will be among the topics dealt with in later chapters.[7]

1.3 Historical Background to the Development of Catalysis and Catalytic Processes

Before proceeding in subsequent chapters to discuss some of the more fundamental aspects of heterogeneous catalysis and then to deal with a number of specific applications in some detail, it is instructive first to examine some of the most significant steps in the development of the subject over the last two centuries (Tables 1.1–1.3). As will be seen, although the foundations for the subject were laid in the 19th century by scientists such as Davy, Faraday, Berzelius, and Sabatier, some of the most rapid developments in the practical use of catalysis occurred in the early 20th century with the introduction of processes such as ammonia synthesis and methanol synthesis. These processes were then followed by the introduction of many other catalytic processes. The following sections will outline some of these developments using the material of Tables 1.1–1.3 as a framework.

1.3.1 Some Milestones in the Early Development of Catalysis

Various examples of catalytic reactions, both heterogeneous and homogeneous (Box 1.4, see also Task 1.1), have been known for many centuries (see Table 1.1). Probably the earliest known example of catalysis was the use of yeast, a material which contains an enzyme which ferments the sugar contained in biological materials (such as grain or grapes), to give ethyl alcohol.

Enzymatic fermentation has been known for some 8000 years: beer was first produced in Ancient Egypt and Mesopotamia (today's Iraq) while the earliest wine production was in Georgia and Iran. The earliest work on what we consider to be heterogeneous catalysis may have been carried out by alchemists in their search for a route from base metals to gold. However, the first formal scientific reports of the use of heterogeneous catalysts were written in 1800 by Joseph Priestley and Martinus van Marum (see Table 1.1); both of these pioneers described work on the dehydrogenation of ethyl alcohol over metal catalysts. Surprisingly, at least to the modern scientist, neither appears to have recognized that the metals were acting as catalysts, seemingly believing that their function was just to supply heat for the reaction.

[7] It should be noted that just as in reaction (1.5), the products of reaction (1.6) are also in equilibrium, the difference being to be found in the reaction conditions; the oxygen is in excess for reaction (1.5) whereas the methane is in excess for reaction (1.6). Both reactions occur at high temperature. As we will see in Chapter 11, reaction (1.6) proceeds in most cases initially by oxidation of the methane followed by reaction of the remaining methane with the water and CO_2 formed in the first step.

Table 1.1: Some early milestones in catalysis

Year/Scientist(s) Involved	Reaction	Catalyst(s)	Applications	Further Details
6000 BC	Fermentation	Yeast	Alcohol production	Chapter 1
1800 Priestley, Van Marum	Ethanol dehydration	Metals	Test reaction	Chapter 1
1813 Thénard	Ammonia decomposition	Heated metals	Test reaction	Chapter 1
1817 Humphrey Davy	Combustion of CO and H_2	Pt wire	Miners' lamp	Chapter 1 Box 1.5
1823 Dobreiner		Pt sponge	Catalytic combustion	
1823 Dulong and Thénard		Pd, Ir, Au, etc.	Catalytic combustion	
1831 Phillips	SO_2 oxidation	Pt	Sulfuric acid production	Chapters 1, 8
1834 Faraday introduces the ideas behind the concept of adsorption				
1835 Berzelius introduces the term "catalysis"				
1838 Kuhlmann	NH_3 oxidation	Pt wire	Fertilizer production	Chapter 1 Box 1.6
1890s Sabatier and Senderens	Hydrogenation	Ni, Co, etc.	Fat hardening, methanation	Boxes 1.2 and 1.3, Chapter 14
1906 Bosch and Claude	Nitrogen fixation	Os	Ammonia synthesis	Chapters 1, 8 Boxes 1.7 and 1.8
1918 I. Langmuir demonstrates the dissociative adsorption of H_2 on metal surfaces and introduces the Langmuir isotherm[a]				
1922 Fischer and Tropsch	CO + H_2 (water gas)	Alkalized Fe filings	Hydrocarbon synthesis	Chapters 1, 12
1923 Bosch and Mittasch, etc.	CO + H_2 (water gas)	ZnO/Cr_2O_3	Methanol synthesis	Chapter 12
1920s–30s Mittasch and Schiller	Steam reforming	Ni catalysts	Syngas production	Chapters 1, 11
1925 H.S. Taylor suggests that preferential adsorption occurs at peaks, fissures, and other crystalline discontinuities on a surface[b]				
1929 A.A. Balandin proposes his multiplet theory: catalytic activity depends on catalyst surface site geometry[c]				
1938 Brunauer, Emmett, and Teller formulate BET equation to describe multilayer adsorption[d]				
1941 Beeck et al. show that the catalytic activities of oriented metallic films depend on the lattice spacing or the d-bond character of the metal concerned[e]				
1943 Schwab and Schwab–Agallidis examined the behavior of metal alloy films and demonstrated that the activity depended on the electron concentration[f]				

[a] I. Langmuir, J. Am. Chem. Soc. 40 (1918) 1361.
[b] H.S. Taylor, Proc. R. Soc., Sect. A108 (1925) 105.
[c] A.A. Balandin, Z. Phys. Chem. 132 (1929) 289.
[d] S. Brunauer, P.H. Emmett, E. Teller, J. Am. Chem. Soc. 60 (1938) 309.
[e] O. Beeck, A.E. Smith, A. Wheeler, Proc. R. Soc., Sect. A177 (1941) 62.
[f] G.M. Schwaab, E. Schwab-Agallidis, Naturwissenschaften, 31 (1943) 322.

BOX 1.4 Homogeneous Versus Heterogeneous Catalysis

A *homogeneous catalytic process* is one in which the catalyst is in the same phase as the reactants and products. A simple example is the use of an acid[a] to catalyze the hydrolysis of an ester (or, alternatively, the reverse esterification process):

$$RCOOR' + H_2O \xrightarrow{H^+} RCOOH + R'OH$$

The rate of reaction in the presence of H^+ ions is very much higher than in their absence and yet the H^+ ion does not enter into the stoichiometric equation and is therefore the catalyst. The reaction proceeds by a series of elementary steps. In the first of these (now for clarity, taking the specific example of the hydrolysis of ethyl acetate), the proton of the acid reacts reversibly with the ester as follows:

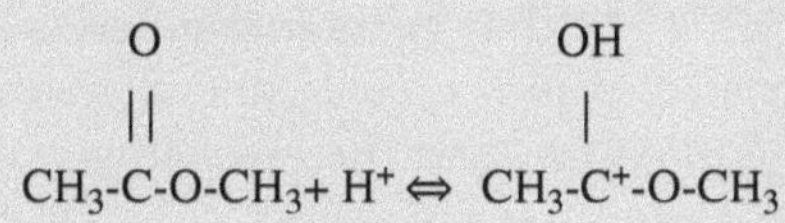

The cationic species that results is much more susceptible to nucleophilic attack by the water molecule than is the original ester molecule:

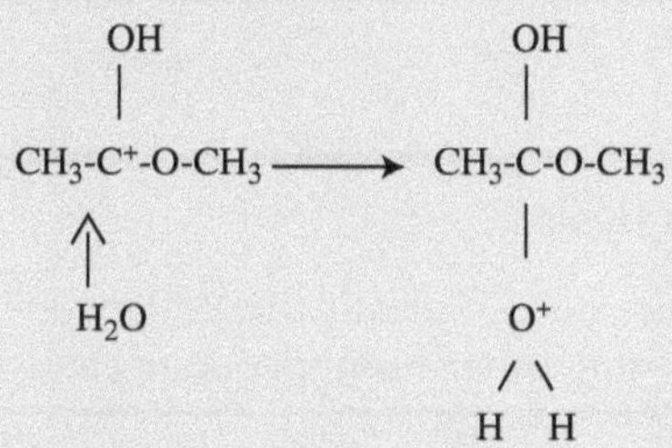

The positively charged species formed by this attack then rearranges by the transfer of a hydrogen atom from the OH group to the methoxy group and elimination of a methanol molecule:

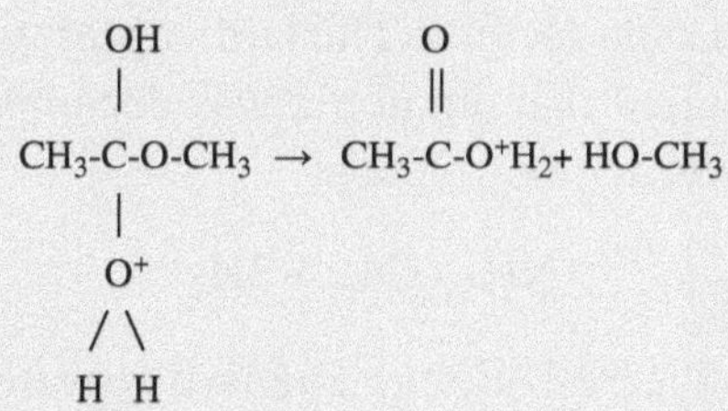

In a final step, a proton (the catalyst) is eliminated and acetic acid is formed:

(Continued)

BOX 1.4 Homogeneous Versus Heterogeneous Catalysis (Continued)

$$CH_3\text{-}\overset{\overset{O}{\|}}{C}\text{-}O^{+}H_2 \longrightarrow CH_3\text{-}\overset{\overset{O}{\|}}{C}\text{-}OH + H^{+}$$

Other homogeneously catalyzed reactions include those in which an inorganic complex is involved. An example of such a species is the Wilkinson catalyst:

$$RhCl(PPh_3)_3 \quad (Ph = \text{phenyl})$$

This catalyst is a chemical species that can be used for the hydrogenation of double bonds and for other reactions in the liquid phase. (http://en.wikipedia.org/wiki/Wilkinson's_catalyst).

In contrast, a *heterogeneous catalyst* is one that exists in a different phase to that of the reactants; the catalyst is usually a solid and the reactants are either gases or liquids. Examples of such catalysts are given in the main text and the mechanism of heterogeneous catalysis is discussed in detail in Chapter 3, How Does a Catalyst Work? It is worth noting that the reactions discussed earlier, the hydrolysis of an ester or the reverse reaction (esterification of an acid), can also be carried out using a solid acid catalyst.

[a]The hydrolysis of an ester can also be catalyzed by an alkali and a similar scheme can be used to show the reaction in the presence of OH − ions.

Task 1.1 Homogeneous Catalysis

Using web entries, find out what you can about the Wilkinson's catalyst and its applications. Can you find any entries concerning other homogeneous catalytic processes?

1.3.1.1 Ammonia decomposition

What seems to have been the first detailed description of the operation of a heterogeneous catalyst was made in 1813 by Louis Jacques Thénard, a professor at the École Polytechnique in Paris. He reported that ammonia was decomposed to hydrogen and nitrogen when it was passed over red-hot metals:

$$NH_3 \rightarrow N_2 + 3H_2 \qquad (1.10)$$

After 10 years, together with Pierre Dulong, Thénard reported that the reaction occurred over iron, copper, silver, gold, and platinum, the rate of reaction decreasing in the order given. This appears to have been the first report of activity patterns in catalysis. (See also Task 1.2.)

BOX 1.5 Activity Patterns in Catalysis

There have been many investigations of the rates of catalytic reactions over a sequence of different catalysts, the aim being to understand the relationships between the catalytic activities observed and the chemical properties of the catalysts involved. For example, much attention was devoted to the so-called d-band theory of catalysis, this being based on the idea that the activity of a catalyst was dependent on the electron concentration in metallic component of the catalyst. This work was initiated by the observation by Beeck and his colleagues that the catalytic behavior of oriented metal films was related to the lattice spacing or the percentage d-character of the metal and by the work of Schwab and Schwab–Agilidis using alloy films (Table 1.1). Other correlations have stressed the importance of the role of surface compounds in determining catalytic activity. For example, Fahrenfort et al.[a] stressed the importance of surface formate species in the decomposition of formic acid over a series of metallic catalysts, showing that the enthalpy of formation of the formate determined the activity of the catalysts. The results gave a so-called volcano curve when the rates of reaction were plotted against the enthalpies of formation of the formates, as shown schematically in the following figure. The most active metals (Pt, Ir, Ru, Pd, and Rh) had intermediate enthalpies of formation, this indicating that the most favorable situation was one in which formic acid is able to adsorb relatively easily but not be held too strongly on the surface to prevent subsequent decomposition occurring.

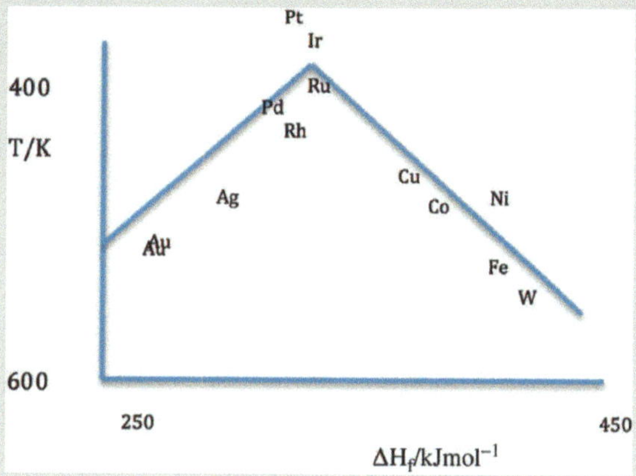

[a]J. Fahrenfort, L.L. van Reijen, W.M.H. Sachtler, The Mechanism of Heterogeneous Catalysis, in: J.H. de Boer et al. (Eds.), Elsevier, Amsterdam, 1960, p. 23.

Task 1.2 Correlations in Catalysis

Using Scopus or a similar citation tool as an entrance point, read about the work of Wolfgang Sachtler and Denis Dowden and their contributions to catalysis. See what you can find on the web about the contributions of these scientists to the theory on the relationships between the catalytic behavior and the physical properties of metals and oxides. If possible, include reference to other contributors in this area.

1.3.1.2 Early work on catalytic oxidation

Shortly after the initial observation on ammonia decomposition had been reported by Thénard, one of the most important early experiments in the development of heterogeneous catalysis as we know it was carried out in the laboratories of the Royal Institution in London in 1817: Humphrey Davy, assisted by a young Michael Faraday, found that a heated platinum wire could bring about the combustion of air and coal gas (largely CO and H_2) without the action of a flame. This was the first reported example of catalytic oxidation. Davy reproduced the result with palladium but failed to do so with copper, silver, iron, gold, or zinc.

The great importance of the surface area of the platinum was demonstrated in 1820 at University College, Cork by Humphrey Davy's cousin, Edmond Davy, who showed that finely divided platinum could oxidize ethyl alcohol at room temperature. Edmond Davy's technique for making high area platinum was improved in 1823 by J.W. Dobreiner, working at the University of Jena; using hydrogen reduction of finely divided platinum, Dobreiner produced a spongy material that caused the combination of hydrogen and oxygen at room temperature. Later that year, Dulong and Thénard showed that the same reaction occurred on palladium and iridium at around room temperature but that higher temperatures were required using cobalt, nickel, rhodium, silver, and gold.[8]

BOX 1.6 The Miners' Lamp

The fact that the platinum brought about reaction of an inflammable gas smoothly, without explosion, and yet with the evolution of heat which caused the wire to glow red-hot, was the basis for the development of the miners' safety lamp. A description of the work carried out by Davy is given by L.B. Hunt in *Platinum Metals Rev.*, 1979, **23**, (l), 29–31(http://www.platinummetalsreview.com/dynamic/article/view/pmr-v23-i1-029-031) and the early work on catalysis is summarized by AJB Robertson in *Platinum Metals Rev.*, 1975, **19**, (2), 64–69 (http://www.platinummetalsreview.com/dynamic/article/view/pmr-v19-i2-064-069). Note that the miners' lamp did not contain Pt but only iron gauze. Note also that the gauze rusted in high water vapor pressures and so there were more, not fewer, explosions.

Task 1.3 Sir Humphrey Davy

Using the references given in Box 1.6, do a literature search on the work of Sir Humphrey Davy and particularly on the history of the miners' lamp. Examine Davy's career and find details of the construction and operation of the lamp.

[8] This appears to have been the first observation of the use of gold as a catalyst. Catalysis by gold has recently become a topic of great interest since highly dispersed gold catalysts (i.e., catalysts with very small gold particles suspended on a support) can have very high activities for catalytic oxidation and other reactions (see Chapter 14: Conversion of Hydrocarbons to Intermediates by Catalytic Hydrogenation or Oxidation).

Task 1.4 Michael Faraday

Having worked as Davy's assistant, Michael Faraday later became Director of the Royal Institution in London. During a very long career, he carried out research on a wide variety of topics and he is recognized as having been one of the most successful practical scientists of all time. Do a web search on Michael Faraday, examining some of the material available on his life and times and paying particular attention to his work on catalytic combustion.

1.3.1.3 Berzelius and the concept of catalysis

As well as being examples illustrating the development of knowledge related to catalytic oxidation, the work outlined earlier shows the way in which new ideas developed internationally, even in the 19th century. There was at that time much discussion on the way in which a catalyst functions. In an article summarizing some of the results outlined earlier that he presented to the Stockholm Academy of Sciences in 1835, J.J. Berzelius was the first to introduce the term "catalysis." The word comes from the Greek word *loosen*. Berzelius proposed that there was a *catalytic force* and that *catalysis* occurred because of the decomposition of bodies by this force: loosening the bonds. The ideas behind the concept of what we now know as *adsorption* had been introduced a year earlier in 1834 by Michael Faraday who had been working on electrolysis and had observed a spontaneous combustion of hydrogen and oxygen occurring on platinum electrodes. He had proposed that the two reacting gases were *condensed* on the surface of the platinum and that "the approximation of the particles to those of the metals may be very great." Faraday's proposition stressed the importance of the surface of the catalyst and also hinted at something that is now generally accepted: that the geometry of the interaction at the interface is also critical.

1.3.2 The Earliest Industrial Catalytic Processes

1.3.2.1 The contact acid process

Examination of the interaction between molecules and surfaces ("chemisorption") inherent in the descriptions given earlier had to wait until techniques were developed that allowed such studies to occur, these being dependent on the perfection of vacuum technologies, as will be discussed later. However, a lack of a full understanding of the importance of chemisorption did not hinder the development of a number of important industrial processes. For example, the "Contact Process" for the production of sulfuric acid was patented by Peregrine Phillips, Junior, in 1831.[9] Phillips brought about the oxidation of sulfur dioxide (SO_2) by drawing it with air through tubes of platinum (or similar material)

[9] The earliest process for the production of sulfuric acid was the Lead Chamber Process (Roebuck, 1746). Although this is often cited as an early catalytic process, it needs to be recognized that the catalyst in the process is gaseous, a mixture of nitrogen oxides, and that the process is thus a homogeneous one. The lead chamber itself, sometimes mentioned as operating as a heterogeneous catalyst, was most probably only an inert reaction vessel providing good heat conductivity. Readers interested in finding out more about the process can find a good description on Wikipedia.

packed with platinum wire, or platinum in any finely divided state, and heating them to a "strong yellow heat" (http://www.platinummetalsreview.com/dynamic/article/view/pmr-v19-i2-064-069). This reaction was not developed as an industrial process until 1875 when Messel introduced the industrial "Contact Acid Process;" this originally operated using a supported platinum catalyst but now uses a vanadia-containing material. http://en.wikipedia.org/wiki/Sulfuric_acid#Manufacture http://www.lenntech.com/Chemistry/Sulphuric-acid.htm

http://www.chemguide.co.uk/physical/equilibria/contact.html

The types of reactor used in the contact process will be discussed in Chapter 8: Mass and Heat Transfer Limitations and Other Aspects of the Use of Large-Scale Catalytic Reactors.

1.3.2.2 Oxidation of ammonia

Fundamental work on another important catalytic oxidation process, the oxidation of ammonia over platinum to give nitric oxide (a step in the formation of nitric acid and hence nitrate fertilizers) was reported by Kuhlmann in 1838.[10] Kuhlmann recognized that the process (using ammonia from coke-oven gas) would in due course become important even though at that time it was uncompetitive due to the cheap and readily available supplies of nitrogenous fertilizer in the form of Chile saltpetre. Toward the end of the 19th century, fears began to emerge that it would become impossible to feed the world's population using the naturally occurring sources of nitrate fertilizers that then existed. However, it was only with the threatened approach of war that scientists in various countries started to examine the oxidation of ammonia (then derived from coal gas) in detail. Hunt describes the work carried out by Wilhelm Ostwald and his assistant Eberhard Brauer at the University of Leipzig in the first decade of the 20th century. Starting by using platinized asbestos, they progressed to using a reactor containing a corrugated platinum strip. It fell to Karl Kaiser of the Technische Hochschule, Charlottenburg to develop in 1909 the gauze-type catalysts of the type used until the present day. (See also Task 1.5.)

BOX 1.7 Ammonia Oxidation

It is well established that the Pt-Rh gauzes used for ammonia oxidation gradually become roughened and develop new structures during use, even though the chemical composition of the catalyst does not change appreciably. An article by F. Sperner and W. Hohmann in *Platinum Metals Review* [20 (1976) 12] includes a series of electron micrographs of gauzes from different parts of the reactor before and after use. These micrographs show very clearly that enormous physical change has occurred in these metallic catalysts during use.

[10] L.B. Hunt, Platinum Met. Rev. 2 (1958)129.

Task 1.5 Ammonia Oxidation Catalysts

The catalysts currently used for ammonia oxidation comprise of a set of gauzes of Pt/Rh alloy. The reaction is highly exothermic and the rate of reaction is determined by external diffusion limitations. Although little or no Pt or Rh is lost in the process, the catalyst changes its appearance quite drastically. Starting with the paper by L.B. Hunt, http://www.platinummetalsreview.com/pdf/pmr-v2-i4-129-134.pdf, examine the development of the ammonia oxidation process and catalyst and how the latter changes during use. [Note: while the general scientific literature contains detailed papers on the subject, much of the material relevant to this search can be found online in *Platinum Metals Review* (http://www.platinummetalsreview.com)].

1.3.2.3 Ammonia synthesis

It was recognized at the turn of the century that the existing supplies of ammonia derived from coal gas would be inadequate as a basis for a fertilizer industry; for example, in his presidential address to the British association in 1898, Sir William Crookes appealed to the scientific community to develop a method of fixing atmospheric nitrogen to solve what was then termed the "nitrogen problem".

Le Chatelier, aware that the decomposition of ammonia should be reversible (see Box 1.8), did experiments in 1901 attempting to bring about the synthesis of ammonia from the elements at elevated pressures. He carried out his experiments in a bomb at 200 atm and at a temperature of 600°C using a reduced iron catalyst heated by a platinum spiral. However, he terminated his experiments after contamination of the reaction mixture with air led to an explosion. Le Chatelier is quoted as saying, near the end of his life, "I let the discovery of the ammonia synthesis slip through my hands. It was the greatest blunder of my scientific career."[11]

Inspired by Le Chatelier's account of his experiments, Haber and Claude succeeded less than 5 years later in producing ammonia. Their work required the development of high-pressure equipment (with the help of Le Rossignol) and also the search for effective catalysts.[12] They found that they could obtain significant yields of ammonia using catalysts based on finely divided osmium, a very rare metal used at that time in light filaments.[13] Shortly after Haber's first patent in March 1909 describing these results with osmium as a

[11] A.J.B. Robertson, http://www.platinummetalsreview.com/pdf/pmr-v27-i1-031-039.pdf.

[12] A detailed account of the development of the Haber–Bosch Process and the activities of the scientists and technologists involved is to be found in the book: Enriching the Earth by Fritz Haber and Carl Bosch; and the Transformation of World Food Production by Vaclav Smil (MIT Press, 2004) ISBN 026293135, 9780262693134.

[13] The total world supply of osmium at that time was 100 kg and BASF acquired all of that for the sum of 400,000 marks; even then, BASF calculated that this quantity was only enough to produce 750 tons of ammonia per year.

catalyst, he obtained a further patent using uranium as catalyst; although uranium was also rare, more of this metal was available. In July of 1909, Mittasch and Kranz of BASF witnessed the operation of Haber's laboratory system with Os as catalyst. In consequence, BASF initiated a crash program to develop the process. Bosch (who had been present in Haber's lab prior to the successful demonstration but had to leave before the critical experiment was carried out) developed the high-pressure equipment required for the reactor system while Alvin Mittasch, together with his colleagues Wolf and Stern, carried out a search for a cheaper and more effective catalyst. (See also Task 1.6.)

BOX 1.8 Le Chatelier's Principle and Ammonia Synthesis

At the beginning of the 20th century, chemists became engrossed by the principle of reversibility. The concept of reversibility was enshrined in Le Chatelier's Principle: that "a reaction equilibrium will move to oppose the constraints placed upon it"; in other words, the position of a chemical equilibrium can be affected by the temperature or pressure of the system.

http://www.woodrow.org/teachers/ci/1992/LeChatelier.html

This principle was applied particularly to the decomposition and formation of ammonia:

$$N_2 + 3H_2 \rightleftarrows NH_3; \quad \Delta H^\circ = -93.3\ \text{kJ}$$

This reaction will proceed in either direction, depending on the conditions applied. At low temperatures and high pressures, it will proceed from left to right; at higher temperatures and lower pressures, ammonia decomposition to the elements will occur, as discussed earlier. It should be recognized that a catalyst that will bring about the forward reaction also catalyzes the reverse reaction.

Task 1.6 Fritz Haber

Fritz Haber, in common with Sir Humphrey Davey (Task 1.3) and Michael Faraday (Task 1.4) and many other well-known scientists of the 19th and 20th centuries, had a career that embraced many different scientific topics. While at the University of Karlsruhe, he worked with BASF on the development of the Haber—Bosch Process for the synthesis of ammonia and later, while at the Kaiser Wilhelm Institute for Physical Chemistry and Electrochemistry in Berlin, he worked on combustion reactions, electrochemistry, and free radical chemistry. During the First World War, he carried out research on chlorine and other poisonous gases. He won the Nobel Prize for Chemistry in 1918 for his work on ammonia synthesis.

Carry out a web search on the history of the ammonia synthesis process and the development of the Haber—Bosch Process. Include in your search the names of Nernst, Le Chatelier, and Bosch and use sources such as Wikipedia to obtain historical details of the main players in the saga. Examine also the great contribution which the synthesis of ammonia has made to the development of agriculture. (Further information on this subject can be obtained, e.g., from the book by Vaclav Smil; see footnote 6). When you have completed your task, you should have assembled a great deal more information than is contained in this text.

The search for improved catalysts (Box 1.9) was concentrated on the more abundant and cheaper elements of the periodic table, particularly on iron. Initially, none of the samples gave yields as high as those obtained with Os or U. However, in November 1909, Wolf found that one of his iron-containing samples, magnetite (Fe_3O_4) from a mine in Gällivare, Northern Sweden, gave very high yields. Mittasch soon realized that the greatly improved properties were due to the presence of impurities; the first patent claiming the beneficial effects of added NaOH and KOH was issued in January 1910. Soon it was also found that MgO and Al_2O_3 were valuable promoters. The final formulation achieved, and generally still used, was based on magnetite containing 2.5%–4% Al_2O_3, 0.5%–1.2% K_2O, 2.0%–3.5% CaO, and 0.0%–1.0% MgO (together with 0.2%–0.5% Si present as an impurity). The only metal not examined by Mittasch and his colleagues was ruthenium as it had not yet been isolated. (See also Task 1.7.)

BOX 1.9 Testing of Ammonia Synthesis Catalysts

In order to commercialize the Haber–Bosch Process, BASF urgently needed a cheaper and more abundant catalyst than the Os catalyst developed by Haber. Mittasch initiated a crash program of the type never before attempted to find a suitable material using a series of 30 parallel high pressure mini-reactors developed by his colleague Stern. Each of these had an easily replacable cylinder containing 2 g of catalyst. This set of systems enabled the Mittasch team to test many thousand samples of catalyst over a period of 2 years and the work led to the development of the promoted iron catalysts of the type that are still used today. It is interesting to note that the element Ru was tested for the first time only in the 1970s by Ozaki and Aika [A. Ozaki and K. Aika, Kinetic and isotope effect of ammonia synthesis over ruthenium, J. Catal. 16 (1970) 97–101.] These scientists found that Ru gave an activity 20–50 times greater than the iron catalysts previously used. According to Bartholomew and Farrauto, Ru catalysts have been used commercially since the late 1990s.

Task 1.7 Ruthenium as an Ammonia Synthesis Catalyst

Using the reference to the work of Ozaki and Aika given in Box 1.9, search the literature regarding the application of ruthenium as a commercial catalyst for ammonia synthesis. Note any other applications of Ru catalysts that you may encounter.

The first experimental pilot plant for the industrial synthesis of ammonia using the Haber–Bosch Process (Bosch being included in the name because of his crucial contribution in the scale-up of the process and the introduction of the associated ground-breaking high-pressure technologies) was built in Appau and began operation in August 1910, producing 18 kg of ammonia per day by January 1911. The first commercial plant in Appau began operating in September 1913, giving 10 tons of ammonia per day. Today's worldwide production is more than 160 million metric tons per year, large plants each producing 1000 tons or more per day.[14] Without the supplies of ammonia and hence of the fertilizers produced from the ammonia, the world's agriculture would be incapable of sustaining the production of foodstuffs necessary for its current population. Smil has estimated that only about half of today's agricultural production would be possible without the Haber–Bosch Process. Haber was awarded the Nobel Prize for chemistry in 1918 and Bosch was awarded the Prize in 1931 (sharing it with Bergius who had been responsible for the development for a process to produce petroleum from coal by hydrogenation).

The development of the Haber–Bosch Process required a combination of a series of parallel developments: a fuller understanding of catalysis; a recognition of the importance of thermodynamic equilibrium and the principle of reversibility; and the development of high pressure and temperature engineering materials and methodologies. The developments occurring during that period have been well summarized in an article by A.J.B. Robertson (Platinum Met. Rev. 19 (1975) 64–69). (See also Task 1.8.)

Task 1.8 Sulfuric Acid and Ammonia Production

Carry out a web search for companies providing plants and catalysts for the synthesis of sulfuric acid and ammonia. If possible, also find out what you can about the worldwide scale of production of these two important chemicals.

1.3.2.4 Methanol synthesis

The award to Bosch of his own Nobel Prize recognized his major contribution in the engineering of the ammonia synthesis process without which it could not have succeeded. BASF continued to capitalize on the expertise gained from the development of the process. In 1923, Bosch, Mittasch, and their colleagues commercialized a process for the production of methanol from water gas (derived from coal) that used a ZnO/Cr_2O_3 catalyst operated at 300–400°C and at 300 atm pressure.[15] We will discuss the production and uses of methanol

[14] We will return in Chapter 8, Mass and Heat Transfer Limitations and Other Aspects of the Use of Large-Scale Catalytic Reactors, to a description of the plants used for the production of ammonia.

[15] Only in 1966 was the alternative low-pressure process (220–300°C and 50–100 atm) introduced by ICI Fertilizer Division in Billingham (United Kingdom), now part of Johnson Matthey (see Table 1.3).

in more detail in Chapter 12, Catalytic Reactions Involving Syngas, Hydrogen, or Carbon Monoxide for the Production of Intermediates and Chemicals.

1.3.2.5 Hydrocarbon synthesis by the Fischer–Tropsch process

Using the high pressure technology developed by Bosch and his colleagues, Fischer and Tropsch, working at the Keiser Wilhelm Institute for Coal Research at Mülheim an der Ruhr, showed in 1922 that alkalized iron filings could convert water gas, at a temperature of 400°C and a pressure of more than 100 atm, to a high molecular weight oxidized hydrocarbon mixture which they called "Synthol." Later developments led to the use of iron and cobalt catalysts for the production of liquid hydrocarbons from coal-derived syngas using the so-called Fischer–Tropsch synthesis process (abbreviated later as FTS); this was commercialized during the years of the Second World War by Ruhrchemie in a plant located at Holten. After the war, the main use of the FTS process was in South Africa where Sasol produced synthetic fuels from coal using the technology. Relatively recently, there has been a significant revival of interest in the process, largely as a step in the conversion of natural gas to fuels (GTL, gas to liquids), for example, using the Shell Middle Distillates Process, for application in regions of the world having large reserves of natural gas.[16] The process is also now being applied in China to make use of its large reserves of coal, China having few natural gas or oil resources.[17] Recently, the technology has also been applied to the conversion of biomass. As a result, the technology has been termed "anything-to-liquids technology."[18] The Fischer–Tropsch process will be discussed in more detail in Chapter 12, Catalytic Reactions Involving Syngas, Hydrogen, or Carbon Monoxide for the Production of Intermediates and Chemicals. (See also Task 1.9.)

---

[16] C.H. Bartholomew and R.J. Farrauto ("Fundamentals of Industrial Catalytic Processes" Wiley-Interscience, New Jersey, 2006) have described the development of FTS technologies as having five phases: (1) discovery of Co and Fe catalysts (1902–28); (2) commercial development of the Fischer cobalt-based process (1928–45); (3) age of iron and Sasol (1946–74); (4) rediscovery of FTS and cobalt (1975–90); and (5) birth/growth of the GTL industry (1990–present).

[17] Many of the developments described above took place in Germany before and during the two world wars; much of this was described in patents but there was also much secret know-how. Some of the earlier technology, such as the steam reforming process, was licensed internationally between the wars. Toward the end of the second world war, the allies recognized that there existed in Germany a lot of technology which was not generally available and so teams of technical experts were sent to Germany (and other countries) to obtain details of processes such as the FTS Reaction. A website exists which gives access to a great deal of this material and which also gives fascinating glimpses of the personalities involved: http://www.fischer-tropsch.org.

[18] See, for example, an authoritative review by J. van Loosdrecht, F.G. Botes, I.M. Ciobica, A. Ferreira, P. Gibson, D.J. Moodley, A.M. Saib, J.L. Visgie, C.J. Weststrate, J.W. Niemantsverdriet, "Fischer–Tropsch synthesis: Catalysts and chemistry, Compr. Inorg. Chem. II 7 (2013) 525–557.

Task 1.9 Fischer–Tropsch SynthesisTechnology

Read the article by J.L. Casci, C.M. Lok, and M.D. Shannon [Catal. Today, 145 (2009) 38–44] on the history of the Fischer–Tropsch process and use a citation search on both that article and the references cited therein to create a list of some of the most important papers describing the development of the process.

1.3.2.6 Steam reforming of methane and higher hydrocarbons

Another critical development associated in one way or another with all these processes was the introduction of the steam reforming of methane and higher hydrocarbons, again pioneered by Mittasch and Schiller in BASF in the 1920s and 1930s. As Germany had no natural gas available, the methane needed for the experiments on methane reforming was synthesized from CO and hydrogen. Schiller succeeded in reforming this methane using a nickel catalyst. The first publication in the open literature of work on the conversion of methane and steam was a paper by Neumann and Jacob[19] that appeared in 1924; this showed that the gas products were close to equilibrium with respect to the steam reforming and water-gas shift reactions. A licensing agreement transferred the BASF technology to Standard Oil of New Jersey who began to produce hydrogen for use in its Baton Rouge refinery in 1931.[20] A steam reformer was commissioned by ICI in Billingham (United Kingdom) in about 1936. While natural gas continued to be the feedstock of choice in the United States, light distillate naphthas became the feedstock used in Europe during the 1950s, this situation only changing when natural gas supplies were located in the North Sea and elsewhere in the 1960s. The development of the steam reforming of naphthas coincided with the development of new steels which could be used in the construction of plants operating at high temperatures and pressures; new catalysts were also developed which had the capability of withstanding these high temperatures and of resisting carbon lay-down. Two tubular reformers capable of operating at around 15 atm were commissioned by ICI in 1962 and one operating at 40 atm was commissioned by Topsøe less than 5 years later. Steam reforming is probably still one of the most critical processes taking place in the chemical industry and improvements in the technology are continuously being reported, the aim being to improve the efficiency and hence lower the cost of hydrogen production. (See Box 1.10.) We will return in more detail to the topic of steam reforming in Chapter 11, Production of Hydrogen and Syngas From Methane and Some Other Reactions of Methane.

[19] B. Neumann, K. Jacob, Z. Electrochem. 30 (1924) 557 (referred to by J.R. Rostrup Nielsen in Catalysis Science and Technology, Vol. 5, in: J.A. Anderson, M. Boudart (Eds.), Springer Verlag, Berlin, Heidelberg, New York, Tokyo, 1984).

[20] Two years earlier, Shell produced hydrogen from methane but by thermal cracking of the methane.

BOX 1.10 Steam Reforming of Methane

The steam reforming of natural gas occurs according to the equation:

$$CH_4 + H_2O \rightarrow CO + 3H_2; \quad \Delta H^{\circ} = +206 \text{ kJ}$$

and this is followed by the water-gas shift reaction:

$$CO + H_2O \rightarrow CO_2 + H_2; \quad \Delta H^{\circ} = -41 \text{ kJ}$$

The allover reaction therefore requires an energy input of about 165 kJ per mol of methane. This energy is supplied by the combustion of excess methane:

$$CH_4 + 2O_2 \rightarrow CO_2 + 2H_2O; \quad \Delta H^{\circ} = -889.5 \text{ kJ}$$

Hence, approximately 0.2 mol of methane must be burned to bring about the reforming of one mole of methane. It should be noted that the requirement for extra methane will be significantly greater, partly due to inefficiencies in heat transfer, etc., and partly due to the fact that the enthalpies given earlier are for the reactions at 298 K.

1.3.3 Some Parallel Advances on the Basics of Catalysis

Much of the early industrial development work associated with the catalytic processes discussed earlier was related to more basic studies in universities and research institutes that were aimed at gaining a greater understanding of the science behind these processes. In some cases, the academic work preceded the industrial development, in other cases, the industrial development came before a build-up of academic interest. (Indeed, the industrial development was quite frequently preceded by fundamental work in the industrial laboratory concerned but the work was not published in the open literature until after the process or product has been commercialized.) A number of names are associated with these developments in the first half of the 20th century: Langmuir for his work on adsorption, Taylor for his development of the idea of the active site, Emmett for his work on physical adsorption, Rideal, Tempkin, Boudart, and many others for their work on the kinetics of catalytic reactions. These names will recur frequently, with others, throughout the remainder of this book, and the interested reader should carry out searches on the internet to find out more about the people behind the names.

1.3.3.1 Irving Langmuir and the development of ideas on chemisorption

Irving Langmuir showed in 1918 that hydrogen dissociates on a metal and he introduced the Langmuir isotherm to describe chemisorption as well as some cases of physical adsorption (see Chapter 2: Surfaces and Adsorption). This led to the development of the Langmuir–Hinshelwood equation to describe the rate of a heterogeneous catalytic reaction (see Chapter 7: The Kinetics and Mechanisms of Catalytic Reactions). While Langmuir's derivation depended on all the sites on a catalyst surface being equivalent, it was recognized by Hugh Taylor in 1925 that preferential adsorption occurred on sites on the surface which are irregular: peaks, fissures, and other crystalline discontinuities. Balandin proposed his multiplet theory in 1929, this postulating that the active sites for a catalytic reaction depend on the geometric arrangement of the catalyst surface. This was followed by the introduction of ideas about the possible importance of the electronic structure of the metallic catalysts by Beeck and by Schwab (see Table 1.1). As a result of these conflicting views, there ensued a long period of controversy with many papers being written supporting either the geometric or electronic theories. As will be seen in later sections, current thinking generally favors the geometric approach and extends this to consider the importance of compounds formed between the reactants and the catalyst surface: chemistry in two dimensions. Task 1.10 covers some of the early discussions of this topic at meetings of the Faraday Society.

Task 1.10 Faraday Discussions

Some of the most important meetings in the history of the development of catalysis have taken place in the framework of the so-called Faraday Discussions. Early meetings in this series were organized by the Faraday Society but they are now held under the umbrella of the Royal Society of Chemistry; the proceedings, which included details of the discussions which ensued, were originally published separately as "Discussions of the Faraday Society."[a] The discussions ranged over a wide range of subjects related to physical chemistry and each covered a separate topic. Several Discussions were devoted to catalysis, these involving many of those active in the field at the time. A notable example was the Discussion held in 1950 that was attended by Langmuir and Sir Cyril Hinshelwood at which emerged the concept of the Langmuir–Hinshelwood mechanism of catalysis (Chapter 7: The Kinetics and Mechanisms of Catalytic Reactions). Search for the contents of the 1950 Faraday Discussion and read the relevant papers and discussion. Identify other Discussions that have dealt with catalysis.

[a]The Faraday Society was merged with the Royal Society of Chemistry and the Discussions have since 1972 continued under the title Faraday Discussions of the Chemical Society.

1.3.4 Developments Related to the Use of Oil as a Fuel or Feedstock (1937–60)

Table 1.2 lists some of the important milestones in catalysis research and development from the late 1930s up until about 1960. All the examples shown are linked in some way or another to the increased use of oil either as a fuel for combustion engines or as a raw material for chemical processes. (Prior to this period, the main feedstock for many chemical processes was coal and, to a lesser extent, biomass in the form of creosote oil and similar materials.) Several sources give details of the historical development of catalysis during part or all of this period. Among these are a list of landmarks in catalysis published online by Burt Davis (http://www.crtc.caer.uku.edu/history) who has also produced a number of

Table 1.2: Development of the use of catalysis for petroleum conversion (1937–60)

Year/Scientist(s) Involved	Process	Catalyst(s)	Applications	Further Details
1937 E. Houdry	Houdry process	Alumino-silicate	Petroleum catalytic cracking	
1942 Standard oil	Fluid catalytic cracking	Alumino-silicate	Continuous cracking process	
1940s–50s R.M. Milton, D.W. Breck, E. Flanegan, R. Barrer	Commercial synthesis of zeolites	A-, X-, and Y-type zeolites	Adsorbents, catalyst supports	Chapter 2
1956 E. Houdry	Developed monolithic catalysts		Exhaust-gas treatment (oxidation of CO and HCs)	Chapter 13
1949 V. Haensel	Naphtha reforming	$Pt-Cl-Al_2O_3$	UOP platforming process. Later developments (1967) led to incorporation of Re etc.	Chapter 10
1953 K. Ziegler	Polymerization of C_2H_4		Polyethylene production. Led to development of UNIPOL Process by Union Carbide and Shell in 1980s giving linear low-density polyethylene	
1954 G. Natta	Polymerization of C_3H_8		Stereospecific polypropylene	
1954 John P. Hogan and R.L. Banks	Polymerization of C_2H_4	Cr_2O_3	Phillips polyethylene process	
1957 James D. Idol	Ammoxidation of propylene	Bi–P–Mo catalyst	Sohio process (Plant completed 1960)	Chapter 14
1959 R.L. Banks and G.C. Bailey	Olefin metathesis	$Mo(CO)_6$, $W(CO)_6$ on alumina	For example, $2C_3^{=} \text{->} C_2^{=} + C_4^{=}$	
1960 P. Weisz, V. Filette, W. Haag	Shape selectivity	Zeolites	Aromatics production	
1960 UOP	Hydrar process	Pt on nonacidic support	Benzene to cyclohexane for adipic acid production	

videos of talks by distinguished researchers in the field of catalysis who were active over the same period.[21] Other very useful sources include reviews by Armor and by Védrine and colleagues.[22] The following sections give brief summaries of some of the most important developments occurring during this period, at least in the context of this book. The reader is encouraged to do more detailed literature studies on a number of these and some key references are included to provide starting points for such searches.

1.3.4.1 Catalytic cracking

Crude oil is a mixture of hydrocarbon-related molecules, these ranging from simple aliphatic species to complex aromatic compounds, many of which contain sulfur and/or nitrogen. These are separated into various fractions by distillation. The fraction with the lowest boiling point range (other than C1–C4 gaseous components) is termed naphtha; this can be used as a fuel for combustion processes but not directly in internal combustion engines. In the early stages of the development of the oil industry as we know it, the naphtha fraction was cracked homogeneously to give a more suitable fraction for use in automobiles. However, this product had a very low octane number. Perhaps the most significant advance was the development by Houdry of the catalytic cracking process. After moving to the United States from France in 1930 (Box 1.11), he started to work on the conversion of heavy liquid tars using acid catalysts and developed a regeneration process to remove the coke deposited on the catalyst. As a result, the first Houdry plant was commissioned by Sun Oil in 1937. This then led to the introduction in 1942 by Standard Oil of New Jersey of fluid catalytic cracking (see Chapter 10: The Production of Energy Carriers From Oil); this process allows the continuous regeneration of the zeolite catalyst used by recirculating it through a second reactor. The zeolite catalysts used for the process have since been continuously improved and this process is one of the most important in the modern petroleum industry. (For further information, carry out Task 1.11.)

1.3.4.2 Other important developments

Other significant processes developed in the same period, 1937–60 (see Table 1.2), included those for the use of the ethylene and propylene formed by the (homogeneous) cracking of naphtha; Ziegler–Natta polymerization to give polyethylene and polypropylene (Task 1.12); and selective oxidation or ammoxidation to give acrolein and acrylic acid or acrylonitrile, respectively. Other developments (Table 1.2) were the introduction of hydrodesulphurization catalysis for the purification of crude oil; the work of Banks and Bailey on olefin metathesis; and the development of zeolite catalysis, particularly in relation to shape selective catalysis and aromatics production. Early work on photocatalysis was

---

[21] This listing covers the years 1949–99 so that some of the work that Davis describes is also included in the following section.

[22] J.N. Armor, Catal. Today 163 (2011) 3–9; I. Fechete, Y. Wang, J.C. Védrine, Catal. Today 189 (2012) 2–7.

BOX 1.11 Eugène Houdry and Catalytic Cracking

Eugène Houdry (1892–1962) was a French mechanical engineer and racing driver who started, while still living in France, to look for ways to convert brown coal to a racing fuel. When he then moved to the United States in 1930, he started to experiment with the use of heavy oils (tars) as raw material. Using acid catalysts to promote the cracking of these species, he recognized that they were being deactivated rapidly due to "coke" formation. He developed a discontinuous reactor system in which the catalyst could be regenerated. This was first used in 1937 in the Sun Oil refinery in Marcus Hook, Pennsylvania, and many more such units were built in the early years of the Second World War. The process contributed to the production of a high-octane aviation fuel that was used in allied aircraft during the war, giving them an advantage over German aircraft using fuel produced by the Fischer-Tropsch and Bergius processes. In 1942, the process of coke removal was much simplified by the development of fluid catalytic cracking using zeolite catalysts; this advance was achieved by two MIT engineers, Warren K. Lewis and Edwin R. Gilliland, working under contract from Standard Oil of New Jersey (now Exxon Mobil).

Later, Houdry also became interested in the problem of automobile emissions and was responsible in 1956 for the first catalytic converter. However, this development was before its time since all the automobile fuel in use at the time contained tetraethyl lead to increase its octane number; this compound proved to be a poison for Houdry's catalyst. It was only because lead was gradually phased out for environmental reasons in the 1980s that catalytic converters were finally used; see Chapter 13, Environmental Catalysis.

Task 1.11 Eugène Houdry and Vladimir Ipatieff

Using Wikipedia and other sources, examine the life and careers of Houdry and Ipatieff, paying particular attention to their work on the production of high-octane fuels by catalytic cracking, alkylation, and isomerization.

also carried out in the 1930s. Many of these topics will receive more attention in later chapters)

Task 1.12 Ziegler–Natta Catalysis

Examine the development of polymerization catalysis, including the work of Ziegler and Natta (leading to the award of the Nobel Prize in Chemistry in 1963) and also the development of the Phillips catalyst (chromia) for ethylene polymerization.

1.3.5 Developments Since 1963

Table 1.3 lists some of the developments in catalysis that have occurred since about 1963. Many of these will be treated in much more detail in subsequent chapters and will therefore only be noted here. Of particular significance has been the gradual switch in interest from the use of naphtha as raw material to the use of natural gas, this being particularly applied to the production of syngas used for methanol synthesis or of hydrogen to be used for ammonia synthesis. Much attention has also been devoted to the use of catalysis in environmental applications such as the cleanup of car exhausts and power station emissions. There have also been significant advances in zeolite chemistry and the use of zeolites in catalysis. One of the holy grails of catalysis has been the conversion of methane, ethane, and propane to more valuable chemicals such as ethylene, methanol, or formaldehyde and, more recently, syngas for use in other processes. Also worthy of note at this stage is the extensive work now in progress on the conversion of biomass to useful chemicals as a means of utilizing "renewables" and minimizing the emission of fossil-derived CO_2. In parallel to this has been the development of the area of "green catalysis" involving processes that minimize the production of undesirable by-products. And this section would not be complete without mention of the fascinating developments in catalysis by gold, a metal previously thought to be chemically (and thus catalytically) inert.

1.3.5.1 Surface science

Langmuir's work on ultra-high vacuum (UHV) technology (see also Box 1.12) led in the 1960s to the development of important surface techniques such as field emission spectroscopy, field ion spectroscopy, low energy electron diffraction, Auger spectroscopy, and X-ray photoelectron spectroscopy.[23] The availability of commercial equipment based on these methods led to a period of intense activity on "Surface Science,"[24] the investigation of the reactions of molecules with atomically clean solid surfaces using these modern physical techniques. Surface science equipment (see, for example, http://www.vgscienta.com/), requiring the use of reproducible ultra-high vacua in which surfaces may be maintained in a clean state over significant periods of time, is very expensive to purchase and also to operate. Hence, surface science has been accused of hogging the research budgets of many research councils over several decades! (Further discussion of these methods is to be found in Chapter 5: Catalyst Characterization.) More recently, with

[23] The 1981 Nobel Prize for Physics was awarded to the Swedish physicist, Kai M. Siegbahn (http://nobelprize.org/nobel_prizes/physics/laureates/1981/siegbahn-cv.html) for developing this technique for studying the species adsorbed on surfaces by monitoring the secondary electrons emitted when the surface is exposed to electromagnetic radiation.

[24] As opposed to "surface chemistry," a term that is applied more to the gas–liquid and solid–liquid interfaces, although it is sometimes also applied to the solid–gas interface.

Table 1.3: Some new processes developed since 1963

Year/Scientist(s) Involved	Process	Catalyst(s)	Reactions Involved	
1963 Steam reforming	ICI process (high T)	Ni/cement/K	Naphtha to syngas and H_2	Chapter 11
	British Gas CRG process	$Ni/Al_2O_3/K$	Naphtha to methane	
1966 Low-temperature methanol synthesis process	ICI	$Cu-Zn-Al_2O_3$	Syngas to methanol using a recycle reactor	Chapter 8
1975 UOP Purzaust exhaust system	Car exhaust cleanup		Installed in Chrysler 1975 cars. This was followed by developments by other catalyst manufacturers (e.g., Engelhard and Johnson Matthey) and car companies (e.g., Ford and General Motors).	Chapter 13
1976 onward: wider application of exhaust catalysts				
1976 Mobil MTG process	Methanol to gasoline	ZSM-5 zeolite	Plant built in New Zealand	Chapter 12
1978 Mitsubishi	Selective Catalytic Reduction (SCR) of NO_x $NO_x + NH_3$ -> $N_2 + H_2O$	V_2O_5/TiO_2	Cleanup of power station emissions. First plant installed by IHI Corporation	Chapter 13
1980s onward	Applications of catalysis in biomass conversion		Biomass-derived esters Conversion of cellulose Gasification and cleanup	Chapter 15
1982 onward Keller and Bhasin, Lunsford, Baerns, etc.	Methane coupling	Pb, Li/MgO, etc.	Ethane, ethylene	Chapter 11
1983 Enichem (Taramasso, Perego, and Notari)	Olefin epoxidation with H_2O_2	Titanium silicalite (TS-1)		Chapter 14
1983 Panov (Boreskov Institute) with Monsanto (now Solutia)	Benzene to phenol	Fe-zeolite	N_2O as oxidant; N_2O formed in adipic acid production	Chapter 14
Late 1980s UOP/BP Cyclar Process	Propane and butane to aromatics	Ga/zeolite	Aromatics from various different saturated hydrocarbon feedstocks	
Late 1980s—present	Partial oxidation	Ni	Production of feed gas for methanol and Fisher—Tropsch synthesis	Chapter 1
Methane conversion	Dry reforming	Noble metals		Chapter 11
1987 Masatake Haruta	Gold catalysis	Au on various supports	CO oxidation, water-gas shift selective oxidation, etc.	Chapter 14
1992 Mobil and other companies	Synthesis of MCM-41 (Mobil Composition of matter No. 41), etc.	Mesoporous materials	Uses as novel supports with high surface areas and very well defined mesopore sizes	
1995 Toyota	NO_x storage catalysts for lean-burn engines	BaO addition to monolith catalyst	Removes NO_x under oxidizing conditions	Chapter 13

BOX 1.12 Irving Langmuir and the Development of Modern Surface Science Techniques

As an example of the contributions made by some of the scientists to the development of the subject, the work of Irving Langmuir, a Nobel awardee in Chemistry in 1932 (*Nobel Lectures, Chemistry* 1922–41, Elsevier Publishing Company, Amsterdam, 1966), will now be discussed briefly. Langmuir graduated in metallurgy from Columbia University in 1903. He then proceeded to work with Nernst at the University of Göttingen and obtained his MA and PhD from there in 1906; his thesis was concerned the "Nernst Glower," the electric light bulb previously invented by Nernst. He returned to the United States and briefly took up an academic position in New Jersey (until 1909) before joining the laboratories of the General Electric Company at Schenectady, New York, ending up as Associate Director[a]. Much of his work in General Electric stemmed from his earlier involvement in vacuum physics related to incandescent lamps and vacuum discharges (developing for that work the modern diffusion pump) and he went on, inter alia, to work on the behavior of gases at low pressure.[b] He showed that hydrogen was dissociated into atoms on the incandescent filament of a light bulb and that it was adsorbed on the surface of the bulb as a monomolecular layer. This led to the so-called Langmuir Isotherm to describe chemisorption on surfaces (see Chapter 2) and also ultimately to the kinetic description of bimolecular catalytic process, the so-called Langmuir–Hinshelwood kinetics (Chapter 7). The work of Langmuir and others in that period led to an understanding of the importance of chemisorbed species in catalytic processes and the concepts of surface intermediates and active sites followed soon after.

[a]Langmuir has been described as being the father of basic research in industry.

[b]Langmuir's name is associated with many other developments such as the Langmuir–Blodgett trough used to study surface films on liquids, the Langmuir probe used to measure charged particles in discharge tubes, the so-called Langmuir waves in plasmas, and the phenomenon of seeding of clouds to bring about precipitation. He was also responsible for the development of the gas-filled incandescent lamp and the discovery of atomic hydrogen.

the development of approaches for using these techniques for studying reactions at surfaces at realistic pressures, more emphasis has been placed on the investigations of catalytic reactions. Scientists whose work is notable in this field, particularly on the relationship between surface science and catalysis, are Gabor Somorjai and Gerhard Ertl. The former has been described as the "Father of Modern Surface Chemistry" (see http://berkeley.edu/news/media/releases/2002/05/10_somorjai.html); the latter was awarded the Nobel Prize in Chemistry in 2007, in particular for having applied his approach to a fundamental examination of the mechanism of the Haber synthesis of ammonia over iron surfaces (see http://nobelprize.org/nobel_prizes/chemistry/laureates/2007/index.html). (See also Task 1.13.)

Task 1.13 Gabor Somorjai

Examine the publication list of Gabor A. Somorjai and identify some of his publications on the importance of "stepped surfaces" in catalysis. Use references to some of his key papers on the subject to identify related work by other authors.

1.4 The Scientific Literature on Catalysis

When the author started to work on catalysis in the early 1960s, the literature on catalysis was spread over a wide range of different journals. Many of those written in English were general journals devoted to fundamental aspects of physical chemistry (e.g., *Journal of Physical Chemistry and Transactions of the Faraday Society* (now *J. Chem. Soc., Faraday Trans.*)) while articles were also published in the more general journals of national societies (e.g., *Journal of the American Chemical Society*, *Journal of the Chemical Society*, or their equivalents in other countries). Much was also published in the German, French, Japanese, and Russian literature and so a scientist needed either to have a basic reading ability of these languages (if the journals of interest were not translated into English) or to rely wholly on abstracts in Chemical Abstracts. As an example of translation into English, the Russian journal, *Kinetika i Kataliz*, was available as Kinetics and Catalysis, this generally appearing a year or so after the publication in Russian. In order to keep up with the literature, one therefore either had to visit the library regularly to scan the contents pages of one's favorite journals or, as methodology developed, to subscribe to a publication (such as Current Contents, produced by the Institute of Scientific Information) giving the contents pages of the journals of interest.

The first specialized English-language catalysis journal, the *Journal of Catalysis*, was published by Academic Press in 1962 with as editors J.H. de Boer (The Netherlands) and P. W. Selwood (United States). Despite this publication filling an obvious need, there was some initial resistance to its publication: many of the older school of academics did not fully approve of commercial (and thus "profit-making") publication of scientific papers.[25] However, the new journal was generally welcomed by the expanding catalysis community and, as a result, it was not long before it became swamped by submissions, this leading to a corresponding significant delay in publication. Academic Press had also launched in 1948 the annual book series, Advances in Catalysis, a collection of reviews on all aspects of

[25] A similar resistance to the existence of commercial journals has re-emerged relatively recently with the move toward the publication of "open access journals": ones that are freely accessible to any reader without the necessity for an "institutional" subscription to the journal. Such a system relies on the author (or his funding organization) paying for the publication and also requires that the publisher has a robust method of archiving the publications; there is little point in publishing a paper that will not be accessible to future potential readers.

catalysis (specifically excluding biocatalysis). The editors of this series initially were W.G. Frenkenberg (Lancaster, Pa), V.I. Komarewski (Chicago), and E.K. Rideal (London).[26] In 1967, a second serial review publication, *Catalysis Reviews*, was launched with Heinz Heinemann as Editor; this is published by Taylor and Francis (Abingdon, United Kingdom).

By the end of the 1980s, the need for other journals devoted to catalysis emerged. This resulted in the launch in 1981 of *Applied Catalysis*, with Bernard Delmon as Editor-in-Chief, and with David Whan, Jacques Védrine, Laslo Guczi and Jim Roth as Editors. A rapid expansion of the number of papers appearing in *Applied Catalysis* led to the launch in 1987 of *Catalysis Today* (with myself as Editor); the intention was that this new periodical would publish predominantly reviews and the proceedings of smaller meetings in order to relieve the pressure on *Applied Catalysis*. Further expansion of Applied Catalysis led to it being split in 1988 into two separate sections (*Applied Catalysis A (General)* and *Applied Catalysis B (Environmental)*). The same year (1988) saw the launch of Catalysis Letters, edited by John M. Thomas and Gabor Somorjai and published by Springer; this was followed in 1994 by the launch by Springer of a sister journal, *Topics in Catalysis*, the latter having similar aims to those of *Catalysis Today*.

These journals remained the main players in the core literature of heterogeneous catalysis until the end of the 20th century although there were parallel developments in the closely related areas of surface science (e.g., *Surface Science, Langmuir*) and homogeneous catalysis (e.g., *Journal of Molecular Catalysis*, later to be split into many different sections) as well as in areas such as solid-state chemistry. The listing given earlier is now far from exhaustive as many other English-language journals (including some open access titles) that are either devoted to catalysis or contain a large proportion of papers on the topic have more recently been launched both in English-speaking countries and in countries in which English is not the *lingua franca*.[27] (A listing of the journals currently being published that have the word "catalysis" in their titles is given in the Appendix.)

---

[26] It is instructive to read the Preface to the first volume of this publication (W.G. Frenkenberg, V.I. Komarewski, and E.K. Rideal (Adv. Catal. 1 (1948) vii–ix)) as this gives a good historical perspective of the state of the subject at that time.

[27] With so many journals to choose from, it is sometimes difficult for a new entrant to the field to decide to which journal he or she should submit a paper. Although metrics such as "impact factor" give an indication of the importance of a particular journal, it is, in the opinion of this author, advisable to try to publish in a journal that is most often used by one's peers. If in doubt, carry out a citation study for your own specific field and then identify the core journals emerging among the most highly cited papers. Although it may be the aim of every scientist to publish in Nature or Science, it may be more advisable to consider a more specialized journal such as *Journal of Catalysis*, *Applied Catalysis*, *Catalysis Letters*, or *Catalysis Today* for your first papers as they are more likely to be read by those working in the same field.

1.5 Other Developments in the Communication of Scientific Results

Just as in the period of the early development of the subject of heterogeneous catalysis described in the preceding sections (1.2 –1.4), personal contact either by letter or by face-to-face encounters at conferences and specialist workshops has continued to be one of the most important ways of aiding the transmission of new ideas. Of great importance too has been the patent literature, a resource frequently largely ignored by academics working in the field. (Perhaps this is because the task of keeping up with scientific papers on the subject is sufficiently time-consuming.) The patent literature (Box 1.13) is of paramount importance to industrial scientists and engineers working in the field and undoubtedly many

BOX 1.13 The Patent Literature

A patent[a] gives to the inventor(s) of a new product or process a legal right to an exclusive benefit from that invention and excludes others from making any commercial use of the knowledge disclosed in the patent. The conditions associated with a patent vary widely from country to country but, in general, an invention is protected in any country in which a patent applies against any commercial exploitation of the content of the patent by any other organization. Again in general, a patent gives a detailed description of the patented product or process and gives a number of examples of its use. Patents often give useful summaries of the current state of knowledge in the area concerned, the aim being to show that the invention provides new knowledge and involves at least one or two new and non-obvious inventive steps. Someone looking at the patent literature without very specific knowledge often has the problem of discerning which of a whole range of variants of the invention is actually applied in practice. Nevertheless, it is very helpful to someone working in an academic catalysis laboratory to be able to get some idea of what the current areas of concern in industry might be. The most important ability for the reader is to be able to see through smoke screens put up by the patentee(s) in order to try to hide the main aims of the work described!

As shown in Task 1.14, one way of accessing patents is to use Google Scholar. There is also a range of patent abstracting services available for those who might need to be aware of recent patents[b].

[a]There is much useful information on the patenting process available on the web and the reader is advised to look up websites of the patent offices of the countries of interest as the actual conditions depend very much on the country involved. A very readable general item on the history of patents and of their use is given in Wikipedia (https://en.wikipedia.org/wiki/Patent).

[b]One of the most used sources of patent information is the Derwent World Patents Index, now published by Thomson Reuters, who also produce a range of other patent-related products.

Task 1.14 The Patent Literature Related to the Development of Zeolite TS-1

As an example of the use of the patent literature, use Google Scholar to look up the work of Bruno Notari, one of the inventors of TS-1. Google Scholar gives access to the primary US patent for this material [US Patent 4,410,501 (1983)] as well as a range of related patents. Examine these patents (and possibly some of Dr. Notari's papers in the open literature) to learn more about the preparation and uses of TS-1.

of the advances in catalytic processing discussed in the earlier sections (1.2 –1.4) would have occurred much less rapidly had the patent system not existed.

Scientific communication has relatively recently been transformed by the development of the Internet. We all now rely on email, Skype, and social networking sites such as LinkedIn to keep us in touch with one another and to learn of new developments as they appear. Most if not all of the journals discussed earlier are now available online and scientific indexing services such as Web of Science, Scopus, and Google Scholar are central to our daily lives. (The use of such services is central to many of the tasks interspersed in this text and the suggested methodology for using them is outlined in detail in the Appendix.) And to use all of these new developments effectively, one has also to keep in touch with developments in information technology. We thus live in an age of continuous learning processes.

1.6 Conclusions

This chapter has been an attempt to introduce some of the ideas behind the subject of catalysis and to trace some of the important milestones in its development. Some of the topics introduced earlier will be considered in more detail later in the book, others not. In subsequent chapters, the student is encouraged to follow up as many of the topics introduced as possible, making use of the web addresses given and then exploring further the wealth of related material. You, the current-day student, are indeed fortunate to have this vast array of information literally at your fingertips. Make use of it and learn from it. But do not get lost in the process: remember to keep your literature searches well focused!

CHAPTER 2

Surfaces and Adsorption

Chapter Outline

2.1 Introduction

When I was a young academic, it was decided at a faculty meeting that we needed to provide remedial classes covering the range of elementary mathematics that would be needed by our first year chemistry students. When at a subsequent meeting our colleague who had agreed to give these classes was asked how they were going, he replied that he had started at the beginning and had been going backward ever since! Having defined catalysis in Chapter 1, and even having talked about adsorption and adsorbed species in relation to ethylene hydrogenation [Eqs. (1.3a–1.3d)], I am now going to go backward to discuss some of the basic aspects of adsorption and desorption. This is because, as we shall see, adsorption and desorption are essential steps in every catalytic process; but it is also because adsorption is an important topic in its own right.

It is generally considered that there are three states of matter, solid, liquid, and gas, even though each of these can be further refined or subdivided. As far as you as a chemist are concerned, most of the reactions with which you have had experience occur in either the gas or liquid phases, even though there are also many reactions that occur in the solid

Contemporary Catalysis.
DOI: https://doi.org/10.1016/B978-0-444-63474-0.00002-3

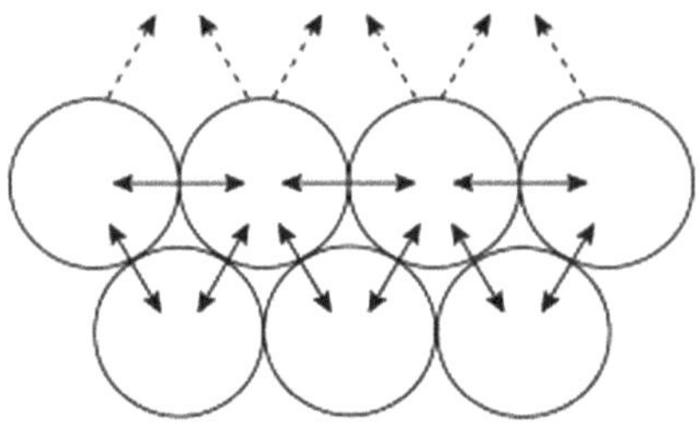

Figure 2.1
Schematic representation of the cross-section of a surface generated by cleaving a metal crystal.

phase. However, further thought should lead you to recognize that you are very familiar with situations occurring at the interface between two and even more phases: the formation of gas bubbles in a liquid, the action of a surfactant at a liquid–solid interface, corrosion at a solid–gas interface (or a solid–liquid–gas) interface, friction at a solid–solid interface. Heterogeneous catalysis is another such phenomenon.

All surfaces are "unsaturated" in that they can formally be thought of as being created by slicing through a bulk phase (see Fig. 2.1). These unsaturated surfaces therefore possess surface energy (or, as is occasionally rather loosely put, it has "dangling valencies" ready and waiting to react with other molecules). The surface energy can contribute to the all-over energy changes occurring when a process occurs at a surface, whether this process is purely a physical interaction or a chemical reaction. The study of the chemical phenomena occurring at such interfaces can be loosely termed "surface chemistry" and, as noted in Chapter 1, the term can also embrace the study of the adsorption and reactions of molecules at solid surfaces (normally adsorbed from the gas phase but sometimes from the liquid phase). When a reaction of a molecule (or molecules) adsorbed at a surface leads to a new surface entity that desorbs to give products and results in the regeneration of the surface, we have a catalytic process and are dealing with the phenomenon of "heterogeneous catalysis." In order to understand heterogeneous catalysis, we therefore have to have a good understanding of the phenomena of adsorption and, to a lesser extent, desorption. This chapter aims to give the student an understanding of the factors determining the behavior of surfaces during adsorption (and desorption) processes and of how one would study these processes.

2.2 Clean Surfaces

In order to study the phenomenon of adsorption experimentally, it is essential first to be in a position to generate clean surfaces. The earliest work on chemisorption coincided with the

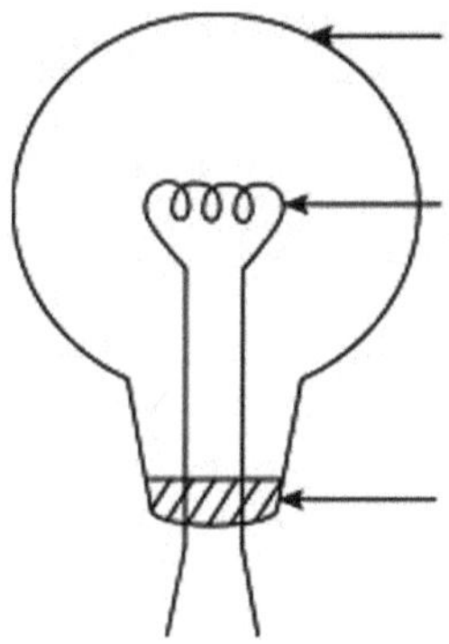

Figure 2.2
Schematic representation of an electric light bulb.

development of the vacuum techniques necessary to generate and maintain clean surfaces (see Chapter 1: An Introduction to Heterogeneous Catalysis—Chemistry in Two Dimensions). The work of Langmuir was driven by the development of the electric light bulb, initiated during his stay in Nernst's laboratory in Göttingen where he learnt of the latter's work on "glowers." When in 1909 Langmuir joined the laboratories of General Electric in Murray Hill, New Jersey, he started to work on the use of incandescent filaments for what we now know as the electric light bulb (Fig. 2.2).

Typically, the filaments in light bulbs are made of thin tungsten wire which, when heated to temperatures of over 2000°C, give out light in the visible wavelength. Tungsten is one of a group of refractory metals that melt at very high temperatures, the melting point of tungsten, T_m, being about 3410°C. The surface atoms of a metal can become mobile at a temperature, the so-called Tamman temperature, T_t, significantly below the melting point, T_m (Table 2.1). The value of the Tamman temperature is given approximately by:

$$T_t \cong 0.6 T_m \tag{2.1}$$

Hence, the surface mobility of tungsten atoms is not expected to occur until c. 2200°C. Tungsten metal can be oxidized in air to give the oxide WO_3. When it is formed at relatively low temperatures, the oxide forms a nonmobile layer that passivates the surface against further oxidation. However, as the temperature is increased further, multiple layers of oxide are first formed and then the oxide layer becomes volatile, desorption of WO_3 taking place.[1] As long as excess oxygen is excluded from the filament and no other gases are present, any surface layer of oxide is desorbed relatively easily, resulting in a clean and stable tungsten surface. The exclusion of oxygen from the filament requires the use of high

[1] Desorption of any surface impurities occurs at the same time; further, some oxygen may desorb as molecular O_2 rather than as WO_3.

Table 2.1: The melting points of some refractory metals

Metal	Melting Point (°C)
Tungsten	3410 ± 20
Molybdenum	2617
Tantalum	2996
Rhenium	3180
Niobium	2468 ± 10
Osmium	3045 ± 30

Source: *Handbook of Chemistry and Physics, CRC Press.*

vacua and it was the development of the methods needed to create such vacua that led to the first successful incandescent light bulbs.

All early vacuum apparatus of the sort used by Langmuir to study the behavior of his tungsten filaments was made of glass and every laboratory using such equipment employed one or more professional glassblowers. (All research students working with such systems became relatively proficient amateur glassblowers! I would number myself among this group.) A typical vacuum system was made up of a rotary pump which would give a vacuum of the order of 10^{-2} mm Hg[2], a mercury diffusion pump which (following suitable outgassing of the system) would give a vacuum as low as 10^{-9} mm Hg[3], various vacuum taps (greased stop-cocks, mercury cut-offs or, rather later, metal valves joined to the system by suitable metal–glass seals) and pressure measuring devices including mercury manometers, McLeod gauges, and ion gauges. In order to achieve the lowest possible pressure, which was dependent on the vacuum components such as the pumps and valves used, elaborate preparation methods, such as bake-out and trapping at low temperatures to exclude mercury vapor, were necessary. Most modern equipment now uses turbomolecular pumps although oil diffusion pumps or ion pumps are also used. (See also Task 2.1.)

Task 2.1 Vacuum Techniques

Carry out a web search on vacuum pumps, pressure gauges, etc., paying particular attention to the design and use of turbomolecular pumps. Why is a forepump necessary for the successful operation of a diffusion pump or turbomolecular pump? Who are the main suppliers of vacuum equipment?

[2] The non-SI unit of mm Hg was a very practical one, the pressure being read directly from a manometer or determined with a McLeod gauge. The term Torr (named after the Italian scientist Torricelli who first used a barometer in 1644) is also often used in its place; 1 Torr or 1 mm Hg is equivalent to 13,341 Pa. For more details on pressure units, search the web for entries on Torr.

[3] Langmuir was responsible for the introduction of a modified mercury diffusion pump that still bears his name.

2.3 Langmuir's Work on Adsorption

During his work on the development of the light bulb, Langmuir discovered that hydrogen could interact at a hot tungsten filament to give hydrogen atoms. He also then recognized that hydrogen could become *adsorbed* on the tungsten surface, this process proceeding through the dissociation of the hydrogen–hydrogen bonds. The word *adsorbed* implies "uptake on" the surface of a solid; when the interaction is chemical, we refer to "chemisorption" but when the interaction is purely physical, we refer to "physical adsorption" (or "physisorption"). As both chemisorption and physical adsorption play very important roles in relation to the phenomenon of heterogeneous catalysis, we will now consider as both of these in some detail.

The phenomenon of chemisorption has received most attention in relation to metal surfaces. Langmuir's early work was carried out using clean metal filaments produced under high vacua. This led to further experimentation using evaporated metal films (generally prepared under essentially what we now recognize as "ultra-high vacuum (UHV)" using laboratory-built glass equipment) and, later, to work using single crystals in purpose-built commercial stainless steel UHV systems.

2.4 The Langmuir Isotherm

Langmuir's pioneering[4] work allowed him to develop the well-known Langmuir adsorption isotherm. This equation can be used to describe the adsorption of a gas on a clean surface, generally treated as being a metal (although the arguments could be modified to apply equally to oxides or other materials).

The simplest form of the Langmuir adsorption isotherm describes the equilibrium adsorption of atoms A from the gas phase on the sites (M) of a metal surface at a constant temperature:

$$M + A_{(g)} \rightleftarrows A - M \tag{2.2}$$

The resultant equation (see Box 2.1 for derivation) is:

$$v_{ads}/v_m = \theta_A = b_A P_A/(1 + b_A P_A) \tag{2.3}$$

[4] In this chapter, references are not supplied to the original literature in relation to material that has become "bookwork." In other words, the material has become so well established that the original references are seldom read. However, the student can perform searches to access the original papers if so desired, remembering that some of them may be in what are nowadays relatively inaccessible journals or in a language other than English.

BOX 2.1 Derivation of the Langmuir Adsorption Isotherm

Consider a surface made up of n_S equivalent surface sites (e.g., metal atoms). If a gas A is in equilibrium contact with that surface and some A is adsorbed on the surface so that the surface concentration is n_A, there will be (n_S-n_A) empty sites.

Assuming that adsorption is not activated, the rate of adsorption of A on the surface, r_{ads}, will be given by:

$$r_{ads} = k_{ads} \cdot P_A \cdot (n_S - n_A) \tag{2.1.1}$$

and the rate of desorption from the filled sites, r_{des}, will be given by:

$$r_{des} = k_{des} \cdot n_A \tag{2.1.2}$$

At equilibrium, $r_{ads} = r_{des}$, and so

$$r_{ads} = k_{ads} \cdot P_A \cdot (n_S - n_A) = r_{des} = k_{des} \cdot n_A \tag{2.1.3}$$

Dividing through by nS and replacing nA/nS by θ_A, we obtain:

$$k_{ads} \cdot P_A(1 - \theta_A) = k_{des} \cdot \theta_A \tag{2.1.4}$$

or

$$\theta_A/(1 - \theta_A) = k_{ads}/k_{des} \cdot P_A = b_A P_A \tag{2.1.5}$$

where b_A is the so-called adsorption constant. The value of b_A depends on the activation energies for adsorption and desorption; in other words, it is directly related to the heat of adsorption.

Rearranging gives:

$$\theta_A = b_A P_A/(1 + b_A P_A) \tag{2.1.6}$$

This is the Langmuir Isotherm describing the adsorption of an atom A on a *uniform* surface.

Eq. (2.3) applies to the adsorption of an undissociated species A on the atoms of a metal surface at a constant temperature, these metal "surface sites" being either empty (M) or occupied by the species A "adsorbed" on the site (A—M). Here, v_{ads} is the volume of A adsorbed, v_m is the volume of A required to give a full monolayer, θ_A is the fractional coverage of the surface by A, b_A is the so-called adsorption constant, and P_A is the pressure of A. (It should be noted that the amounts adsorbed can be expressed in many ways. The volume of gas adsorbed (measured at STP) is frequently used; however, this quantity could equally well be converted to the number of molecules adsorbed n_A or the mass of gas adsorbed m_A.) The derivation of the equation shown in Box 2.1 depends on all the sites being identical; in other words, the heat of adsorption of the hydrogen is unchanged as the surface fills up. As b_A is related to the heat of adsorption [$b_A = a.\exp(\Delta H_{ads}/RT)$], this implies that b_A is independent of the coverage of the surface by A. As we shall see, this is far from the real situation. Nevertheless, the isotherm of great importance as it allows us to gain a better picture of the adsorption/desorption equilibrium that exists and it allows us to proceed to apply the same approach to actual catalytic processes. (See Chapter 7: The Kinetics and Mechanisms of Catalytic Reactions.)

Fig. 2.3 shows the typical form of an isotherm for the adsorption of molecule A on a surface. There is no adsorption at $P_A = 0$ while the volume adsorbed approaches the value v_m asymptotically as the pressure increases. In other words, the sites are gradually filled until there are no empty sites remaining; i.e., once v_{ads} reaches v_m, the so-called monolayer capacity of the surface. In handling data of this sort, it is helpful to be able to linearize the plots in order to obtain a more accurate value of v_m and to allow the constant b to be determined. One such linearization (can you find another?) is to plot the data simply as reciprocal values as shown in Fig. 2.4. The value of v_m is obtained from the y intercept and the "adsorption constant" b is obtained from the negative intercept on the $1/P_A$ axis.

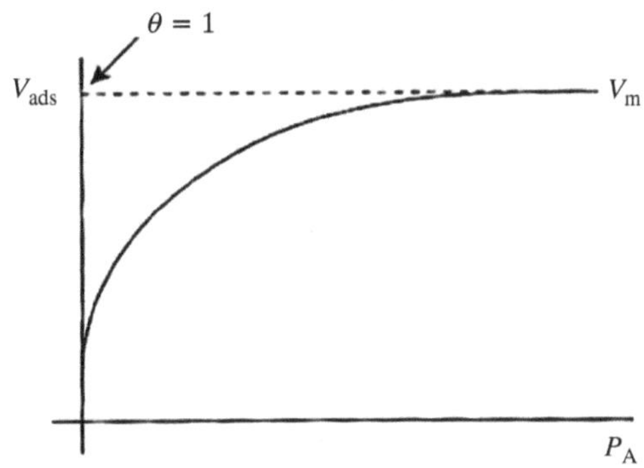

Figure 2.3
Typical adsorption isotherm following the Langmuir equation; when the volume adsorbed (V_{ads}) is equal to the monolayer capacity (V_m), no further adsorption occurs.

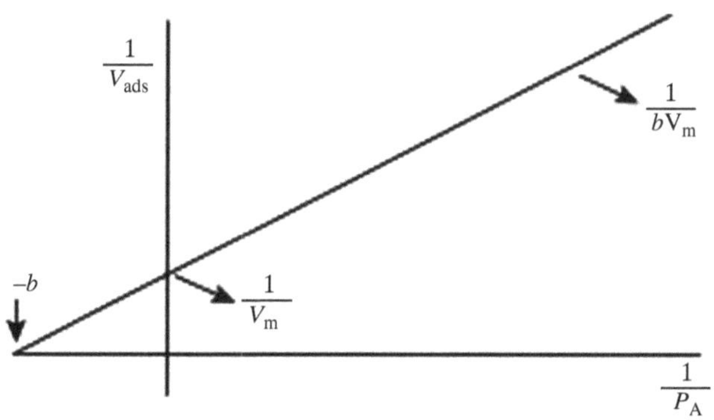

Figure 2.4
Plot of the reciprocal form of the Langmuir equation from which V_m and b_A may be determined.

As was shown in Box 2.1, the constant b_A is given by k_{ads}/k_{des} and is hence related to the heat of adsorption ΔH_{ads}: the higher the value of the heat of adsorption (or the greater the strength of the M–A bond formed), the higher is the value of b_A. The shape of the isotherm is strongly dependent of the value of b_A. For low heats of adsorption, the isotherms have poorly defined "knees" and the asymptotic value of v_m is difficult to determine; in contrast, for high values of the heat of adsorption, the "knee" is well defined and it is possible to determine the value of v_m purely by inspection. The shape of the isotherm is also determined by the adsorption temperature: an isotherm with a diffuse knee obtained for adsorption at a relatively high temperature can become relatively sharp at lower temperatures. We will return to the topic of the temperature of adsorption in Section 2.8 in relation to the subject of the adsorption of more complex molecules.

The isotherm shown in Fig. 2.3 has two important features: at low pressures, the volume adsorbed is proportional to P_A while, at high pressures, the volume adsorbed is independent of pressure ($\theta \approx 1.0$). [This has consequences in relation to the rates of catalyzed reactions (see Chapter 7); a unimolecular catalyzed reaction can have kinetics ranging from first order to zero order, depending on the coverage of the catalyst surface by the reactant.] In practice, it is also often found for intermediate coverages that

$$\theta_A \approx b_A^n P_A^n \tag{2.4}$$

where $0 < n < 1$. This empirical equation is known as the Freundlich Isotherm (see further below). (See also Task 2.2.)

Task 2.2 Information from the Reciprocal Plot of Fig. 2.4

a. The value of b can be determined from the gradient of the plot. How?
b. How can one determine the heat of adsorption of A on the surface of the adsorbent for the data of Fig. 2.3? (Hint: you would need to acquire additional data.)
c. What would be the form of the plots of Figs. 2.3 and 2.4 if the heat of adsorption were to be much lower?

2.5 The Chemisorption of Hydrogen

For the case of the adsorption of hydrogen, the form of the Langmuir isotherm has to be modified to allow for the dissociation of the hydrogen. The equation (see Box 2.1 for derivation) now contains terms for the square root of the hydrogen pressure and the constant c has a different relationship to the heat of adsorption:

$$\theta_H = c_H P_{H_2}^{0.5} / (1 + c_H P_{H_2}^{0.5}) \tag{2.5}$$

For intermediate coverages,

$$\theta_H = c_H^n P_{H_2}^n \tag{2.6}$$

where again $0 < n < 1$ (See also Box 2.2, Task 2.3.)

BOX 2.2 Dissociative Hydrogen Adsorption—The Pressure Function

When a molecule such as hydrogen dissociates on adsorption, we can no longer use the simple Langmuir isotherm to describe the adsorption behavior. As we shall see in Section 2.9, dissociative adsorption takes place following physical adsorption of the hydrogen molecule. It is possible to set up the corresponding Langmuir isotherm using a similar approach to that shown in Box 2.1. However, an alternative method is to consider the existence of a so-called *virtual equilibrium* involving the formation of gas-phase hydrogen atoms in equilibrium with molecular hydrogen in the gas phase as shown in the following scheme:

$$\begin{array}{ccc} H_{2(g)} & \leftrightarrows & 2\,H_{(g)} \\ \downarrow\uparrow & & \downarrow\uparrow \\ H_{2\,(ads)} & \leftrightarrows & 2\,H_{(ads)} \end{array}$$

As long as the system is in equilibrium, the *virtual pressure* of hydrogen atoms so formed determines the adsorption behavior.

For the gas-phase (virtual) equilibrium, we can write:

$$K_{eq} = P_H^2/P_{H2} \tag{2.2.1}$$

from which we obtain:

$$P_H = \left(K_{eq}\cdot P_{H2}\right)^{0.5} \tag{2.2.2}$$

Inserting this expression for P_H in place of P_A in Eq. (2.2) (with $C_H = b_H.K_{eq}{}^{0.5}$) gives us Eq. (2.4). In practice, it is not important that the gas phase equilibrium is unlikely to be achieved as long as the alternative route via adsorbed molecular hydrogen is fully equilibrated.

Task 2.3 Determination of the Metallic Area of a Sample

If the value of V_m determined for the nondissociative adsorption of a species A on a sample of 0.15 g of a metal is 1.7 cm^3 (STP)*, determine the specific surface area of the metal, assuming that one A species occupies 0.65 nm^2 on its surface.

*STP, standard temperature and pressure; one mole of ideal gas occupies 22.4 dm^3 at STP.

2.6 The Chemisorption of More Complex Molecules

The approach outlined earlier (Box 2.2) can also be applied to more complex situations. For example, if one considers that the adsorption of methane occurs on a metal surface according to the following equilibrium reaction:

$$M + CH_{4(g)} \rightleftarrows CH_3 - M + H - M \quad (2.7)$$

it can be shown that:

$$\theta_{CH_3} = kP_{CH_4}/P_{H_2}^{0.5}/(1 + kP_{CH_4}/P_{H_2}^{0.5}) \quad (2.8)$$

If we now use the approximation for intermediate coverages,

$$\theta_{CH_3} = k'(P_{CH_4}/P_{H_2}^{0.5})^n \quad (2.9)$$

where k′ is a new constant and $0 < n < 1$.

The same approach can be applied, for example, to the cases of CH_{2ads} and CH_{ads}. (See also Tasks 2.4 and 2.5.)

Task 2.4 Pressure Functions for Other Adsorbed Species

Work out the pressure functions for the species CH_{2ads} and CH_{ads}. Using these pressure functions, write Langmuir equations expressing the coverages of a surface with each of these adsorbed entities in terms of P_{CH4} and P_{H2}.

Task 2.5 Adsorption Constants for Adsorbed Hydrogen and C_1 Species

The adsorption constants for dissociated molecules [e.g., hydrogen adatoms or the various C_1 species resulting from the adsorption of methane (Task 2.4)] are no longer simply related to the heats of adsorption of the parent molecule as shown earlier for the adsorption of a species A. Work out the relationships for the following adsorbed species: H, CH_3, and CH_2.

2.7 Adsorption of Two Species, A and B

When two species, A and B, compete for the same sites on a catalyst surface, the coverages by each species depend on the relative heats of adsorption and can be written as:

$$\theta_A = b_A P_a/(1 + b_A P_A + b_B P_B) \quad (2.10)$$

$$\theta_B = b_B P_B/(1 + b_A P_A + b_B P_B) \quad (2.11)$$

where b_A and b_B are the adsorption coefficients for the adsorption of A and B, respectively, these constants being related to the heats of adsorption as before. (See also Box 2.1.)

These isotherms will be used later (see Chapter 7) to describe the kinetics of a reaction involving two reacting species. It is instructive to note that if B is very much more strongly adsorbed than A, then $b_BPB >> (1 + b_AP_A)$ and hence:

$$\theta_A \approx bAP_A/b_BP_B \tag{2.12}$$

and

$$\theta_B \Rightarrow 1.0 \tag{2.13}$$

In other words, the adsorption of A will become negligible if B is very strongly adsorbed. One consequence of this is that even a small trace of a strongly adsorbed contaminant (e.g., oxygen or a sulfur-containing molecule such as H_2S or SO_2) can completely poison a surface for the adsorption (or further reaction) of another molecule such as hydrogen. We will return to the subject of poisons in Chapter 7, The Kinetics and Mechanisms of Catalytic Reactions.

Task 2.6 Langmuir Isotherms for Coexisting Adsorbed CH_2 and CH Species

Using the pressure functions you derived in Task 2.4 for CH_2 and CH, set up Langmuir isotherms for each of these species. What would the result be if the CH species were much more strongly adsorbed than the CH_2 species?

2.8 Nonhomogeneous Surfaces

It is well documented that the heat of adsorption of a molecule on a surface seldom remains constant as a function of coverage as required for the Langmuir isotherm to hold; instead, the heat of adsorption almost invariably decreases as coverage increases. Various attempts have been made to describe the adsorption behavior for such cases. Examples of such approaches are the Freundlich and Temkin equations. Although the former equation [Eq. (2.4)] was initially seen as being entirely empirical, Zeldovich showed that it could be derived for a situation in which the heat of adsorption falls logarithmically with coverage, the resulting linear form being:

$$\ln \theta = RT/Q_o \cdot \ln p + \text{constant} \tag{2.14}$$

where Q_o is the heat of adsorption at zero coverage. This reduces to the more familiar form of the Freundlich equation (see earlier):

$$\theta = kp^{1/m}$$

where m is greater than unity, this being equivalent to Eq. (2.3).

The so-called Temkin equation

$$\theta = a \cdot \ln(bp) \tag{2.15}$$

where a and b are constants whose values depend on the enthalpy of adsorption at zero coverage, was first derived by Slygin and Frumkin but was popularized by Temkin and Pyzhev in connection with their work on the decomposition of ammonia over platinum and tungsten surfaces; it is based on the assumption that the heat of adsorption drops linearly with increasing coverage. This assumption is close to the situation met practically in many cases.

Although these two isotherms describe the adsorption behavior of many gases better than does the Langmuir isotherm, the latter is more frequently used in the derivation of rate equations (see Chapter 7: The Kinetics and Mechanisms of Catalytic Reactions).

2.9 Nonequilibrium Adsorption

In all that we have discussed until now, the assumption has been made that we are dealing with an equilibrium situation. After all, the word "isotherm" implies that there is an equilibrium at a constant temperature. However, not all adsorption processes occur at equilibrium and it can be that adsorption gives rise to a species that cannot be removed reversibly. In such a case, the chemisorption process (or its reverse, desorption) is "activated"; in other words, there is an activation energy associated with one or more of the steps leading to chemisorption or desorption and so these processes may not reach equilibrium. In some cases, the adsorbed species can desorb once more under suitable circumstances. However, the adsorbed species often transform to some other species prior to, or during, desorption; this sometimes requires elevated temperatures. An example of such complex behavior is the adsorption of methane on various metals and this will now be discussed.

Experiments carried out on the adsorption of methane on evaporated films of tungsten prepared under UHV (a new film was used for each experiment) showed that methane could adsorb irreversibly at temperatures as low as 173K.[5] (The amount adsorbed corresponded to approximately a single monolayer.) When the surface covered by the adsorbed methane was heated, no gaseous product was found until a temperature of c. 300K, at which temperature hydrogen desorbed. At a temperature of 293K, the average composition of the surface layer was CH_2.

The question remains: was all the hydrogen remaining on the surface at any temperature associated with the carbon (i.e., as CH_{xads}); or was it adsorbed separately on the surface ($C_{ads} + xH_{ads}$); or was there some intermediate situation involving some free hydrogen and

[5] P.G. Wright, P.G. Ashmore, C. Kemball, Trans. Faraday Soc. 54 (1958) 1692; J.R.H. Ross, M.W. Roberts, C. Kemball, J. Chem. Soc. Faraday Trans. I, 68 (1972) 221.

some CH species? The question was answered (at least partially) by carrying out "exchange experiments" with the adsorbed layer: deuterium was added to the system after heating to various different temperatures and the formation of HD and H_2 was monitored. Immediately after adsorption of methane at 193K, one hydrogen atom per adsorbed methane could be rapidly exchanged with deuterium at the same temperature; when the sample was then heated, there was a gradual exchange of more hydrogen until four hydrogen atoms had been exchanged at a surface temperature of about 300K. Assuming that the rapid exchange corresponded to free hydrogen adsorbed on the surface and that the slow exchange corresponded to exchange with a CH_x species, it could be concluded that the methane adsorbed on tungsten at 173K according to the equation:

$$CH_4(g) + 2W \rightarrow CH_3 - W + H - W \quad (2.16)$$

On heating the adsorbed species to a higher temperature, the results indicated that the CH_3-W species decomposed further and it was found that the adsorbed entities at 373K were more probably CH-W species. While these conclusions are only semiquantitative (since they depend on the assumptions relating to the rate of exchange of different types of species and that they also relate to average surface compositions), they indicate quite clearly that the way in which a simple molecule such as methane decomposes when adsorbed on a metal surface is very dependent on the conditions and that kinetic processes are quite important. Further, not all metals behave in the same way. When the experiments were repeated using Pd films, it was found that methane did not adsorb until 400K and that it did so according to the equation:

$$CH_4(g) + Pd \rightarrow CH_3 - Pd + \tfrac{1}{2}H_2(g) \quad (2.17)$$

In the case of Pd, there was no evidence for any further dissociation of the adsorbed CH_3 species. One other example of the adsorption of a "complex" molecule will now be considered: that of H_2S on metals such as Fe and Pd.[6] The H_2S molecule adsorbs at temperatures as low as 193K on both metals without desorption of any hydrogen. However, the behavior of the two metals differs considerably if the temperature is raised to c. 300K. With iron, hydrogen is desorbed whereas, with Pd, little or no hydrogen is found. At even higher temperatures, in the presence of excess H_2S, "sulfidation" of both metals occurs so that the equivalent of multiple layers ("multilayers") of sulfide forms. (Equivalent results are found for the chemisorption of oxygen: bulk oxidation is found to occur if the temperature is above room temperature.) The stoichiometry of the sulfidation reaction with iron follows the equation:

[6] M.W. Roberts, J.R.H. Ross, Trans. Faraday Soc. 62 (1966) 2301–2308; M.W. Roberts, J.R.H. Ross, in: Reactivity of Solids, J.W. Mitchell, R.C. DeVries, R.W. Roberts, P. Cannon (Eds.), Wiley, 1969, pp. 411–423.

$$Fe + H_2S \rightarrow FeS + H_2 \quad (2.18)$$

However, with Pd, yet another complication occurs: some of the hydrogen (in the form of atoms) is "absorbed" in the interstices of the metal to form a hydride rather than desorbing to the gas phase. The degree to which this can occur depends on the thermodynamics of the Pd-H system and to whether or not the hydrogen pressure resulting from the reaction exceeds the equilibrium pressure for the so-called ($\alpha + \beta$) region of the phase diagram. It is, however, interesting to note that the formation of a surface layer of adsorbed sulfur species on the Pd is sufficient to prevent the dissociation of any molecular hydrogen added to the system; hydrogen does not dissociate on the sulfur-covered surface and so cannot enter the hydride phase. Hence, it can be concluded that the twin sites necessary on the metal surface to bring about the dissociation process have all been blocked by the adsorption of the sulfur.[7] (See also Box 2.3.)

BOX 2.3 Metal Hydrides

A fuller discussion of hydrides and their formation is beyond the scope of this text but interested readers should carry out a suitable literature search to find out more on the subject. Relevant web addresses are: http://en.wikipedia.org/wiki/Palladium_hydride/ and http://www.platinummetalsreview.com/pdf/pmr-v15-i1-021-025.pdf. It should be noted in passing that palladium catalysts are often used in reactions involving hydrogen and the possible formation of hydrides during the catalytic process is often not considered. Although the dilute α phase will be formed at all pressures, the more extensive β phase will only form above a specific critical pressure at any temperature.

BOX 2.4 Theoretical Approaches to Adsorption

The approach outlined in the following paragraphs is "classical" in that it only provides a phenomenological description of the adsorption process. Modern computing methods allow much more precise descriptions to be given, not only for adsorption processes but also for relatively simple catalytic processes. With the availability of steadily increasing computer power, the complexity of the systems studied is also increasing steadily. In essence, the methods now available allow detailed prediction of the steps that will occur when a molecule approaches a surface and when it undergoes transformations or catalytic processes. However, details of computational approaches to catalysis are beyond the scope of this text. Interested readers should carry out citation searches on the work of M. Neurock and his group from the University of Virginia and use this as a lead into other work of the same sort.

[7] However, it was found that if the hydrogen was dissociated on an incandescent filament in the proximity of the Pd film, the hydrogen atoms formed could be absorbed in the metal lattice; the hydride equilibrium pressure was achieved when the hot filament was turned off, this indicating that the desorption process was not hindered by the adsorbed sulfur layer.

2.10 The Process of Adsorption

As discussed in Section 2.5, hydrogen adsorbs on a metal surface in the form of atoms. However, atoms are not likely to be present at any appreciable concentration in the gas phase. How then does molecular hydrogen dissociate at a metal surface?

The process of adsorption can be described using the so-called Lennard-Jones potential energy diagram shown in Fig. 2.5. This diagram shows the potential energy of hydrogen molecules and hydrogen atoms as a function of the distance of each of these from the metal surface. Consider first the approach to the surface of a hydrogen molecule. At an infinite distance from the surface, there is no attraction between the molecule and the surface; the molecule is at zero potential energy. However, as the molecule begins to get closer to the surface, van der Waals attractive forces occur and the molecule becomes "physically adsorbed" (i.e., held by physical forces of attraction, $-\Delta H_p$) at a distance from the surface corresponding to the sums of the van der Waals radii of the hydrogen atom and the Ni surface atom, $r_{Me(VdV)} + r_{H2(VdV)}$. (If the metal is Ni, the distance is about 0.32 nm; for the other transition metals, the values will be only slightly different.) If the molecule approaches the surface any more closely, there will be a strong repulsion between the electron clouds of the molecule and the surface.

Now consider the approach of a hydrogen atom. In order to dissociate a hydrogen molecule into two hydrogen atoms, it would be necessary to provide the dissociation energy for the

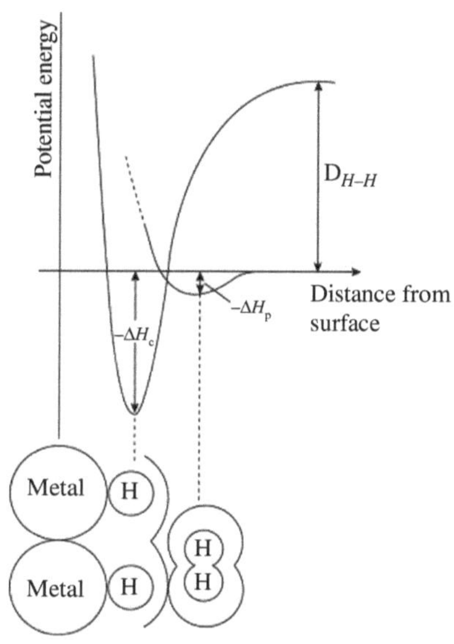

Figure 2.5
Lennard-Jones potential energy diagram for the adsorption of hydrogen on a metal surface; as the hydrogen approaches the surface from the right-hand side, it becomes physically adsorbed and then, following dissociation in the region of the surface, hydrogen atoms are chemisorbed on the surface.

hydrogen molecule, D_{H-H}, of 434 kJ. This dissociation is most unlikely to occur at any appreciable rate in the gas phase. If, however, we consider that a gaseous hydrogen atom if formed will, as it approaches the surface, feel a much greater force of attraction than the hydrogen molecule and will be able to form a metal–hydrogen bond, hence becoming adsorbed with a heat of adsorption $-\Delta H_c$ (chemisorption curve). The equilibrium position for this chemisorbed state is at a distance equal to the sum of the atomic radii of the hydrogen atom and the metal, $r_{Me} + r_H$. (For Ni, the distance is about 1.6 nm but the values for the other transition metals will again be similar.) It can be seen that the chemisorption curve intersects the physical adsorption curve at a value below or very close to the zero energy axis. For the transition metals, the intersection generally occurs, as shown here, below the zero of potential energy and so the chemisorption process is not activated. (In other words, there is no activation energy barrier to be overcome as the hydrogen molecule approaches the surface and it transfers smoothly from the physically adsorbed state to the chemisorbed state.) (See also Box 2.5.)

BOX 2.5 Activated Chemisorption

In early textbooks describing chemisorption, it was often stated that chemisorption can be activated. However, there are few, if any, instances where that is the case for a clean surface. If activated adsorption did occur, the point of intersection of the two curves would be above the zero potential energy line. It is quite possible that the early observations of chemisorption that led to these statements were made on "dirty" surfaces that were contaminated by partial or even complete monolayers of adsorbed oxygen. It was only with the advent of "UHV" experimental adsorption systems that it became possible to maintain clean surfaces prior to adsorption measurements. We will return to the topic of UHV systems in relation to experimental methods of studying catalysts and catalytic reactions (see Chapter 5).

The situation may however change as the surface becomes more and more covered by hydrogen as shown in Fig. 2.6: the minimum of the chemisorption curve decreases in magnitude (dotted curves 1, 2, 3), with the consequence that the curves intersect above the energy minimum. This leads to the occurrence of an activation energy for the adsorption process. We will return to this concept in Chapter 3 when discussing the phenomenon of catalysis. If the adsorbed hydrogen is now to desorb to the gas phase, becoming reassociated with another hydrogen atom to form a hydrogen molecule, the reverse path has to be followed. At low coverages, the activation energy for desorption will be ΔH_c; for higher coverages, it will be ΔH_c plus the value of the activation energy for adsorption.

It is worth considering the use of such curves to describe the phenomenon of absorption discussed earlier in relation to the Pd/H_2 system. Hydrogen can become adsorbed on the surface of the Pd in exactly the same way as described earlier. However, as shown in Fig. 2.7, the curve for the chemisorbed hydrogen atom now intersects with an additional

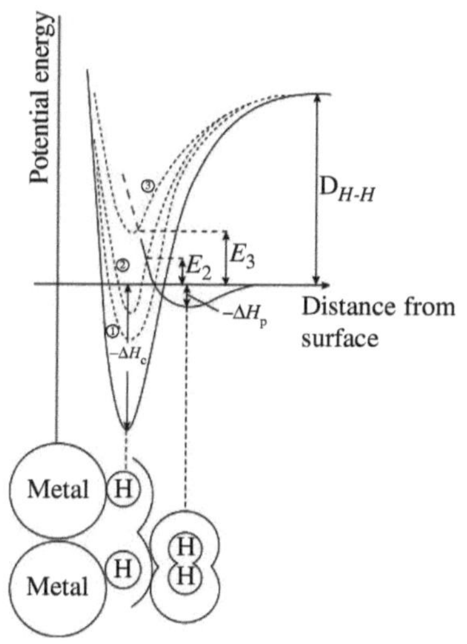

Figure 2.6
Extension of the Lennard-Jones potential energy diagram of Fig. 2.5 showing the effect of increasing the coverage of hydrogen on the surface (curves 1, 2, 3); the decrease of heat of adsorption (ΔH_c) with coverage can result in the occurrence of an activation energy for adsorption. If desorption from the adsorbed state can occur, the activation energy for desorption becomes ($\Delta H_c + E$), the value of this decreasing with increasing coverage.

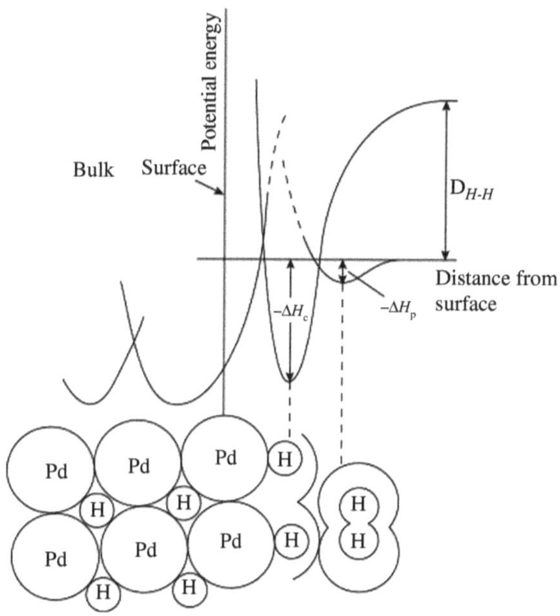

Figure 2.7
Extension of the Lennard-Jones potential energy diagram to show the absorption of hydrogen atoms within a metal such as palladium.

curve representing hydrogen within the bulk of the metal (in interstitial sites within the metal structure); in this case there are a series of minima extending into the bulk of the metal, each representing the most stable position for an interstitial hydrogen atom. As long as the chemisorbed curve intersects with the absorption curve at or near the energy zero, hydrogen from the gas phase can be transferred seamlessly into the Pd bulk.[8] Diffusion into the bulk may itself be activated and the slow step may be either entry into the first layer of the bulk or diffusion through the Pd lattice. (See also Box 2.6.)

BOX 2.6 Precursor States

It is possible that the atom A can adsorb on several different sorts of sites on a single surface. Adsorption of A_2 may occur on one site (Site X) and the adsorbed species may then migrate to a second more stable site (Site Y). In that case, there will be curves in the Lennard-Jones potential energy diagram corresponding to each of these. The adsorbed species at A can then transfer to B. The rate at which transfer from site X to site Y occurs will be related to the energy barrier to be overcome and will be dependent on the temperature. If there is no energy barrier between Site X and Site Y, then all the adsorbed species will immediately go to Site Y as long as adsorption at that is more energetically favored.

2.11 Some Generalizations on Chemisorption

In the preceding sections (2.5, 2.6, 2.8, 2.9), we have considered the adsorption and absorption of hydrogen on metals and have discussed also the adsorption of other molecules such as methane, oxygen, and hydrogen sulfide. It is important to recognize that virtually every molecular species can adsorb on one or another surface and that the surfaces are not restricted to being metals: oxides, sulfides, various forms of carbon, as well as many other types of solid can all chemisorb molecules. While there are some clear regularities in the behavior of different molecules on different solids, each combination must be considered in its own right. The essential factor determining

[8] Originally, it was thought that the hydride of palladium could only be formed electrochemically. That was because it was necessary to achieve atomically clean Pd surfaces before the dissociation of molecular hydrogen, necessary for the subsequent absorption step, could occur; "as received" Pd surfaces are invariably covered by species such as adsorbed oxygen or other impurities. This is consistent with the observation mentioned earlier that an adsorbed layer of sulfur on the surface prevented absorption of hydrogen: the sulfur (or any other contaminant) blocks the surface sites to such an extent that the dissociation of hydrogen cannot occur.

whether or not chemisorption occurs is that the reaction that is occurring has a negative value of ΔG. As ΔG_{ads} is given by the equation:

$$\Delta G_{\mathrm{ads}} = \Delta H_{\mathrm{ads}} + T\Delta S_{\mathrm{ads}} \tag{2.19}$$

and the values of ΔH_{ads} and ΔS_{ads} therefore determine the situation. If we are dealing with the dissociative adsorption of a molecule A_2 at the surface to give $2A_{ads}$, the value of ΔS_{ads} will always be positive since the all-over system gains entropy. Hence, adsorption will only occur if ΔH_{ads} is negative. For nondissociative adsorption (e.g., of CO on a metal), the entropy change could be negative. Hence, in principle, the adsorption of CO could involve an endothermic process. However, in practice, there is no evidence for such a situation.

Relatively early in the history of the study of chemisorption, much effort was devoted to determining the heats of adsorption and attempting to find correlations between the heats measured and other properties of the adsorption system. The most successful approach was one in which the heats of adsorption were found to be relatively well correlated with the heats of formation of the equivalent bulk compounds. An example is the adsorption of oxygen on the transition metals, for which the heat of adsorption was reasonably well correlated with the heat of formation of the most stable oxide (see Fig. 2.8).

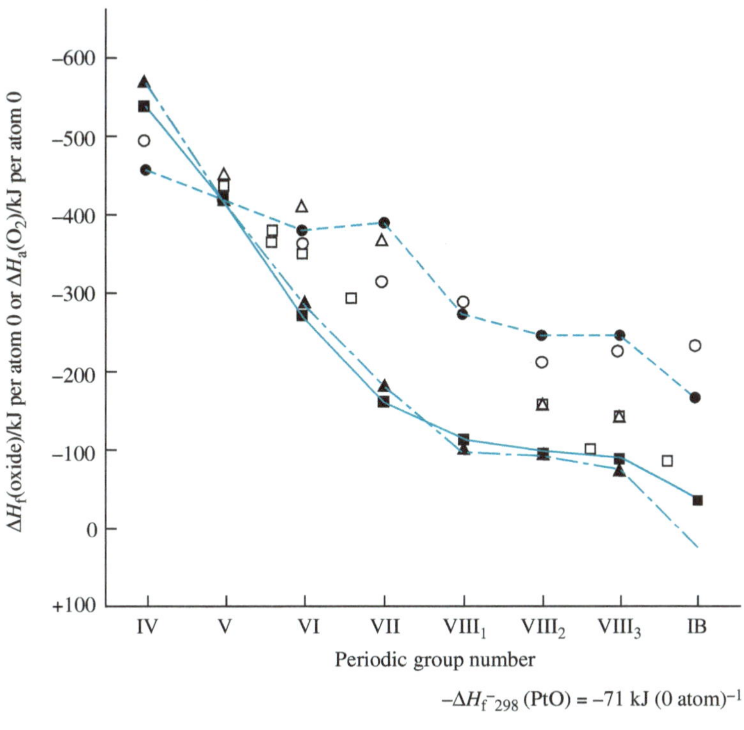

Figure 2.8
Enthalpies of formation of oxides (ΔH_f (oxide): *filled symbols*) and of chemisorption of O_2 on metals (ΔH_a: *open symbols*) as a function of the Periodic Group Number. Symbols: *circles*, first transition series; *squares*, second transition series; *triangles*, third transition series.

Table 2.2: A classification of metals according to their abilities in chemisorptions

Group	Metals	Gases						
		O_2	C_2H_2	C_2H_4	CO	H_2	CO_2	N_2
A	Ti, Zr, Hf, V, Nb, Ta, Cr, Mo, W, Fe, Ru, Os	+	+	+	+	+	+	+
B_1	Ni, Co	+	+	+	+	+	+	−
B_2	Rh, Pd, Pt, Ir	+	+	+	+	+	−	−
B_3	Mn, Cu	+	+	+	+	±	−	−
C	Al, Au	+	+	+	+	−	−	−
D	Li, Na, K	+	+	−	−	−	−	−
E	Mg, Ag, Zn, Cd, In, Si, Ge, Sn, Ph, As, Sb, Bi	+	−	−	−	−	−	−

+ means strong chemisorption occurs; ± means it is week; − means unobservable.

Once a reasonable corpus of data had been accumulated, a "chemistry" of chemisorption was established. For example, B.M.W. Trapnell listed the metals which would adsorb various different molecules; see Table 2.2. It is interesting to note that gold, generally considered as being very inert, is listed as chemisorbing O_2, acetylene, and ethylene but not H_2, CO_2 or N_2. The catalytic properties of gold are currently receiving much attention. We will return to that subject in a later chapter (Chapter 14).

As mentioned in Box 2.4, computational methods are now successfully applied to investigating the way in which a molecule interacts with a surface and to calculating heats of adsorption. However, a full treatment of the subject is beyond the scope of this book.

2.12 Physical Adsorption

Physical adsorption[9] is closely related to the phenomenon of the condensation of a vapor on a surface to form a liquid layer. A first layer of molecules adsorbs on the surface (as in chemisorption) and then additional layers of molecules adsorb on top of this first layer in a manner similar to that observed in the condensation of a liquid. Physical adsorption is frequently studied at temperatures not far from, or equal to, the boiling point of the

[9] Physical adsorption is often discussed in detail in textbooks on heterogeneous catalysis before any mention is made of the phenomenon of chemisorption. Apart from mentioning earlier that physical adsorption is a precursor to chemisorption, I have chosen to introduce it after chemisorption for the reason that a detailed understanding of physical adsorption is less essential in understanding heterogeneous catalysis. Nevertheless, physical adsorption is important as it allows one to check the total area of a catalyst and also to find out about its pore structure. Your instructor may prefer to handle this section in connection with the material of Chapter 5 on catalyst characterization.

corresponding liquid. The most commonly encountered example encountered in work related to catalysis is the physical adsorption of molecular nitrogen on a high surface-area solid at the temperature of liquid nitrogen (78K), its boiling point in an open vessel at atmospheric pressure. However, other molecules have also be used in specific cases, examples being the use of the adsorption of Kr or Xe at 78K to determine low surface area values. The physical adsorption of CO_2 has also been used to determine the surface areas of zeolites and other very porous solids.

The most widely used mathematical description of physical adsorption was published in a paper in 1938 by Brunnauer, Emmett, and Teller: the so-called BET equation. (See also Box 2.7.) Emmett and his research group had for some time been studying iron catalysts for ammonia synthesis and had started to develop physical adsorption methods to distinguish between one catalyst formulation and another. The adsorption of gases such as nitrogen on surfaces at low temperature had been recognized for some time. However, because the isotherms obtained for different solids had very different shapes (see Fig. 2.9), it was difficult to compare one isotherm with another. The BET theory allows a comparison to be made in a much more quantitative way than had hitherto been possible since it allows one to determine, for any solid, a quantity v_m, the so-called monolayer capacity, and hence its (total) surface area.

BOX 2.7 Brunnauer, Emmett, and Teller

Stephen Brunnauer, a Hungarian born in Budapest in 1903, was first a student and then a postdoc with Paul Emmett. After wartime work on explosives, he spent some years as manager of basic research at Portland Cement before becoming Chair of the Chemistry Department of Clarkson College (see http://www.caer.uky.edu/energeia/PDF/vol9-1.pdf).

Paul Emmett (1900–85) became chairman of the Department of Chemical Engineering of John Hopkins University in 1937; in 1943, he joined the Manhattan Project (where he developed a method of enriching U^{235}) and then spent several years and the Mellon Institute of Industrial Research before returning to John Hopkins as W.R. Grace Professor in the Chemistry Department. Following his retirement in 1971, he became a Research Professor at Portland State University (see http://en.wikipedia.org/wiki/Paul_H._Emmett/).

Edward Teller (1908–2003), also a Hungarian and later a very controversial figure, was then Professor of Physics at George Washington University. He was persuaded to collaborate on the project by Brunnauer. He later became a member of Manhattan Project, becoming known as Father of Hydrogen Bomb. For more details, see the extensive Wikipedia article: http://en.wikipedia.org/wiki/Edward_Teller

The BET theory is derived in a similar way to that used for the Langmuir theory outlined earlier (that a molecule is adsorbed on a free surface site and that a monolayer of such species can be achieved, this having a constant heat of adsorption, ΔH_1) with the extension that additional layers can be adsorbed on top of the first layer with heats of adsorption equivalent to the latent heat of vaporization of the gas (ΔH_L); see Fig. 2.10. (These subsequent layers also behave according to the Langmuir model.) In most cases, the second and subsequent layers begin to form before the first layer is complete. The resultant equation is:

$$v_{ads} = \frac{v_m c P}{(P^o - P)\{1 + (c-1)P/P^o\}} \tag{2.20}$$

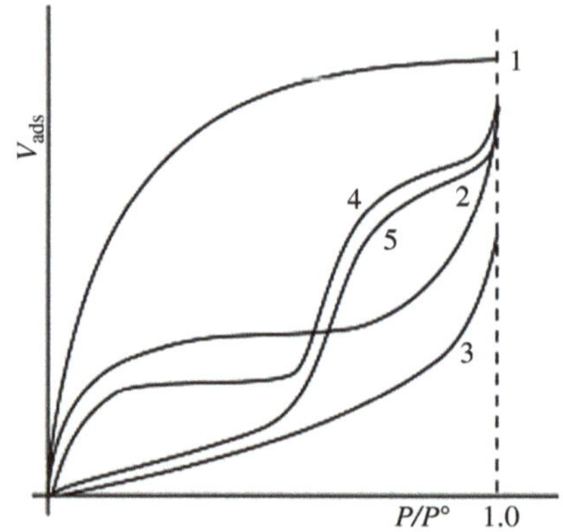

Figure 2.9
Typical physical adsorption isotherms showing five distinct types: (1) monolayer adsorption only; (2) monolayer followed by multilayer; (3) multilayer without prior monolayer formation; (4) monolayer, then pore filling; and finally (5) multilayer without prior monolayer formation.

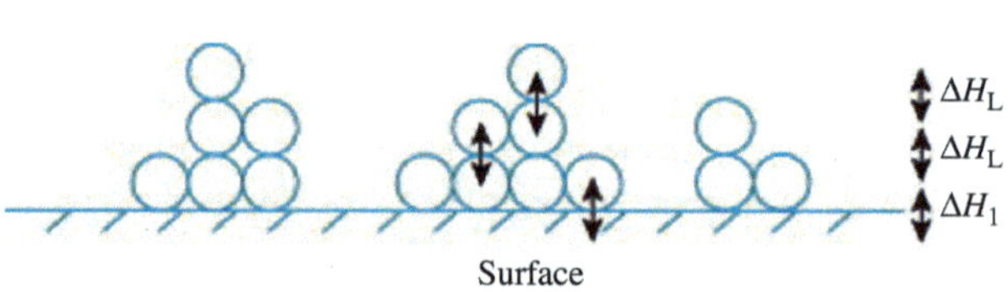

Figure 2.10
Schematic representation of the formation of first and subsequent layers; note that multilayers start to form before the monolayer is complete.

where v_{ads} is the volume of gas adsorbed, v_m is the volume required to give a monolayer of adsorbed molecules, P is the measured pressure of the gas, P^o is the saturated vapor pressure of the gas at the temperature of adsorption, and c is a constant given by the equation:

$$c = \exp\{(\Delta H_1 - \Delta H_L)/RT\} \tag{2.21}$$

Eq. (2.22) can be linearized as:

$$P/\{v(P^o - P)\} = 1/v_m c + (c - 1)P/v_m c P^o \tag{2.22}$$

Hence, a plot of $P/\{v(P^o - P)\}$ (or, alternatively, $P/P^o/\{v(1 - P/P^o)\}$) vs P/P^o will have a y intercept of $1/v_m c$ and a gradient of $(c - 1)/v_m c$. The values of the constant c and of monolayer volume v_m can hence be determined. (Alternatively, the x intercept is $-1/(c - 1)$ and hence the value of c can be determined directly.)

Fig. 2.11 shows a typical plot of an isotherm that obeys the BET equation relatively well. It can be seen that the theoretical curve (*solid line*) follows the experimental points quite closely over a range of $0.1 < P/P^o < 0.3$. However, above the second value, the curve begins to deviate quite significantly from the theoretical one.

Fig. 2.12 shows the equivalent plot of the linearized form of the equation [Eq. (2.16)]. Again, it can be seen that the experimental values lie close to the theoretical curves up until $P/P^o \approx 0.3$ but that there are significant deviations above that value. Much has been written concerning deviations from the BET equation and attempts continue to be made to develop more accurate equations. Nevertheless, the BET equation is firmly established as a tool to

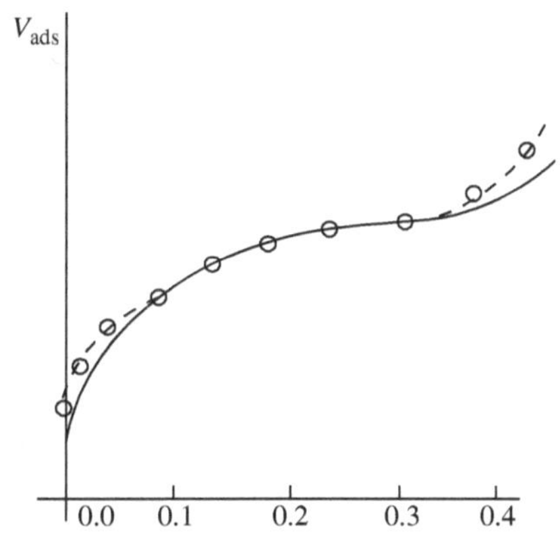

Figure 2.11
Typical results for the adsorption of nitrogen on a solid surface at 78K. The solid curve is drawn to fit the BET equation and shows a reasonable fit to the experimental points, especially at $0.1 < P/P^o < 0.3$.

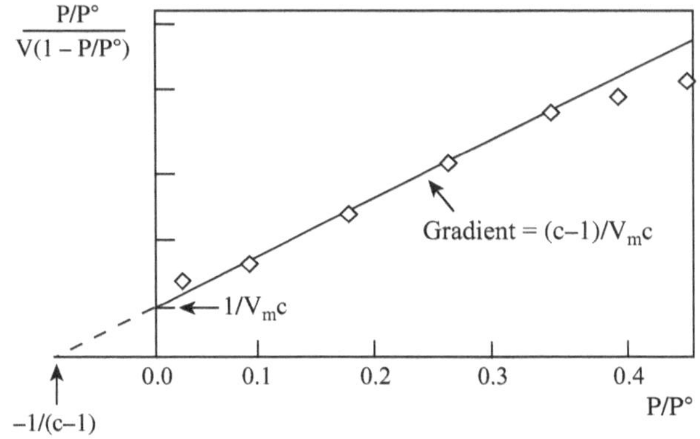

Figure 2.12
Linearized form of the BET equation showing how V_m can be obtained from the X and Y intercepts or from the gradient and Y intercept.

determine the value of V_m and hence the surface area of the adsorbent in question. This is because it allows an unequivocal and reproducible value to be determined for any solid, regardless of the values of ΔH_1 and ΔH_L. The student should beware of applying the equation blindly, for example, by using computerized best-fit methods; this is because the computer program may wrongly attempt to include points that deviate from the linear reciprocal plots (at lower or higher P/P^o values).

It should also be recognized that the method using nitrogen adsorption at 78K, the boiling point of liquid nitrogen ($P^o = 1$ bar, 1014 hPa), works best for relatively high surface areas. This is for the very practical reason that adsorption is most commonly carried out using volumetric or gravimetric methods and that the measurements are therefore of differences in pressure or weight; if the surface area is too low, the accuracy in determining a pressure or weight difference may be insufficient to give a meaningful surface area. This difficulty using a volumetric system can on occasion be overcome by using a different adsorbent such as Kr or even Xe at 78K as these atoms have much lower values of P^o: 2 and 4×10^{-3} hPa, respectively. The experiments therefore have to be carried out using equipment capable of measuring accurately pressures in the region of 0.2 and 4×10^{-4} hPa, respectively, for example, using a McLeod gauge (http://en.wikipedia.org/wiki/McLeod_gauge/); with such equipment, it is possible to measure much smaller uptakes than if nitrogen was used.[10]

Fig. 2.13 shows two typical plots obtained for the adsorption of Kr at 78K on an evaporated metallic film of iron, prepared using high vacuum methods, before (curve 1) and after

[10] M.W. Roberts, J.R.H. Ross, Trans. Faraday Soc. 62 (1966) 2301.

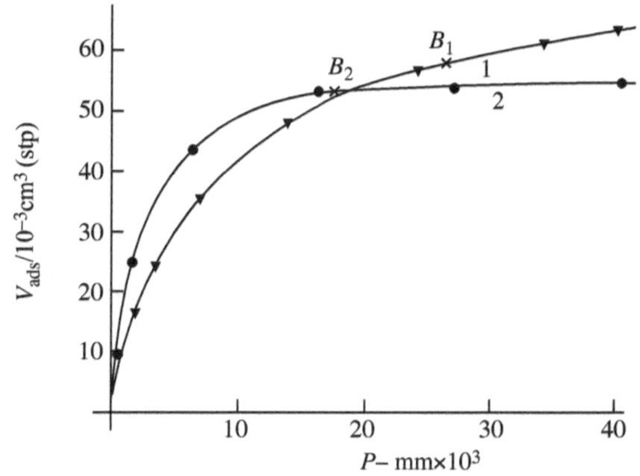

Figure 2.13
Isotherms for Kr adsorption at 78K for an evaporated metal film of iron (1) before and (2) after reaction with H_2S at a temperature of 306K. *Source: The adsorption of hydrogen sulphide and hydrocarbons by metal films, J.R.H. Ross, PhD thesis, QUB, 1966.*

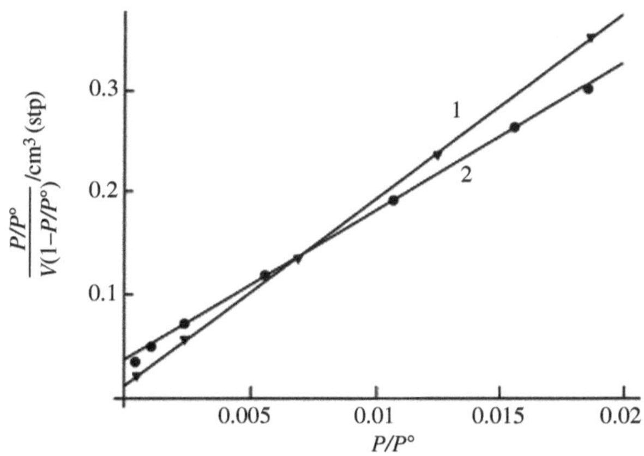

Figure 2.14
Corresponding BET plots for the data of Fig. 2.13.

(curve 2) exposure to H_2S at 306K. The corresponding BET plots are shown in Fig. 2.14, from which it can be calculated that the V_m values were 69.1 and 55.5 cm^3 (stp).[11]

While visual inspection of the curves of Fig. 2.13 allows one to estimate very approximate monolayer values corresponding to the values of "Point B" (B_1 and B_2 shown) of 58.0 and 53.5 cm^3 (stp), these values are lower than those obtained by the BET method, particularly for the fresh Fe sample for which the "knee" in the adsorption

[11] In this example, the amount of gas adsorbed is expressed in units of volume adsorbed; however, numbers of molecules or weights could also be used.

isotherm is more diffuse. The different shapes of the isotherms are related to different free energies of adsorption of Kr on the fresh and sulfided surfaces, these being related to very different values of the adsorption constant c (401 and 1511, respectively). It therefore appears that the Kr is more strongly adsorbed on the sulfided surface than on the uncovered metal surface.

It is generally accepted that the cross-sectional area occupied by a Kr atom under these circumstances is about 0.195 nm^2. Hence, it can be calculated that the areas of the film before and after reaction with H_2S were 362 and 290 cm^2, respectively. As the area of the surface of the pyrex vessel onto which the film had been evaporated was approximately 150 cm^2, the iron film had an area higher than the geometric value but the area was decreased by the adsorption of H_2S.

2.13 Behavior of Physical Adsorption Isotherms at Values of $P/P^o \geq 0.3$

Fig. 2.12 showed that there was significant deviation from the BET equation at values of P/P^o greater than about 0.3. This deviation is in part due to a failure of the BET equation to take into account changes in the adsorption behavior as the thickness of the adsorbed layer increases. However, a more important contribution to the observed deviation at higher values of P/P^o is the onset of so-called pore condensation. It is beyond the scope of this text to examine pore condensation in very great detail; readers interested in expanding their knowledge of the subject should refer to one of the many texts dealing with adsorption and pore condensation.

Every solid exhibiting high surface area consists of small particles, often in the nanometer size range, stacked on one another in either a random or an ordered way.[12] The result is that there will be gaps between the particles and these gaps tend to be relatively uniform in shape and size. These gaps are more often referred to as "pores." This term must not be taken to mean that these pores are always cylindrical or totally uniform in appearance: they can be slit-shaped; they can have the form of ink-bottles, with narrow openings and wider interiors; they can have the form of gaps between platelets; or they can be perfectly uniform as is the case of pores in zeolite crystals (see Fig. 2.15). However, for the present purposes, we will assume that all the pores are cylindrical.

When physical adsorption occurs on a solid with cylindrical pores, a monolayer initially forms on the external and the internal surfaces at relatively low values of P/P^o. As the relative pressure increases, multilayer formation begins to occur on all the surfaces. If we had only external surfaces that approximated to flat bodies, we would start to get condensation of a liquid only as P/P^o approached 1.0; in other words, liquefaction would have

[12] The preparation of typical catalyst materials is discussed in more detail in Chapter 4, Catalyst Preparation.

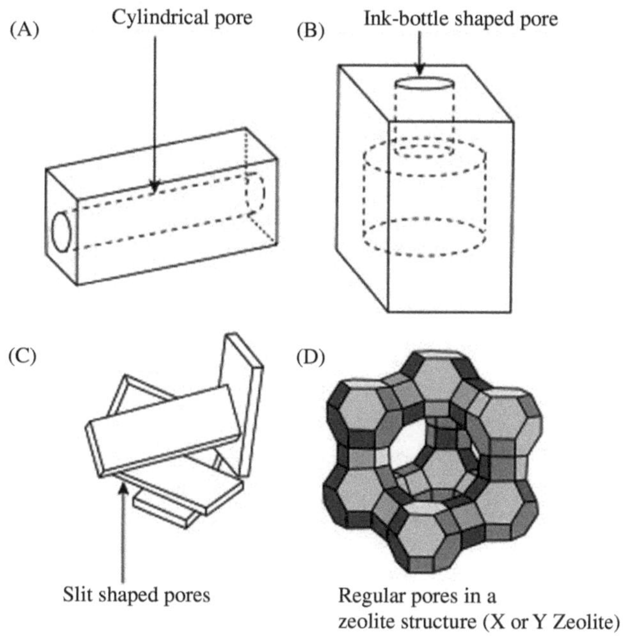

Figure 2.15
Some typical pore structures: (A) idealized cylindrical pore in a solid matrix; (B) an ink-bottle shaped pore in a solid matrix; (C) slit-shaped pores formed between platelets of a solid; and (D) regular pores in an X or Y zeolite structure.

occurred at a pressure equal to the saturation vapor pressure of the adsorbate. (Reminder: for nitrogen at 78K, this occurs at 1 atmosphere pressure.) Hence, at P/P^o approaching 1.0, the volume adsorbed rises rapidly and tends to ∞. However, if pores are present, they begin to fill (see Fig. 2.16) at values that are much below P^o as the pressure is increased, the corresponding value of P/P^o being determined by the pore radius or radii (see the following text). Once the pores have filled, multilayer adsorption occurs on the external surfaces of the solid as the pressure tends to P^o. If the pressure is then decreased, the isotherm follows the adsorption branch for a short range of P/P^o, this corresponding to the removal of the multilayers on the external surfaces. However, the isotherm then continues to a much lower level of P/P^o before desorption from the pores occurs. The "desorption branch" of the isotherm is generally parallel to the "adsorption branch" but this is not necessarily the case.

We will now give a brief explanation of how this adsorption/desorption behavior occurs. If the solid adsorbent contains uniform cylindrical pores of radius r (see Fig. 2.17), the effective saturation vapor pressure at a curved liquid surface within these pores is reduced, the new value P' being given by the Kelvin equation:

$$\ln P^o/P' = 2\gamma V.\cos\varphi/rRT \qquad (2.23)$$

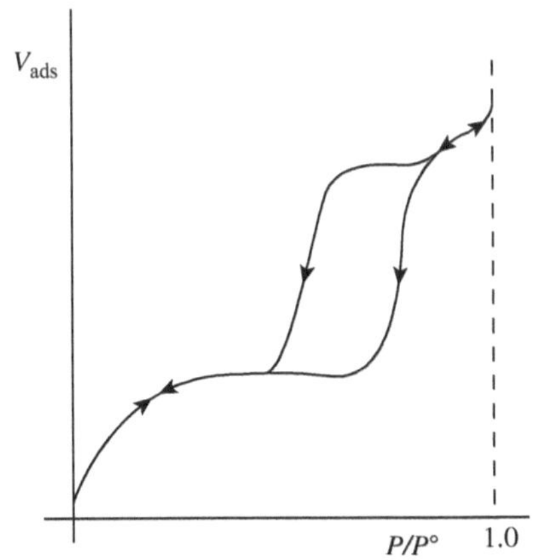

Figure 2.16
Schematic representation of an isotherm exhibiting hysteresis, showing filling and emptying of the pores.

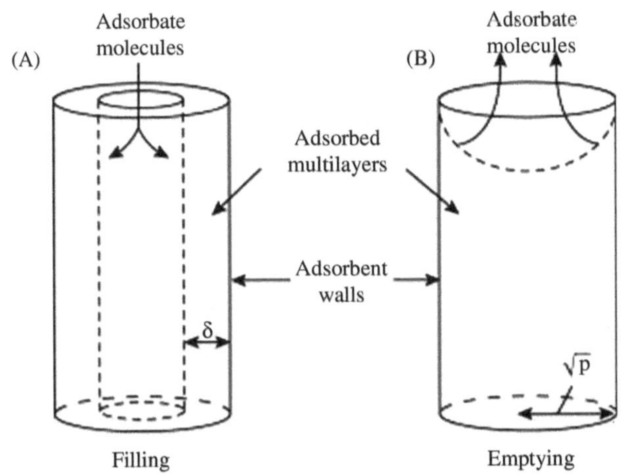

Figure 2.17
Schematic representation of the filling and emptying of cylindrical pores of radius r. The pores fill as shown in (A) by the build-up of molecules in layers within the pores until the latter are filled; however, as they empty (B), desorption occurs from a spherical interface that gradually moves inward within the pores.

where γ is the surface tension of the liquid [8.72 mN/m for N_2)], V is its molar volume (34.68 cm^3/mol for N_2), φ is the contact angle of the liquid with the gas (assumed to be zero for liquid N_2) and r is the radius of curvature of the liquid surface.

If there are two radii of curvature, r_1 and r_2, then the equation is:

$$\ln P^o/P' = \gamma V.\cos\varphi(1/r_1RT + 1/r_2RT) \tag{2.24}$$

When, as a result of an increase in the pressure, a linear cylindrical pore begins to fill due to condensation within the pore, this condensation occurs on a layer of the adsorbate that is already present (see Fig. 2.17A). The pore surface can be thought of as having two radii, one being (r−δ), where δ is the thickness of the layer within the pore, the other being ∞ as a result of the linear nature of the pore. Inserting these values in Eq. (2.24) gives:

$$\ln P^o/P' = \gamma V\cos\varphi/(r-\delta)RT \qquad (2.25)$$

If one then assumes that the angle of contact of the liquid in the pore is zero (complete wetting), cosφ is 1.0 and we obtain:

$$\ln P^o/P' = \gamma V/(r-\delta)RT \qquad (2.26)$$

If the cylindrical pores are entirely uniform, the volume adsorbed increases rapidly at a single value of P′ until the pores are completely filled; however, if there is a range of pore sizes, the volume adsorbed increases less sharply as a function of pressure as it is increased. The data can be used to determine the pore diameter (or the pore size distribution if the pores are not all the same) as long as the thickness δ is known; the value of δ can be calculated from the amount adsorbed at the value of P/P^o corresponding to the onset of capillary condensation.

Once the pores are completely filled, all that remains to be covered is the external surface and the isotherm may show a final corresponding slight increase in the volume adsorbed. This condensation on the exterior of the sample will continue until the vessel containing the solid is completely filled. The total volume adsorbed up until the onset of this last section of the isotherm is equivalent to the total volume of the pores in the solid, an important parameter in relation to catalysis. We will return to the subject of catalyst pores in Chapter 8 (Mass and Heat Transfer Limitations and Other Aspects of the Use of Large Scale Catalytic Reactors) in connection with diffusion limitations. (Task 2.7 allows the reader to gain further information on pore structure.).

Task 2.7 Surface Area and Pore Structure

For a good starting point to study the scientific literature on adsorption methods in order to determine the surface area and pore structure of a catalyst or other porous material, see a relatively recent article by K.S.W. Sing in Adv. Colloid Interface Sci. 76 (1998) 3. Using Scopus or another citation index, examine other more recent articles citing Sing's review and tabulate the most interesting of these.

If, following the filling of the pores, the system is then evacuated in a controlled way (in other words, the adsorbed gas is removed in known steps), a "desorption branch" to the isotherm may also be obtained (see Fig. 2.16). Desorption occurs from the pores in a different way to the way in which adsorption occurred, as discussed earlier: the adsorbate is now desorbed from a hemispherical liquid surface in each pore. The results are now be described by Eq. (2.20) since r_1 and r_2 are both equal to the actual pore radius, r, and there is no adsorbed material remaining on the pore wall (see Fig. 2.17B). As shown in Fig. 2.16, the "desorption branch" of the isotherm is significantly different to the adsorption branch. Values of r calculated from both branches should agree but that is not always the case.

A full discussion of pore filling and the shapes of pore systems and of more advanced treatments of adsorption data (for example, the BJH method[13]) is beyond the scope of this book. It is recommended that the interested reader should study the subject in more detail using one of the standard texts on the subject. One of the earliest such books, a classic, was "Adsorption, Surface Area and Porosity" by S.J. Gregg and K.S.W. Sing, (2nd ed.) Academic Press, NY and London, 1982. A more recent book is "Characterization of Porous Solids and Powders: Surface Area, Porosity and Density," by S. Lowell, J.E. Shields, M.A. Thomas, M. Thommes, Kluwer Academic Publishers, 2004. (See also Task 2.8.)

Task 2.8 Adsorption by Porous Solids

Adsorption by porous solids is frequently used in practice. One well-known example is the use of silica gel to remove water vapor from a container. A recent application of adsorption in well-structured solids is the use of so-called metal organic frameworks (MOFs) for the adsorption of CO_2, H_2, and other gases such as methane. Recent work has shown that it is possible to adsorb sufficiently high quantities of these gases to enable the use of them to extract CO_2 from the atmosphere or to store hydrogen or methane for transport purposes. Carry out a literature study for work on MOF structures for use as adsorbents using the following open-access article as a starting point for your work:

D. Alezi, Y. Belmabkhout, M. Suyetin, P.S. Bhatt, L.J. Weselinski, V. Solovyeva, K. Adil, I. Spanopoulos, P.N. Trikalitis, A.-H. Emwas, M. Eddaoudi, J. Am. Chem. Soc. 137 (2015) 13308.

---

[13] E.P. Barrett, L.G. Joyner, P.P. Halenda, The determination of pore volume and area distribution in porous substances. I. Computations from nitrogen isotherms, J. Am. Chem. Soc. 73 (1951) 373.

CHAPTER 3

How Does a Catalyst Work?

Chapter Outline

3.1 Introduction

In Chapter 1, An Introduction to Heterogeneous Catalysis and Its Development Through the Centuries—Chemistry in Two Dimensions, we discussed the phenomenon of catalysis and traced the history of the development and understanding of catalysis. Then, in Chapter 2, Surfaces and Adsorption, we examined the phenomena of adsorption, both chemical and physical, that are so important in the functioning of a catalyst. In this chapter, we will discuss in more detail the way in which a heterogeneous catalyst functions, outlining the importance of the application of both kinetic and thermodynamic principles to catalysis. We then go on to discuss in a general fashion a number of examples of the various types of heterogeneous catalysts used in practice.

Contemporary Catalysis.
DOI: https://doi.org/10.1016/B978-0-444-63474-0.00003-5

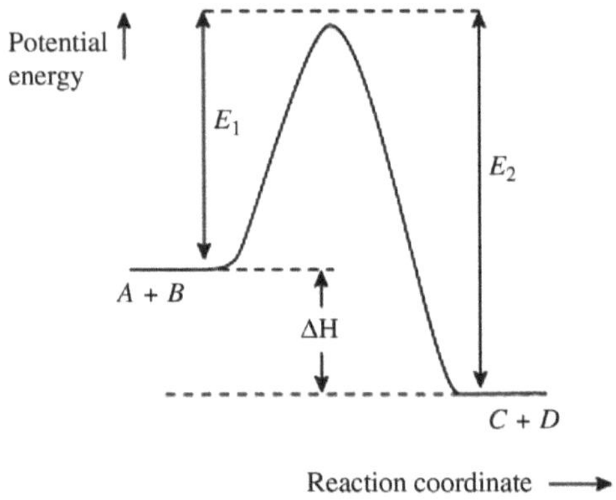

Figure 3.1
Energetics of the reaction: A + B→C + D

3.2 The Catalytic Process

3.2.1 Bimolecular Catalytic Processes

In Chapter 1, we considered the abstract gas-phase reaction:

$$\mathrm{A + B \rightarrow C + D} \tag{3.1}$$

As written, this equation represents an equilibrium situation in which the species A reacts reversibly with the species B to give the new species C and D. The equilibrium position for the reaction is determined by thermodynamics and, once equilibrium is achieved, the rates of the forward and reverse reactions are equal. The energetics of the process are shown schematically in Fig. 3.1.

In this figure, the reaction is shown as being exothermic (ΔH −ve) but it could equally well have been shown as being endothermic. In order to transform into the products C and D, the reactants A and B must come together to form an activated (or transition state) complex $\text{A-B}^{\ddagger}$ which then decomposes to form the products.[1] The activation energy for the formation of the complex is E_1 and this has a numerical value well above the average thermal energy of the molecules A or B. (As the reaction occurs in the gas phase, the acquisition of the activation energy is most easily visualized by considering that the

[1] The formation of transition state complexes is handled in standard textbooks covering reaction kinetics. The subject has also been extended to consider reactions at surfaces (see, e.g., S. Glasstone, K.J. Laidler, and H. Eyring, The Theory of Rate Processes, McGraw Hill, New York, 1941) but this topic will not be considered further in this text.

molecules pick up the energy by a series of gas-phase collisions.[2]) Once C and D have formed, they can transfer back to A and B by the reverse process, the activation energy for that change being E_2. When the rate of the reverse reaction r_2 becomes equal to that of the forward reaction r_1, the all-over reaction has reached a dynamic equilibrium. If the forward and reverse reactions are both elementary processes,

$$r_1 = k_1[A]\cdot[B] \text{ and } r_2 = k_2[C][D] \tag{3.2}$$

from which we can derive that:

$$r_1/r_2 = k_1/k_2[C][D]/[A][B] = K_{eq} \tag{3.3}$$

Task 3.1 Enthalpy Change for Reaction 3.1

Show that $E_1 - E_2 = \Delta H$. Remember that $k_1 = A\ \exp - E_1/RT$. (This is the Arrhenius equation in which A is the Arrhenius constant and E_1 is the activation energy for the forward reaction.)

We now consider the situation in which A and B are reacted in the presence of a solid surface that will adsorb both A and B. As shown in Eq. (3.4), the adsorbed species A_{ads} and B_{ads} may then transform on the surface to give C_{ads} and D_{ads}. These can in turn desorb to give the gaseous products C and D:

$$A + B \xrightarrow{1} A_{ads} + B_{ads} \xrightarrow{2} (AB)^{\neq}_{ads} \xrightarrow{3} C_{ads} + D_{ads} \xrightarrow{4} C + D \tag{3.4}$$

Reaction (3.4) will occur on the surface in preference to the gas-phase bimolecular process as long as the rates of the constituent steps [adsorption (1), surface reaction (2 and 3), and desorption (4)] are all more rapid than the allover rate of the gas-phase process. The situation can be represented schematically as in Fig. 3.2; in this, as a simplification, single energy troughs are shown for A_{ads} and B_{ads} and also for C_{ads} and D_{ads} and a single step is shown for the inter-conversion to the adsorbed products.[3]

[2] Collision theory explains the acquisition of the activation energy required for the gas-phase process by considering that it occurs by random gas-phase collisions between A and B (and also collisions with other atoms or molecules, including the products). The excess energy acquired by the products is then dissipated by further gas-phase collisions with all the molecules present in the system.

[3] This representation is unrealistic since there will inevitably be different heats of adsorption for A and B as well as for C and D. It is also probable that a number of steps will be involved in the surface transition rather than the single step shown. Further, if the adsorption of each of A and/or B involves an activation energy, these values will be different. The same will be the case for the desorption process. The diagram should therefore be seen only as a means of gaining a picture of how the catalyzed reaction occurs.

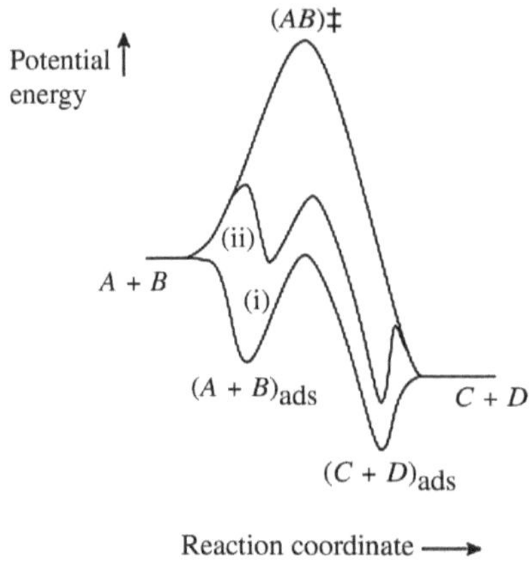

Figure 3.2
Schematic representation of the energetics of the catalyzed reaction: A + B→C + D

Two curves are shown in Fig. 3.2 for possible surface reactions. Curve (i) represents the situation in which both A and B adsorb without any activation energy for the adsorption step while curve (ii) represents a situation in which there is an activation energy for the adsorption of A and B. Once A and B are adsorbed, they can transfer, via an adsorbed complex $(AB)^{\neq}_{ads}$, to give adsorbed C and D and these can then desorb to give products.[4] The most important message to get from a very schematic diagram of this type is that the catalyst surface provides an alternative route for the reaction and that this route is energetically much more favored than is the gas-phase reaction.[5] In the schematic representation shown in Fig. 3.2, the surface reaction is rate determining and the apparent activation energy for the all-over reaction is associated with the barrier for the surface reaction. The second message to gain from the diagram and discussion is that the catalyzed reaction may have any number of steps, any one of which (or even a combination of them) might be rate determining.[6]

4 Many other possible sets of curves could be imagined. For example, the adsorption of A could have a high activation energy while the adsorption of B might not be activated. Alternatively, desorption of C or D might have a high activation energy. Furthermore, as A and B may adsorb preferentially on two different types of site that are some distance apart, there might also be a step involving activated surface diffusion of A and/or B over the catalyst surface to allow them to reach that position on the surface at which reaction can occur.

5 Compare this situation with the water analogy discussed in Chapter 1, An Introduction to Heterogeneous Catalysis—Chemistry in Two Dimensions.

6 In order to distinguish between the multiple feasible reaction sequences for a catalytic reaction, one of the most well-established methods is to carry out careful kinetic measurements of the all-over process (discussed in more detail in Chapter 6, Catalytic Reactors and the Measurement of Catalytic Behavior). However, many other methods have been used to distinguish possible surface reaction mechanisms. Many of these will be featured in subsequent chapters.

BOX 3.1 Hydrogen Exchange Reactions

The early developments of the theories of bimolecular reactions on catalyst surfaces were based on the reactions of hydrogen species: ortho (*o*-) to para (*p*-) hydrogen exchange or H_2/D_2 exchange. Ortho hydrogen is a molecule of hydrogen in which the nuclear spins are parallel (i.e., in the same direction) while para hydrogen has spins that are antiparallel. The para modification is favored at low temperatures (c.80K) and the transfer to this form may be catalyzed by graphite. Its transfer back to the ortho modification has been examined over a wide range of different catalytic surfaces, for example evaporated films of transition metals. It has been shown that the reaction occurs in a sequence involving dissociative adsorption of the hydrogen on the surface followed by recombination of hydrogen atoms in a desorption step:

$$p\text{-}H_2 \rightarrow 2H_{ads} \rightarrow o\text{-}H_2$$

A similar mechanism applies to the hydrogen–deuterium exchange reaction:

$$H_2 + D_2 \rightarrow 2H_{ads} + 2D_{ads} \rightarrow 2HD$$

While the *o*- to *p*-H_2 exchange reaction requires the use of experimentally difficult thermal conductivity measurements to distinguish the two forms, it is relatively simple to study the hydrogen–deuterium reaction using mass spectrometry. Examination of these two reactions led to the development of two well-established mechanisms of surface reaction: the Langmuir–Hinshelwood and Eley–Rideal mechanisms. These are discussed in Chapter 7, The Kinetics and Mechanisms of Catalytic Reactions.

3.2.2 *Unimolecular Catalytic Processes*

In the same way that a bimolecular process may be catalyzed by a surface, a unimolecular decomposition reaction may also be catalyzed. Fig. 3.3 shows the energetics of a simple model reaction, the decomposition of molecule A to give products.

The upper curve in Fig. 3.3 represents the uncatalyzed reaction.[7] As with the bimolecular reaction discussed earlier (Figs. 3.1 and 3.2), the reaction is taken to be exothermic but the arguments apply equally to an endothermic reaction. For the reaction in the gas phase, the activation energy for the forward reaction is shown as E_1 and that for the reverse reaction is

[7] For a gas-phase unimolecular reaction, Lindemann developed a modified collision theory to explain how the necessary activation energy for the reaction is obtained. This is based on collisions between the reactant A and other molecules present in the gas phase: other molecules of A, product molecules, P, or additional inert molecules, M. In these collisions, A acquires energy by collisions with M, A, or P, with the result that the activated molecule $A^{\neq}$ is formed; $A^{\neq}$ and the product molecules then loose their excess energy by transfer to M or to the other molecules (or even the surface of the reaction vessel). However, for a surface reaction, the transfer of energy will be mostly from and to the surface on which reaction occurs.

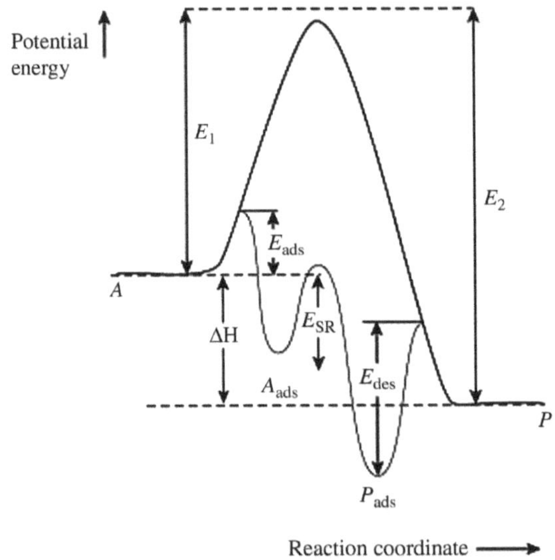

Figure 3.3
Schematic diagram of the energy profile for a reaction A → B without and with a catalyst.

shown as E_2. The enthalpy of the reaction ΔH is therefore $E_1 - E_2$; in the figure as drawn, E_2 is greater than E_1 and so the value of ΔH is negative.

When the reactant A can adsorb on a surface, a new species, A_{ads}, is formed, this having a heat of adsorption, $\Delta H_{A(ads)}$. Depending on the system, there may or may not be an activation energy E_{ads} for this adsorption step. Once the species A_{ads} is formed, it may be transformed into P_{ads} by a surface reaction process with an activation energy E_{SR}. The product species P is shown as being adsorbed with an adsorption energy of $\Delta H_{P(ads)}$. Finally, P is desorbed with an activation energy E_{des}. The initial and final states for the uncatalyzed and catalyzed reactions are exactly the same: all that the catalyst has done is to change the allover energetics, providing an alternative route from reactant to product. This is a fundamental tenet of catalysis: *A catalyst can bring about a change in the rate of a chemical reaction but has no effect on the allover thermodynamics of the reaction, that is, the equilibrium position.*

3.2.3 Reversibility in Catalytic Processes

An important tenet of catalysis is illustrated by examination of Fig. 3.3: that the catalyst for the forward reaction will also, under suitable conditions, catalyze the reverse reaction and that the step or steps which dictate the rate for the forward reaction will also dictate the rate

BOX 3.2 Decomposition of Formic Acid

An example of a unimolecular catalytic reaction is the decomposition of formic acid[a]:

$$HCOOH \rightarrow H_2 + CO_2$$

This is a slightly exothermic reaction, with a value of $\Delta H^\circ = -32$ kJ mol^{-1}. It can be catalyzed by a wide variety of different metals. As there are two separate products, the representation shown in Fig. 3.3 (as for the bimolecular reaction discussed earlier) is not strictly correct since the desorption paths for the two products will be different. There is a further complication: alternative products, CO and H_2O, can be formed, depending on the catalyst used and the conditions of the reaction, and so the "selectivity" of the catalyst (see Section 3.2.4) can become important. In many cases of "kinetic selectivity" such as this, the products of the reaction depend on the ease of formation of the different intermediate surface species and the ease with which they can transfer into other species.

[a]Strictly speaking, this is a reversible reaction and it should therefore in principle be possible (see Section 3.2.3 and Task 3.2) to make formic acid by the reaction of hydrogen with CO_2; although such a reaction would be very attractive as a method of sequestering CO_2, it is not thermodynamically feasible to produce more than an insignificant quantity of formic acid in this way. In practice, routes for the synthesis of formic acid all rely on the production of another more thermodynamically favored product that is then converted to the acid in a second step. (See also Task 3.2.) The decomposition of formic acid to produce hydrogen is discussed in more detail in Chapter 15.

Task 3.2 The Catalytic Synthesis of Formic Acid

Do a web search on formic acid and examine the various indirect routes used for its production. One of these routes requires first the formation of methyl formate and then the conversion of the methyl formate to formic acid. What are the steps involved and what are the catalysts that can be used for each? See what you can find out about the thermodynamics of each step.

for the reverse reaction. (Similar arguments apply also to the bimolecular reactions discussed in Section 3.2.1.) For example, if it is found that the surface reaction:

$$A_{ads} \rightarrow P_{ads} \tag{3.5}$$

is the rate determining step (rds) for the forward reaction, then the reverse surface reaction:

$$P_{ads} \rightarrow A_{ads} \tag{3.6}$$

is the rds for the reaction in the other direction.

3.2.4 Selectivity in Catalytic Processes

Another essential aspect of catalysis can be illustrated by an extension of the example shown in Fig. 3.3. If A can react to give two different products, P and Q, the presence of the catalyst can have a very significant effect on the "selectivity" of the reaction as shown in Fig. 3.4.

Consider first the uncatalyzed reaction. A can now give the two products, P and Q, with rates determined by the activation energies E_1^P and E_1^Q; in other words, as long as there are no significant differences in the Arrhenius constants for the two reactions, P will be formed more rapidly than Q and therefore P will be the product that is formed selectively. [In Fig. 3.4, the lower activation energy is associated with the reaction to give the product (P) for which the enthalpy of formation is more negative, but that is not necessarily the case.] If a catalyst is now introduced and A adsorbs to give A_{ads} and this species can, in turn, decompose to give either P or Q, the relative values of the activation energies for the two processes may be reversed. The molecule Q in the example shown in Fig. 3.4 will now be the selective product even though P may be thermodynamically favored, again as long as the pre-exponential factors do not differ greatly.

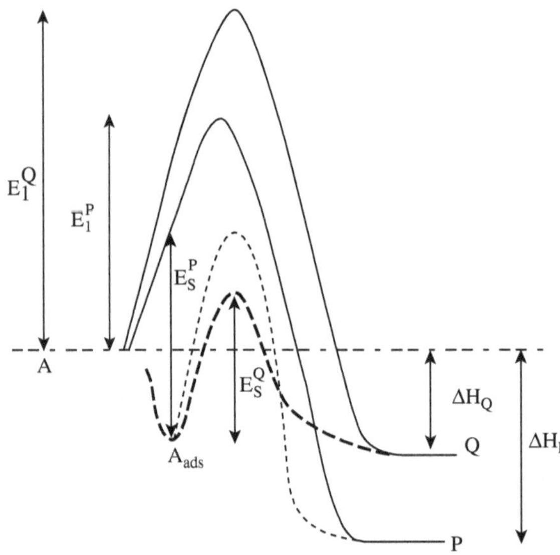

Figure 3.4
Schematic representation of the energetics of two parallel reactions.

$$A \rightarrow P \text{ and } A \rightarrow Q$$

showing that a catalyst can change the selectivity of a reaction

It is the ability of a catalyst to affect the selectivity of a reaction, often producing products that are not found in the equivalent noncatalytic systems, that makes catalysis so important in practice. One such application is the selective oxidation of hydrocarbon molecules. Noncatalytic oxidation of hydrocarbons (i.e., combustion) typically produces only the two most thermodynamically favored products, CO_2 and H_2O; however, selective catalytic systems are known which can produce either molecules such as the corresponding aldehydes or ketones (by selective insertion of oxygen in the molecule being oxidized) or the corresponding olefins (by the selective removal of hydrogen) in so-called "oxidative dehydrogenation" (ODH) reactions. We will return to the subject of catalytic oxidation in much more detail in Chapter 14.

3.2.5 Kinetics and Mechanism

In outlining the ways in which adsorbed molecules react on a catalyst surface to form products, we have concentrated on the energetics of the allover catalytic reaction and on how the presence of the surface affects the reaction, sometimes even changing the product formed. Energetics and kinetics are inevitably closely linked and we have already mentioned the importance of the activation energy barriers for the various steps in a catalytic process. However, we will reserve a full discussion of the most commonly encountered types of kinetic expressions for catalytic processes to Chapter 7 and 8 where we discuss topics such as Langmuir–Hinshelwood, Eley–Rideal, and Mars–van Krevelen mechanisms and will also deal with the importance of diffusion rates in determining the rates of catalytic reactions. Before doing that, we will discuss in this chapter the different types of catalyst in common use, giving some examples of some practical catalysts.

3.3 The Catalyst and the Catalytic Site

So far, apart from mentioning the allover compositions of the catalysts used in various processes discussed from an historical perspective in Chapter 1 and the phenomenon of adsorption in Chapter 2, we have said little about the nature of either the catalyst or the catalytic site. We will now give a brief general description of the most common types of catalyst currently in use, with particular attention being given to the nature of the surfaces of these materials. We will return in detail to the topics of catalyst preparation and characterization in Chapters 4, Catalyst Preparation, and Chapter 5, Catalyst Characterization.

There are many different types of catalyst currently in use in a wide variety of chemical processes. The most commonly encountered catalyst types are materials containing metals, oxides, or sulfides. However, although each of these may exist as single component materials, the active species is often accompanied by other components that are often classified as either "supports" or "promoters."[8] Further, a catalyst may have several different active components; for example, one of these may be a metal and the other an oxide. It is important to recognize that the active form of a catalyst can be very dependent on the reaction mixture to which it is exposed and also to the way in which it is activated: whether the activation and reaction conditions are reducing or oxidizing or even sulfiding. (See also Box 3.3.) We will now discuss briefly some examples of active metals, oxides, and sulfides and their combinations.

BOX 3.3 Oxidizing Versus Reducing Conditions

Let us consider as an example the complete oxidation of a simple molecule such as methane:

$$CH_4 + O_2 \rightarrow CO_2 + 2\ H_2O$$

This reaction can be catalyzed by most metals. However, whether or not the operating catalyst will be actually in the metallic or oxidic form will depend on the thermodynamics of the oxidation of the metal, M:

$$M + 1/2\ O_2 \rightleftarrows MO$$

If this reaction is considered as an equilibrium as shown, the equilibrium constant, Kp is given by the equation:

$$K_p = a_{MO}/a_M \cdot P^{e^{1/2}}_{O_2}$$

where a_M and a_{MO} are the activities of M and MO, respectively, and $P^e{}_{O2}$ is the "equilibrium pressure" of O_2.

Since the activities of both solids are unity, the expression simplifies to:

$$K_p = P^{e^{-1/2}}_{O_2}$$

Hence, the value of the equilibrium pressure $P^e_{O_2}$ determines whether or not a metal will be oxidized in the presence of a partial pressure of oxygen (assuming of course that there is sufficient oxygen available at that pressure to oxidize the metal completely). Above that pressure, complete oxidation occurs; below that pressure, the metal will be the thermodynamically favored state. In practice, this means that noble metals such as Pt, Ir or Au will remain in their metallic states while operating under the conditions used for methane oxidation. However, Pd lies on the thermodynamic borderline and there is strong evidence that the active phase for methane oxidation is the oxide.

[8] Generally, the support of a catalyst is thought of as being inert, just supplying a high area on which the active component is located, common examples being alumina or silica. A promoter, on the other hand, is a component that is added to the catalyst to improve the activity or selectivity of the active component. An example of a promoter that has already been mentioned (Chapter 2, Surfaces and Adsorption) is the K present in the iron ammonia synthesis catalyst. It should be noted however that even apparently inert supports can under certain circumstances contribute to catalytic behavior and it is now recognized that there is often migration of reactant species between the surface of the active component and that of the support; whether or not such migration contributes to the all-over behavior of the catalyst depends strongly on the relative rates of the various surface processes that occur.

If a metal M is to be used for the steam reforming of methane:

$$CH_4 + H_2O \rightarrow CO + 3H_2 \tag{3.7}$$

the state of oxidation of the metal will depend on the equilibrium:

$$M + H_2O \rightleftarrows MO + H_2 \tag{3.8}$$

for which the (temperature dependent) equilibrium constant *Kp* is given by:

$$Kp = P_{H_2}/P_{H_2O} \tag{3.9}$$

The most commonly used metal for steam reforming is nickel (Chapter 11) for which the thermodynamics are such that the metallic state is favored at most temperatures as long as the partial pressure of hydrogen in the mixture is above about 1% of the total mixture. In practice (see later), the catalyst will be prereduced in a hydrogen-rich gas and it can be maintained in its reduced state during the steam reforming reaction by the addition of small quantities of hydrogen to the reactor feed (or by recirculation of some of the product of the reaction).

Task 3.3 Metal or Oxide Under Oxidizing Conditions Used for VOC Control?

Trace volatile organic species (VOCs) can be removed from gaseous mixtures using catalytic oxidation and catalysts for this purpose may contain either metals or oxides as the active phases. Search the web for suitable information on types of catalysts used for VOC control and the conditions used for reaction. Determine whether noble metals such as Pt and Pd will maintain their metallic state under typical VOC control conditions. Show that a metal such as Ni or Mo would be oxidized under these conditions.

The same types of argument apply to many other situations in which changes of the catalyst surface are possible under reaction conditions. For example, the gradual "sulfidation" of a metal oxide may occur under conditions where H_2S is present in the reaction mixture:

$$MO + H_2S \rightarrow MS + H_2O \tag{3.10}$$

the rate at which that sulfidation occurs depending on the reaction conditions (temperature and partial pressure of H_2S). While such sulfidation is sometimes desirable (e.g., to promote the reaction), irreversible poisoning of a metal or metal oxide by trace quantities of sulfur (originating from H_2S or from some other sulfur-containing molecule present in the feed) can also occur; the extent to which this occurs will now be determined by the total quantity of the poison present in the reaction system over time rather than on the kinetics of the poisoning reaction. We will return to the topic of catalyst poisoning at a later stage.

While the use of bulk thermodynamic quantities can frequently give adequate information about the possible formation of new phases, it has to be recognized that the quantities involved in the thermodynamics of surface and bulk processes are not necessarily the same: they will differ due to the fact that a metal atom in the surface of a catalyst will not have exactly the same environment as that in the bulk of the same metal.

3.4 Catalysis by Metals

3.4.1 Introductory Remarks

Table 3.1 lists some examples of metallic catalysts used in commercial processes. Although nearly all metals in the periodic table can be used as catalysts, those most commonly encountered in common catalyst formulations are the transition metals as these tend to be more reactive as a result of their unfilled d-bands.

While historically much attention was given to attempts to correlate the nature of the unfilled d-bands to the catalytic properties of metals, much more attention is now given to the nature of the chemical bonding between the metal and the reactant species (two-dimensional chemistry). Similar remarks apply to catalysis by oxides, sulphides, etc.

A number of catalytic processes can be carried out over pure metals. However, as discussed earlier, metals are most commonly "supported" on oxides or other high-area materials; furthermore, more and more examples are being reported in which the oxide is not just an inert "support" but instead contributes to the all-over activity and selectivity of the catalyst.

Table 3.1: Some catalysts used in commercial processes

Process	Catalyst	Comments
Ammonia synthesis	Promoted iron	Contains K and other promoters
Ammonia oxidation	Pt/Rh gauze	Gauze becomes severely pitted during use
Fischer–Tropsch synthesis	Fe or Co on support	Support originally kieselguhr; includes various promoters such as K
Steam reforming of methane	Ni on support (typically Al_2O_3)	Variety of supports and promoters used
Methanol synthesis	$Cu/ZnO/Al_2O_3$	
Methanol oxidation	Unsupported silver FeMo oxide	Ag uses excess methanol; Fe/Mo uses excess air
Reforming of hydrocarbons	Pt/Re/support	
Automobile exhaust treatment	$Pt/Pd/CeO_2/Al_2O_3$	Prepared by wash-coating monolith support
Selective reduction of NO_x in flue-gas	V_2O_5/TiO_2/matrix support	Reducing agent commonly ammonia or urea
Ethylene oxidation	$Ag/\alpha\text{-}Al_2O_3$	
Selective oxidation and ammoxidation of olefins	Bi molybdate	
Fat hardening	Ni/Al_2O_3	Prereduced catalyst stored in vegetable oil

We will encounter in later chapters many examples of the importance of the support used. In most of the examples to be discussed, the metal surface area is the most important parameter since the rate of any catalytic reaction is under most circumstances proportional to the number of metal atoms exposed. However, there are exceptions; for example, the catalytic properties of gold catalysts are very dependent on the way in which the catalysts are made and are only active if gold nanoparticles are present on the support. (See also Box 3.4 and Task 3.4.)

BOX 3.4 Catalysis by Gold

Although it was shown by Trapnell (see Table 2.2) that gold would adsorb molecules such as O_2, CO, C_2H_2, and C_2H_4, gold has long been considered to be relatively inactive as a catalyst. However, two groups of researchers led by Haruta[a] and Hutchings[b] have relatively recently shown that carefully prepared gold nanoparticles supported on a wide variety of supports (particularly the oxides of the first row of the transition series such as Fe_2O_3) are active for various reactions including CO oxidation. An early application of these catalysts pioneered by Haruta and his colleagues was the use of such catalysts for the destruction of the odors from lavatories. The field of catalysis by gold has now become a hot topic and a series of conferences discussing the topic have been instituted.[c] A book by Bond, Louis and Thompson gives a very good overview of the subject.[d] These authors point out that although small gold particles can be made on almost any type of support, the conditions of preparation and storage are critical to avoid sintering of the gold particles. The topic of catalysis by gold receives further attention in Chapter 14.

[a]M. Haruta, Catal. Today 36 (1997) 393; M. Haruta, Gold Bulletin, 37 (2004) 27.

[b]B. Nkosi, N.J. Colville, G.J. Hutchings, J. Chem. Soc., Chem. Commun. (1988) 71; G.J. Hutchings, Gold Bulletin, 29 (1996) 123.

[c]The World Gold Council has sponsored a number of meetings on the applications of gold, these having had a strong bias toward its catalytic applications; see, for example, Catal. Today, 122, Issue 3/4.

[d] G.C. Bond, C. Louis, D.T. Thompson, Catalysis by Gold, Catalytic Science Series, vol. 6, Imperial College Press, 2006.

Task 3.4 Selectivity in Gold Catalysis

Using the references in Box 3.4 as starting point, examine the many reactions that have been shown to have high selectivity over gold catalysts. List these reactions and give details of the preparation of the catalysts.[a]

[a] The second half of this task precedes the content of Chapter 4, Catalyst Preparation, and it would therefore be advisable if the reader without background knowledge was to defer this task until the contents of Chapter 4, Catalyst Preparation, have been studied.

Many metal-catalyzed reactions are ones that are carried out in their entirety under reducing conditions, the best known of these being processes such as ammonia synthesis or paraffin reforming (Table 3.1) in which neither oxygen or sulfur is a component of the reaction mixtures. Under these conditions, all the metals will be thermodynamically stable and many of them are potential candidates as catalysts. However, other parameters are also important: will the metal be susceptible to sintering as a result of the diffusion of surface metal atoms under reaction conditions (low Tamman temperature); will the metal be susceptible to other reactions such as carbon deposition; will it catalyze undesired side-reactions? If the reaction mixture contains oxygen or sulfur, then the situation is different and the metal must be chosen from a more limited group that is not susceptible to oxidation or sulfidation under the reaction conditions to be used. (See also Box 3.3.)

3.4.2 Unsupported Metal Catalysts

Most metallic catalysts consist of metal particles supported on a high-area oxide material in order to provide as high a metallic surface area as possible. However, there are several examples of the use of unsupported metals. One well-known example is the use of finely woven Pt/Rh gauzes for the oxidation of ammonia:

$$4NH_3 + 7O_2 \rightarrow 4NO_2 + 6H_2O \tag{3.11}$$

This reaction occurs in two steps:

$$4NH_3 + 5O_2 \rightarrow 4NO + 6H_2O \tag{3.12}$$

followed by:

$$2NO + O_2 \rightarrow 2\,NO_2 \tag{3.13}$$

The NO_2 is then absorbed by water to form nitric acid:

$$3NO_2 + H_2O \rightarrow 2HNO_3 + NO \tag{3.14}$$

The gauzes used gradually change in form during use as result of surface roughening until they break and have to be replaced.[9] The reason for this gradual change is not fully known but its progress can be slowed by inclusion of about 10% Rh in the gauze material.[10] The ammonia oxidation reaction is carried out with very short contact times (less than 10^{-3} s) at temperatures above 1000 K and the rate of reaction is controlled by the rate of mass-transfer of the reactants to the surface of the gauze; as the allover reaction is very

[9] See, for example, F. Sperner, W. Hohmann, Platinum Metals Rev. 20 (1976) 12; free access to this useful in-house journal is available at http://www.platinummetalsreview.com/dynamic/advanced search.

[10] Some unreported experiments carried out in our laboratory in about 1968 showed that small amounts of a volatile Pt oxide (probably PtO) could be determined mass-spectrometrically in the gas phase directly above a Pt filament heated in a low pressure of oxygen at a temperature about 1273 K.

exothermic, the process is self-sustaining once it has started. Pt gauze materials have also been used for the catalytic removal of volatile organic compounds (VOCs) from gas mixtures by means of total oxidation. (See also Task 3.3.)

Another example of the use of an unsupported metal is that of unsupported silver (electrochemically produced granules arranged in a thin bed about 1 cm in depth) for the selective oxidation of methanol to give formaldehyde:

$$CH_3OH + 1/2O_2 \rightarrow HCHO + H_2O \tag{3.15}$$

About half of the formaldehyde produced worldwide comes from the silver process.[11] This reaction, first observed by August Wilhelm von Hofmann, uses an excess of methanol and hence some dehydrogenation also occurs in parallel:

$$CH_3OH \rightarrow HCHO + H_2 \tag{3.16}$$

The other process used commercially for formaldehyde production employs excess oxygen with Fe/Mo/O or Fe/V/O catalysts (FORMOX process); http://en.wikipedia.org/wiki/Formaldehyde.

As unsupported metallic catalysts are often susceptible to sintering due to relatively high temperatures of use, the metal particles have to be stabilized in order to give as high a surface area as possible. As sintering often occurs by a surface transport mechanism, this stabilization may be achieved by adding a "promoter" that helps to anchor the metal atoms on the surface and prevent surface migration and/or particle coalescence. In the case of the platinum gauzes used for ammonia oxidation referred earlier, the addition of Rh slows down surface rearrangement. Another example of stabilization of metal particles is found in the case of Raney nickel (see Box 3.5) the skeletal structure of which is stabilized by small quantities of alumina on the surface of the Ni lattice. (See also Task 3.5.)

3.4.3 Supported Metal Catalysts

In many of the examples of unsupported metal catalysts discussed earlier, the metallic surface area is relatively unimportant as the rate of the reaction for which these metals is used is frequently dictated by the external metal area, the rate of reaction being related to the rate of supply of reactant to the external surface of the metal (e.g., in ammonia oxidation, the rate of supply of the ammonia to the Pt surface by gaseous diffusion is rate determining; see Chapter 8). For reactions in which the rate is determined by the rate of surface reaction, the total metallic area of the catalyst (i.e., both the external area and that within the pores of the support) becomes the most important factor. Hence, it is important

[11] Although Pt and Ag are very expensive metals, their use in reactions such as these can fully be justified because the metal can easily be recovered and recycled.

BOX 3.5 Stabilization of Ni Particles Used in Fat Hydrogenation—Raney Nickel

At the beginning of the 20th century, unsupported Ni powders were used for the hydrogenation of vegetable oils and these suffered from serious sintering problems. In 1926, the American scientist and engineer Murray Raney discovered that it was possible to produce a more stable Ni-based catalyst from Ni/Al alloys by leaching out the majority of the aluminum using concentrated sodium hydroxide, this leaving a "skeletal" Ni structure which has a high activity for the hydrogenation reaction. According to a Wikipedia article discussing Raney nickel, the activated material contains both Ni metal and traces of an Ni/Al alloy that helps to stabilize the Ni particles. (An alternative explanation, mentioned later, is that some alumina particles are formed during the leaching process and that these help to stabilize the Ni.) These Raney Ni catalysts (and Cu, Co, and Ru equivalents have also been made) are not only used for fat hydrogenation but also for hydrogenation of nitro-compounds and similar reactions. Raney Ni has also been used commercially for the production of bipyridyl, a precursor used in the production of pesticides, by the combination of two molecules of pyridine:

$$2\ C_5H_5N \rightarrow (C_5H_5N)_2$$

However, it has been shown that there is no particular advantage of using Raney Ni for this reaction as better conversions and catalyst stabilities can be obtained using prereduced coprecipitated Ni/Al catalysts.

The particular properties of Raney nickel which make it superior to unsupported finely divided Ni in reactions such as fat hydrogenation have been associated with the "skeletal" structure formed during the leaching process. Whether this skeletal structure is related to the existence of a residual phase of unleached Ni/Al alloy or with an amorphous alumina phase formed in the leaching process is not clear; in the latter case, the alumina particles would sit between the larger Ni crystallites and prevent them from sintering. A similar model has been put forward for the stabilization of the reduced coprecipitated catalysts referred to earlier.[a]

[a]L.E. Alzamora, J.R.H. Ross, E.C. Kruissink, and L.L. van Reijen, Coprecipitated nickel—alumina catalysts for methanation at high temperatures, Part 2, Variation of total and metallic areas as a function of sample composition and method of pretreatment, J. Chem. Soc., Faraday Trans. 177 (1981) 665—681.

Task 3.5 Preparation and Uses of Raney Nickel

Carry out a web-based search on Raney nickel to obtain more detail (e.g., from Wikipedia) on how it is prepared and used. What can you find on the structure of such materials? Can you find any information on the production of bipyridyl using Raney nickel?

to make and stabilize the highest possible and easily accessible metal surface area. Although finely divided metallic powders can be made with high surface areas, these are very unstable under reaction conditions. Supported metal catalysts are therefore used.

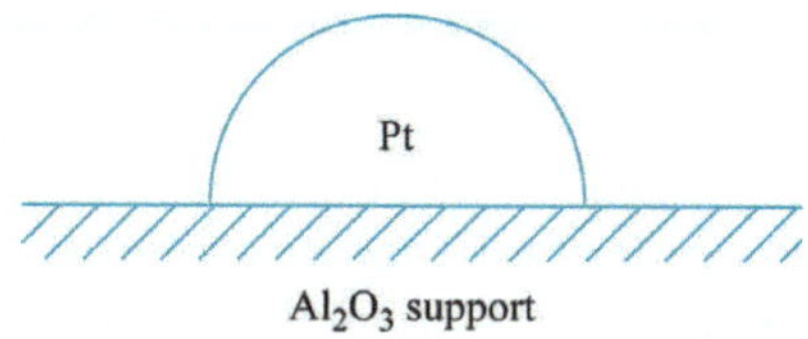

Figure 3.5
Schematic representation of a supported catalyst.

As the name implies, a supported metal is one in which the metal crystallites exist on the surface of a support. Methods have been developed (see Chapter 4: Catalyst Preparation) to prepare very small metal crystallites on high-area supports, the particles having diameters with values of the order of nanometers.[12] The most commonly used supports are single oxides such as alumina or silica, complex oxides such as silica-alumina or zeolites and active carbons. We will return in Chapter 4, to a more detailed discussion of supports and catalyst preparation while in Chapter 5 we will consider catalyst characterization methods. For the present purposes, all we need say is that the metallic crystallites of a supported metal catalyst are generally located within the pores of the support and that these are large enough to allow the diffusion of the reactants to, and of the products away from, the metal surface. A typical supported metal catalyst, for example, Pt/alumina, can be depicted schematically as shown in Fig. 3.5.

A well-known example of the use of supported Pt catalysts is in automobile exhaust catalysis.[13] In typical exhaust catalysts, the Pt is supported on a complex mixture of oxides, including ZrO_2, CeO_2, and Al_2O_3, which in turn is supported on a honeycomb structure, generally in the form of a refractory oxide such as cordierite; alternatively, a metallic structure made of a special alloy such as FeCrAlloy (containing Fe, Cr, Al, and Y and fabricated into a honeycomb arrangement) is sometimes used.

Many other examples of the importance of supported metal catalysts will be discussed in subsequent chapters, either in relation to methods of catalyst preparation or in relation to catalysts used in particular processes. It should be recognized that the support itself may not only be an inert material on which the metal is positioned but it may also contribute to the activity, selectivity, and stability of the catalyst under operation conditions. (See also Box 3.6.) As indicated earlier, the catalyst formulation may also include modifiers or promoters of various types. We will return to these topics later.

[12] Many of the methods used for the preparation of these nanosized particles were developed long before the term nanotechnology was introduced. Nanotechnology can be defined as "the study of controlling matter on an atomic and molecular scale" (en.wikipedia.org/wiki/Nanotechnology) and so the preparation of supported metal catalysts clearly falls within this definition.

[13] An equally important application of supported Pt catalysts is in hydrocarbon reforming; see Chapter 10.

Task 3.6 Car Exhaust Catalysts*

Examine the literature describing the technical development of car exhaust catalysts. As we will return later (Chapter 13: Environmental Catalysis) to the topic to discuss the various reactions occurring in the "catalytic muffler," your study should concentrate on the technical developments from the initial introduction of the first emission control catalysts for conventional petrol and diesel engines to more recent developments in relation to catalysts for use in lean-burn engines. You might also investigate other technologies for reducing automobile and other emissions such as the use of alternative fuels (e.g., biofuels) or hydrogen fuel cells. As starting points for your study, you might examine the websites of catalyst manufacturers such as Johnson Matthey or BASF or of car manufacturers such as Ford, General Motors, or Toyota.

The following website gives a useful overview of the types of catalytic system available commercially: http://www.dcl-inc.com/.

*This topic is dealt with in more detail in Chapter 13, Environmental Catalysis.

BOX 3.6 Evidence for the Participation of Oxygen of a Catalyst Support

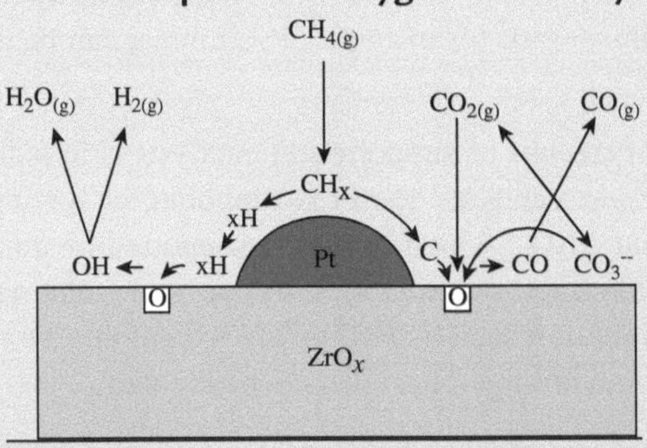

Experiments carried out using diffuse reflectance Fourier transform infrared spectrometry (DRIFTS) have shown that the mechanism of the CO_2 reforming of methane over a Pt/ZrO_2 catalyst probably takes place according to the scheme shown earlier. Carbon formed by the decomposition of the methane on the Pt sites reacts with oxygen from the support in the proximity of the Pt crystallite, this giving partial reduction of the support. The oxygen is then replenished by reaction with CO_2 from the gas phase. [A.M. O'Connor, F.C. Meunier, J.R.H. Ross, Stud. Surf. Sci. Catal. 119 (1998) 819.]

In an earlier publication from the author's laboratory [J.R.H. Ross, M.C.F. Steel, A. Zeini-Isfahani, J. Catal., 52 (1978) 280], it was shown by tracer studies using D_2O or H_2O^{18} that the oxygen originating from the surface of an alumina support participates in the steam reforming reaction of methane over supported Ni/γ—alumina catalysts, possibly through the formation of nickel aluminate sites on the surface of the alumina.

3.5 Oxides

In just the same way that a metal is most commonly used under reducing conditions, oxides are generally used as active catalytic components either in oxidation reactions under oxidizing conditions or in reactions in which reduction of the oxide is prevented by kinetic or thermodynamic limitations.

In selective oxidation reactions, discussed in more detail in Chapter 14, the catalytically active oxide frequently acts according to the so-called Mars–Van Krevelen mechanism (see Chapter 7) in which an oxidation-reduction sequence occurs on the oxide surface. This generally requires that the metal ions of the oxide which constitutes the active phase has a variety of possible oxidation states. Examples of the oxides found in such catalysts include V_2O_5 (reducible to +4 and +3 states) and MoO_3 (reducible to +5 and +4 states). Mixed oxides that are also commonly found in use in catalysts, for example the BiMoOxide used for the selective oxidation of propylene (propene) to acrolein. An example of the use of vanadia-containing catalysts is given in Box 3.7.

BOX 3.7 Selective Reduction of NO_x Over V_2O_5/TiO_2 Catalysts

Selective catalytic reduction (SCR) of the NO_x emissions[a] in the oxygen-containing stack gases from power stations using ammonia as reductant is generally carried out using V_2O_5/TiO_2 catalysts. It has been shown[b] that the reaction probably involves reduction of surface V^{5+} sites in steps (i) and (ii) and reoxidation in steps (iii) and (iv):

$$V{=}O + NH_3 + NO \rightarrow V{-}OH + N_2 + H_2O \qquad \text{(i)}$$

$$3V{=}O + 2NH_3 \rightarrow 3V\square + N_2 + 3H_2O \qquad \text{(ii)}$$

$$V\square + 1/2\ O_2 \rightarrow V{=}O \qquad \text{(iii)}$$

$$V\square + NO \rightarrow V{=}O + 1/2\ N_2 \qquad \text{(iv)}$$

where $V\square$ represents a surface vacancy or a V-OH group. A more detailed mechanistic description based on detailed isotopic exchange experiments has been reported elsewhere.[c]

[a]NO_x is a shorthand for a mixture of the oxides of nitrogen, predominantly NO. The whole topic of SCR is summarized in a comprehensive review by H. Bosch, F.J.J.G. Janssen, Catal. Today 2 (1988) 369.

[b]H. Bosch, F.J.J.G. Janssen, F.M.G. van den Kerkhof, J. Oldenziel, J.G. van Ommen, J.R.H. Ross, Appl. Catal. 25 (1986) 239.

[c]F.J.J.G. Janssen, F.M.G. van den Kerkhof, H. Bosch, J.R.H. Ross, J. Phys. Chem. 91 (1987) 5921, 6633.

Nonreducible oxides generally function as inert supports that help to stabilize metal particles (see earlier), examples being alumina (Al_2O_3), silica (SiO_2), magnesia (MgO), and zirconia (ZrO_2). However, even with these relatively stable oxides, there is some evidence that the interface between the metal and the support provides a special type of site and that the catalytic reactions occurring on the supported catalysts depend on sites on both the metal and the oxide surface near the interface, even utilizing oxygen ions of the support in the reaction. (See the example given in Box 3.6.)

Nonreducible oxides can also function as solid acids or bases. For example, alumina, especially when calcined at relatively low temperatures, has some acidic properties, while MgO, CaO, and BaO have basic properties. Silica-alumina, which can be produced by co-gelling the individual components followed by washing and calcination, has relatively strong acidic properties and has been used as a hydrocarbon isomerization catalyst. In fluid catalytic cracking, the catalysts now used are zeolites.

A recent application of alkaline earth oxides is in the trapping of SO_2 from automobile exhausts. For example, BaO is added to the catalyst formulation in the so-called Toyota system; the noble metal component of the catalyst, Pt, catalyzes the oxidation of SO_2 to SO_3 and the BaO then adsorbs the SO_3 to give $BaSO_4$. A more detailed treatment of the various aspects of exhaust gas catalysis is given in Chapter 13, Environmental Catalysis.

3.6 Sulfides

The most common use of sulfides is in reactions in which a compound of sulfur is present, the reaction conditions favoring the formation of the sulfide from the metal or oxide originally present in the catalyst. The most common example is the $Co/Mo/Al_2O_3$ catalyst used for hydrodesulfurization:

$$RSH + H_2 \rightarrow RH + H_2S$$

The catalyst as prepared consists of a mixture of CoO and MoO_3 supported on a high-area alumina such as γ-Al_2O_3. However, under the hydrodesulfurization conditions, the Co and Mo surface species transform to give a stable sulfide entities. Much has been written about the relative positions of the Co and Mo species but it is now generally accepted that the Co ions sit on the edges of MoS_2 platelets. We will return to a discussion of the process in Chapter 10.

Task 3.7 Hydrodesulfurization Catalysts*

The company Haldor Topsøe A/S of Lyngby, Denmark (http://www.topsoe.com/), is at the forefront in the application of hydrodesulfurization catalysts as well as in other technologies for the chemical industry. Haldor Topsøe is also an increasingly rather unique example of a company which, in addition to marketing commercial catalysts and processes, also carries out important basic research underlying the processes which it markets.

Study the Haldor Topsøe website and follow up on a number of the processes and catalysts that Haldor Topsøe markets. Attempt to identify some of the companies competing with Haldor Topsøe in relation to these. Some of the important developments of Haldor Topsøe in the field of hydrotreating catalysis are based on the work of Henrik Topsøe, son of the founder of the company. Examine some of the publications of Henrik Topsoe and outline the science behind the "Brim" Technology (http://www.topsoe.com/research/BRIM_story.aspx).

*This topic is dealt with in more detail in Chapter 10.

3.7 Conclusions

We have seen in this chapter how a catalyst works and we have examined some aspects of the use of metals, oxides, and sulfides as catalysts. We will now proceed in subsequent chapters to examine in more detail how a catalyst is made, characterized, and tested.

CHAPTER 4

Catalyst Preparation

Chapter Outline

4.1 Importance of Active Surface Area and of Catalyst Structure

We have seen in Chapter 3, How Does a Catalyst Work?, that a heterogeneous catalytic reaction proceeds on the surface of the catalytically active material by the making and breaking of bonds between the reacting species and the atoms of the catalyst surface. In discussing heterogeneous catalysis, we are therefore essentially dealing with *chemical processes occurring in two dimensions*, with the added complication that the reacting and product molecules have to approach and leave the surface in a three-dimensional process. In order to optimize the rate of the reaction, the surface area of the catalyst must be high; if the surface is nonuniform in composition (e.g., with a supported metal, there are metallic sites as well as inert, or relatively inert, support sites), it must contain as many positions as possible at which the molecules involved in the process can be adsorbed. Further, as will be discussed in more detail in Chapter 7, The Kinetics and Mechanisms of Catalytic Reactions, all of the surface, external and internal, must be accessible to the reactant molecules

Contemporary Catalysis.
DOI: https://doi.org/10.1016/B978-0-444-63474-0.00004-7

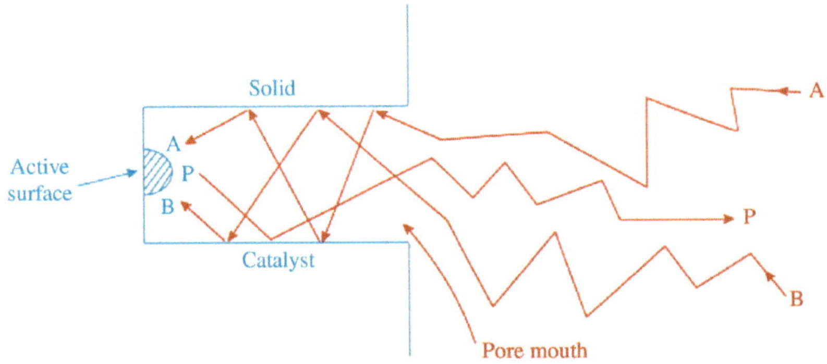

Figure 4.1
Schematic representation of molecules A and B approaching and entering a catalyst pore and being adsorbed on the active surface within the pore, this being followed by the desorption of product P and its diffusion out of the pore.

approaching it from the gas phase (or from the liquid phase) and it must be possible for the products to get away from the surface. The situation for the reaction:

$$A + B \rightarrow P \tag{4.1}$$

(see Chapter 3, where the reaction was given as A + B → C + D) is shown schematically in Fig. 4.1.

Each of the molecules A and B approaches the solid catalyst material from the gas phase, making a series of gas-phase collisions on the way. These collisions can be with other A molecules or with the species B, the product P or even an added inert gas. Once the molecule A reaches the solid, it then has to get to the active surface which might be, for example, a metal crystallite embedded within a pore. A therefore has to diffuse down the pore and will make collisions with the walls of the pore as well as with gas-phase molecules; the relative number of surface collisions relative to gas-phase collisions will then depend on the pore diameter as well as on the reaction conditions (pressure, temperature). Once adsorbed on the active surface, the adsorbed species derived from molecule A can react with an equivalent species resulting from the diffusion and adsorption of molecule B to give adsorbed product P; P then desorbs and has to diffuse out of the pore and away from the surface. The rate of the reaction may still be determined by the rate of adsorption of A or B, by the surface reaction of A with B, or by desorption of product P; however, the rate may also be affected by the rates of diffusion to or from the surface and also into or out of the pore. Hence, in addition to having a high active catalytic surface, the catalyst must have easily accessible outer and inner surfaces; and, as we shall see later, not all of the solid surface needs of necessity to be active.

The following sections (4.2–4.4) will consider the creation of catalytic materials with high active surface areas in a form suitable for practical application and will also discuss the importance of the texture of the catalyst so formed. Chapter 5 then discusses some of the more important methods of characterizing the catalysts prepared.

4.2 Catalyst Preparation

4.2.1 The Active Surface

An effective catalyst may have very many different types of active surface species and each will be capable of catalyzing either a single reaction or a whole range of related reactions. The most commonly encountered active species are metals, oxides and sulfides, as discussed in the previous chapter (Chapter 3), but many other materials have also been shown to have catalytic activities. The choice of the active species is determined very largely by the reaction to be catalyzed; some reactions can be catalyzed by many different catalytic materials, others by only single and unique catalysts. For example, as discussed previously, metal species are most commonly used for reactions of hydrocarbons, examples being hydrocarbon reforming or steam reforming of natural gas. Oxides are used for different types of reaction, typically selective oxidation reactions. However, as the oxides can exhibit acidic or even basic properties, they are also used for reactions requiring either acids or bases, a typical example being catalytic isomerization of hydrocarbons. And sulfides can be used for reactions involving sulfur-containing molecules such a hydrodesulfurization. However, a word of caution is necessary here: all rules can be broken, especially in such a complex subject as catalysis, and there are many exceptions to these generalizations. We will now consider some of the most important methods of making and characterizing high surface area metals, oxides and sulfides and will consider how these materials are likely to be used in practice.

In order to obtain a high surface area and hence increase the rate at which a reaction occurs at the resultant surface, it is desirable to obtain a highly dispersed material (e.g., metal, oxide, or sulfide). Fig. 4.2 shows schematically what happens when a solid of cubic form and with side of length 1 cm is divided into smaller cubes of side 0.5 cm. The original cube has a surface area of 6 cm^2 while the eight smaller cubes have a total area of $8 \times 6 \times 0.5^2 = 12\ cm^2$. A further subdivision gives 32 cubes of side 0.25 cm and a total area of 24 cm^2, etc. If the side of the cube is 1 nm, the area of material is now $6 \times 10^7\ cm^2$ or 6000 m^2, an increase of area by a factor of 10^7. The same factor will apply to the increase the rate of a catalytic reaction occurring on the particles, at least as long as the reactant molecules have full access to the surface of each of the small (nano-) particles. (See also Task 4.1.)

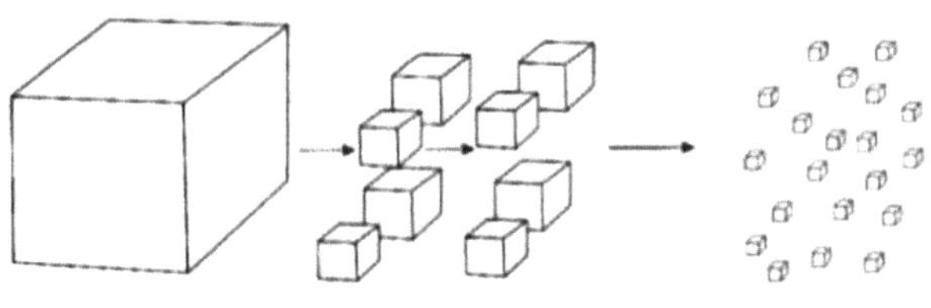

Figure 4.2
Schematic representation of the increase in surface area caused by subdividing a cube of a solid material.

Task 4.1 Increase of Surface Area by Subdivision of a Particle

Work out the area if the dimension of the cube formed in Fig. 4.2 is 5 nm (50 Å).

All materials have surface energies, these being equivalent to the surface tension of a liquid. Just as with a liquid, the surface area will tend to try to minimize itself in order to reduce the total surface energy. This means that the particles will have a tendency to coalesce to reform the original cube unless some way is found to stop that happening; this coalescence of the particles will be particularly important if the temperature is near the Tamman temperature for the solid, the temperature at which atoms of that solid become mobile. The coalescence process is known as "sintering." While sintering is desirable in some applications, for example, in ceramic manufacture, it must be minimized in the case of a heterogeneous catalyst since any sintering will cause loss of catalytic activity. (See also Box 4.1 and Task 4.2.)

BOX 4.1 Minimization of Surface Energy

A liquid will have a minimum surface energy if it exists as spherical drops as these will have the minimum surface area for any given drop size. Two separate drops will tend to coalesce to form a single one, hence minimizing the total surface energy, a good example being the coalescence of two drops of mercury sitting on a plane surface. Similarly, two soap bubbles brought into contact with one another will coalesce into a single bubble.

Task 4.2 Decrease of Surface Energy of Two Connected Bubbles

Two soap bubbles are connected to one another by a cylindrical tube.

Will the bubbles remain the same size? If not, what will happen? (Note that the bubbles will each be approximately spherical, any deviation being due to the effect of gravity.)

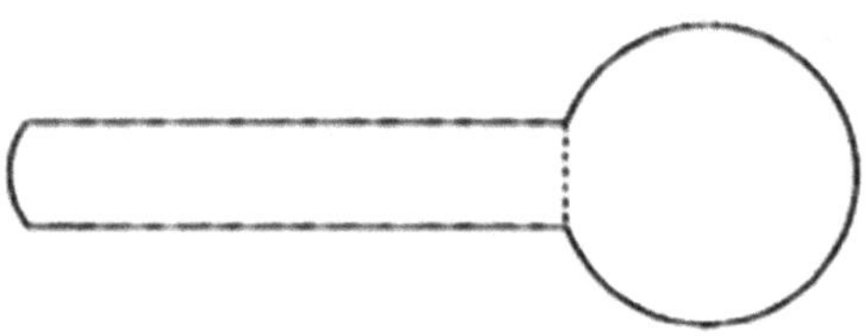

Answer: one will get smaller, the other larger, until the surface areas of the two bubbles added together have reached a minimum.

What will determine the diameter of the resultant single spherical bubble?

One way to prevent the sintering of a catalytically active component is to fix the particles of that component (which are not necessarily a single metal or oxide) onto another high-area surface ("support") in such a way that the particles cannot come into contact with one another and that surface migration of the component atoms of the active particles across the support surface cannot occur (See Box 4.2.). This is the principle of the "supported catalyst" in which the active component is situated on the surface of an inert and stable support material (see Chapter 3: How Does a Catalyst Work?).

BOX 4.2 Support Effects

The support may or may not be considered to be totally inert. Frequently, there is some sort of interaction between the active component and the support; at one extreme, the active component is "glued" to what is otherwise an unreactive material, in the other extreme, there is a "cooperation" between the active material and the support.

In another form of catalyst, the active component is the majority phase present and sintering is prevented by the presence of another minority phase that prevents the particles from coming into contact with one another and hence from coalescing. An example is the coprecipitated Ni-alumina catalyst with high Ni content that can be used for steam reforming or fat hardening; in this, alumina particles sit between the Ni crystallites (see Section 4.4.2).

We will now consider in turn the preparation of some typical supports and will then discuss the preparation (and, where necessary, pretreatment) of some typical supported and coprecipitated catalysts. Some of the methods that can be used to characterize these catalysts will be discussed in Chapter 5, Catalyst Characterization.

4.3 Catalyst Supports

4.3.1 Introductory Remarks

No attempt will be made in this book to discuss the preparation and characterization of each and every type of known catalyst support. Instead, several typical supports will be considered and a number of generalizations likely to be applicable to all types of support will be discussed.[1]

[1] For a more detailed discussion of catalyst preparation, the reader is referred to texts such as "Principles of Catalyst Development" by James T. Richardson (Plenum Press, New York and London, 1989), "Catalyst Preparation, Science and Engineering", Edited by John Regalbuto, (CRC Press, Boca Raton, London, New York) 2007, and the proceedings of the series "International Symposia on the Scientific Bases for the Preparation of Heterogeneous Catalysts" held in Louvain la Neuve, Belgium and published by Elsevier Science Publishers in the Studies in Surface Science and Catalysis series, all accessible through Science Direct and indexed in Scopus.

Table 4.1: Values of the melting points and Tamman temperatures for a series of different oxides commonly used as catalyst supports. Several other oxides are also shown which may in some cases be added as "promoters"

Oxide	Melting Point (K)	Tamman Temperature (K)	Oxide Type/ Comments
γ-Al_2O_3	(2318)	(1159)	Acidic Transfers to α-Al_2O_3 on heating (see text)
α-Al_2O_3	2318	1159	Amphoteric
SiO_2	1973	987	Acidic
SiO_2–Al_2O_3	1818	909	Very acidic
MgO	3073	1537	Basic
CaO	2853	1427	Basic
$CaSiO_4$	2407	1204	Basic
ZrO_2	2988	1494	Amphoteric
Cr_2O_3	2708	1354	Amphoteric
CeO_2	2873	1437	Amphoteric
La_2O_3	2588	1294	Amphoteric
$MgAl_2O_4$	2408	1204	Neutral
$ZnAl_2O_4$	2100	1050	Neutral

Note: The values shown in Table 3.1 are not precise and different sources give slightly different data.
Source: *J.T. Richardson, Principles of Catalyst Development, Plenum, New York and London, 1989. The values agree reasonably well, but not entirely, with those give in the Handbook of Chemistry and Physics, Chemical Rubber Company, 1974. However, the absolute values are not important in this discussion, only the orders of magnitude.*

The most commonly used support materials are refractory oxides; these are metal oxides that have high melting points and are therefore very stable under most operating conditions. The constituent ions of an oxide become mobile at temperatures around the Tamman temperature (Table 4.1) that is at approximately half the temperature of the melting point (in K). As can be seen from Table 3.1, the melting points of typical oxides used as supports range from about 1500 to about 2700°C and so all these oxides will be usable up to temperatures of about 640°C (silica–alumina) or higher. The examples that we will consider in detail later are Al_2O_3, SiO_2, and ZrO_2, for which the values of the Tamman temperature are 886, 714, and 1221°C, respectively. (See also Box 4.3.)

Some oxide systems are complicated further by the possibility of phase transition under operating conditions. A number of oxides can be formed in several different phase modifications, some of which are stable over a wide range of temperatures, others of which are not. (Further, a phase that may be stable under one set of operating conditions can change when subjected to other conditions, for example, exposed to water vapor.) For example, TiO_2 can be made in three different forms: anatase, rutile, and brookite. Anatase can have areas up to 80 m^2g^{-1} while rutile, which forms at temperatures above about 550°C, has a much lower area. The Tamman temperature for TiO_2 is around 800°C (Table 3) and so phase transition in this case does not involve the bulk diffusion process

BOX 4.3 Tamman Temperatures of Oxides

The Tamman temperature has also been used as an indication of the mobility of oxygen ions in selective oxidation catalysts. For example, based on the following data, Chen et al. have argued that vanadia is more likely to have mobile lattice oxygen ions than would the molybdenum or tungsten oxides since it has the lowest Tamman temperature.

Melting Point and Tamman Temperature of Various Metal Oxides

Metal Oxide	Melting Point (K)	Tamman Temperature (K)
V_2O_5	963	482
MoO_3	1068	534
Bi_2O_3	1098	549
CuO	1599	780
WO_3	1745	873
Fe_2O_3	1838	919
TiO_2	2128	1064
ZnO	2248	1124
NiO	2257	1129
Cr_2O_3	2538	1269

K. Chen, A.T. Bell, E. Iglesia, Kinetics and mechanism of oxidative dehydrogenation of propane on vanadium, molybdenum, and tungsten oxides, J. Phys. Chem. B 104 (2000) 1292–1299.

http://iglesia.cchem.berkeley.edu/JPhysChemB_104_1292_2000.pdf

associated with sintering as discussed earlier. A consequence is that each material used as a catalyst support must be examined in detail to make sure that it does not sinter or suffer from phase transitions under the conditions that will be encountered should that material be used in a particular catalytic reaction.

Another important parameter in determining which materials (oxides or other compounds) are used as supports is the texture of the solid. It is important that the support has a relatively high surface area and also that it has an accessible porous structure with reasonably large and regularly shaped pores. The surface area is of importance since the material to be supported also must have a high surface area to give it acceptable catalytic activities. The porous structure is also an important parameter since the molecules taking part in the catalytic reaction must be able to approach the active catalytic surface by diffusion through the pores and the products must also be able to diffuse out of the pores. (If the reaction only occurs on the outermost surface of the catalyst, then its "effectiveness factor" is well below unity and the consequence is that much of the active component in the catalyst is unused; if the active component is an expensive element such as Pt, this is highly undesirable.)

We will now examine some of the most important aspects related to the preparation and use of several of the most commonly encountered catalyst supports: alumina, silica, titania, zirconia, and carbon.

4.3.2 Alumina, Al_2O_3

Alumina is one of the supports most commonly used in the preparation of heterogeneous catalysts. This is largely because it is relatively inexpensive and also because it can be prepared easily in forms that have high surface areas, coupled with readily accessible pore structures. However, it is also of importance because it is relatively inert and does not easily form compounds with the supported components (see later). In certain applications, the fact that the transition aluminas (see later) are acidic has some importance. It should be recognized that the preparation routes used for alumina as a catalyst support are also used in the preparation of the alumina that is smelted to give aluminum metal; as the aluminum industry is very large, there is therefore a ready supply of alumina for other purposes.

The raw material used for alumina preparation is bauxite. This impure mineral is treated in the so-called Bayer Process in which the aluminum is dissolved in NaOH and separated from the impurities present in the original ore. The resultant sodium aluminate solution is then treated to precipitate aluminum hydroxide (bayerite) from which alumina is subsequently made by calcination. (http://www.eaa.net/en/about-aluminium/production-process/) The alumina so prepared is relatively pure but may contain sodium ions as trace impurities. For use in catalysis, the alumina has to be particularly pure and is generally prepared from a solution of a pure salt such as the nitrate[2]. The salt solution is treated with an alkali[3], e.g., ammonia, to give a precipitate of one of the hydroxides (gibbsite, bayerite, or nordstrandite) or oxyhydroxides (boehmite or diaspore[4]) of aluminum, the actual structure produced depending on the conditions used for the precipitation step and the aging of the precipitate.

The hydroxide or oxyhydroxide is then "calcined" to bring about dehydration and to form an alumina. Again, the conditions used for the drying and calcination steps can be crucial in determining the structure of the resultant alumina. In particular, the temperature of calcination has a very significant effect on the form achieved and its surface area.

[2] If another soluble salt such as the sulfate or chloride is used, the resultant alumina will contain trace sulfate or chloride ions and these may be detrimental to any catalyst made with such materials.

[3] If an alkali such as NaOH or Na_2CO_3 is used, the resultant alumina may contain traces of the associated cation and this is again undesirable.

[4] Diaspore is also found as a naturally occurring mineral.

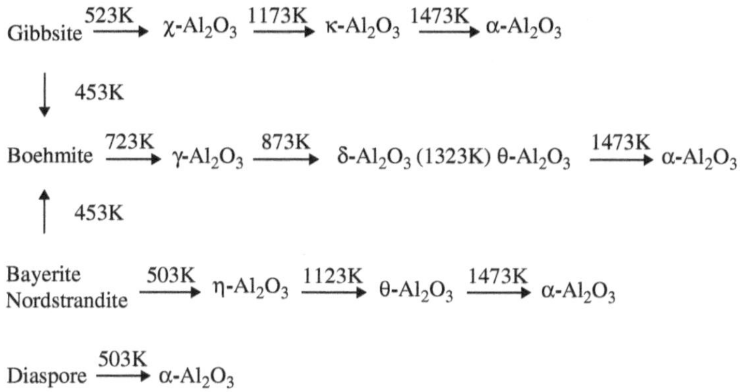

Figure 4.3
The preparation of the transition aluminas. *From B.C. Lippens (Thesis, T.U. Delft, 1961; B.C. Lippens, J.J. Steggerda, in B.G. Linsen (Ed.), The Physical and Chemical Aspects of Adsorbents and Catalysts, Academic Press, 1970).*

There have been many studies of the precipitation, drying, and calcinations steps, one of the most definitive being that by B.C. Lippens (Thesis, T.U. Delft, 1961; B.C. Lippens, J.J. Steggerda, The Physical and Chemical Aspects of Adsorbents and Catalysts, B.G. Linsen (Ed.), Academic Press, 1970.) The work of Lippens, which helped to clear up much of the confusion of nomenclature that had existed in the earlier literature, showed that the sequence for the production of the various transition aluminas and ultimately α-Al_2O_3 are as shown in the scheme of Fig. 4.3. (See also Task 4.3.)

Task 4.3 Alumina Supports

Examine the literature on aluminas using Scopus or the Web of Science. The thesis of Lippens is not cited in Scopus, nor is the Lippens and Steggerda book chapter. Hence, one needs to start by searching for other papers by B.C. Lippens (noting that his son, who later published several papers on catalytic subjects, had the same initials), particularly a series of papers in J. Catal. [3 (1964) 32–37, 38–43 and 44–49 and 4 (1965) 319–323.] Another useful lead is a more recent paper by C. Monterra and G.T. Magnacca, Catal. Today 27 (1996) 497–532. Papers by J.B. Peri on the use of spectroscopic methods to study oxide surfaces (see later) are also relevant.

As indicated earlier, which of the hydroxide or oxyhydroxide precursors is formed is very dependent on the preparation conditions. However, if the desired product is γ-Al_2O_3, then boehmite, an oxyhydroxide, is first formed by hydrothermal treatment of one of the trihydrates and this is then calcined at about 723K. Of the so-called transition alumina, γ-Al_2O_3 is the most commonly used as a catalyst support; the χ- and η-forms have also been used as supports but the results obtained seldom differ too significantly from those obtained using γ-Al_2O_3, apart from the fact that these two forms appear to be stable up to rather higher temperatures.

All of the transition aluminas transfer into other modifications at temperatures higher than about 873K and all are transformed into α-Al_2O_3 (corundum) at temperatures around 1473K; the relative rates of transition depend markedly on the degree of crystallinity of the precursors used in the preparation. Corundum has a well-crystallized spinel structure while the transition aluminas are described as having deformed pseudospinel structures. Lippens and other authors are in agreement that the structures of the transition aluminas differ mainly in the relative occupation of the octahedral and tetrahedral sites.[5]

The transition aluminas all have relatively high surface areas, typically several hundred m^2g^{-1}, and are used for the preparation of catalysts that in general are not used at temperatures much above 500°C. They all have acidic surfaces characterized by the presence on their surfaces of significant concentrations of hydroxyl groups formed either during the decomposition of the precursor or by interaction subsequently with water from the atmosphere:

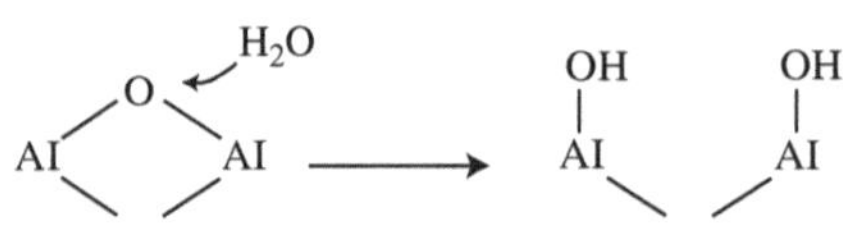

(4.2)

The surface AlOH species so formed have Brønsted acidic character. In other words, they give rise to protons on dissociation and hence act as solid acids. The dehydrated surface shown as the precursor of the reaction shown in Eq. (4.2) is a Lewis acid. If the surface is not fully hydrated, or becomes partially dehydrated on calcination, the surface has a mixture of Lewis and Brønsted acid character. It is this acidity which makes alumina particularly suitable as a support for the preparation of heterogeneous catalysts: in addition to providing "sites" at which precursors of catalytic components may be adsorbed, as described later, these OH groups have acidic character which may contribute to the catalytic behavior of the final catalyst. An important paper on the behavior of alumina surfaces upon dehydration by J.B. Peri reports an investigation of the dehydration of γ-Al_2O_3 using infrared and gravimetric methods.[6] Peri showed that OH species persist on specific sites on the surface of alumina even after dehydration at high temperature.

---

[5] A comprehensive study of the structure and properties of γ-Al_2O_3 has been carried out by G. Paglia. The work is described in detail in his PhD thesis available online at http://adt.curtin.edu.au/theses/available/adt-WCU20040621.123301/unrestricted/01front.pdf. The thesis also contains a list of the papers based on the work.

[6] J.B. Peri, Infrared and gravimetric study of the surface hydration of γ- Al_2O_3, J. Phys. Chem. 69 (1965) 211.

The forms of alumina other than γ-, χ-, and η-Al_2O_3 are seldom used as catalyst supports as their surface areas are generally much lower than those of the transition aluminas. (There is of course the chance that some of these crystal structures may be formed in a catalyst which has been exposed to operation at higher temperatures, this usually resulting in deactivation; see later.) Corrundum (α-Al_2O_3) is a very hard, refractory material with a very low surface area that generally finds application as a ceramic. However, porosity can be built into the structure and such materials have been used as matrices in some catalyst formulations (see Section 4.4.2.3).

Before proceeding to describe the preparation of catalysts using high-area supports, we will first discuss briefly some other support materials commonly used in catalyst formulations.

4.3.3 Silica, SiO_2

Silica has many applications, these including use in glass and whitewear ceramics, as a dessicant, as a component in toothpaste, scouring powders and similar preparations, and as an addition to certain types of food (e.g., as a thickener in soups[7]), as well as being used as a catalyst support. (http://en.wikipedia.org/wiki/Silicon_dioxide). The types of silica most used as catalyst supports are generally the same as those used as dessicants since they have high surface areas and significant pore volumes.[8]

Typical dessicant silica is made by dissolving silica sand (quartz) in hot caustic soda:

$$SiO_2 + 2NaOH \rightarrow Na_2SiO_3 + H_2O \tag{4.3}$$

The sodium silicate is then reacted with acid (usually sulfuric acid) to form silicic acid and sodium sulfate:

$$Na_2SiO_3 + H_2SO_4 \rightarrow H_2SiO_3 + Na_2SO_4 \tag{4.4}$$

The silicic acid formed then polymerizes to form long silicate chains by a polymeric chain growth mechanism involving further condensation steps:

$$\begin{aligned} H_2SiO_3 + H_2SiO_3 \rightarrow\ & HOSi(=O)_2\text{-}O\text{-}Si(=O)_2OH + H_2O \\ & \downarrow \\ & HO\{Si(=O)_2O\}nH \end{aligned} \tag{4.5}$$

[7] Silica is one of the main components of nettle, also commonly used as a thickener for soups.

[8] Fumed silicas, formed by burning $SiCl_4$ in an air/hydrogen flame: $SiCl_4 + 2H_2 + O_2 \rightarrow SiO_2 + 4HCl$, have also been used.

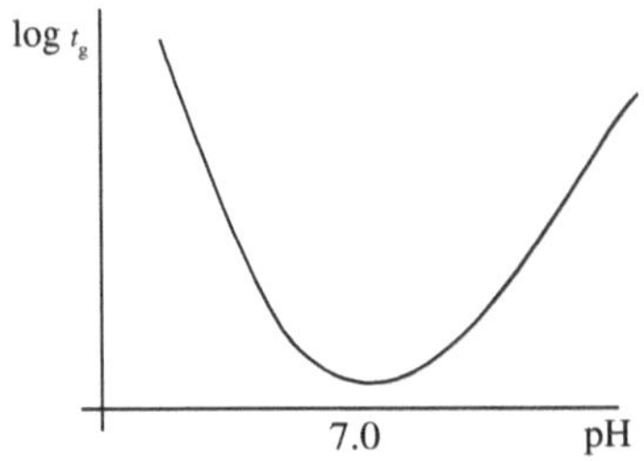

Figure 4.4
Effect of pH on gelation time for a silica sol.

The chains interact with one another forming sols and these then solidify to give hard gels.[9] The higher the concentration of the silica sols, the harder the resultant gels. Particles of these hard gels are those used in the many domestic cleaning preparations mentioned earlier.

The time taken for polymerization and subsequent gel formation (gelation) depend strongly on the conditions: pH, concentration, and temperature. In the industrial process, in which highly concentrated solutions are used, the mixing of the acid and the sodium silicate have to be carried out extremely rapidly in order to avoid conditions under which the gelation process occurs simultaneously in the mixer, a potential disaster if not controlled properly since the gel is extremely hard and unreactive and could block the reactor. This can be understood by reference to Fig. 4.4 that shows schematically the gelation time (t_g, on a logarithmic scale) versus the pH of the gel. It can be seen that if the pH of the gel is about 7, the gelation time is less than a minute. However, if the pH is maintained close to 1, the gelation time will be many hours. The pH of the sodium silicate used for the preparation is high, exact value depending on the concentration of alkali added to dissolve the silica sand, Eq. (4.3). If only enough acid is added to neutralize the sodium silicate, the pH will be close to the minimum of Fig. 4.4 and gelation will occur almost instantaneously; if the concentrations are high, the resultant gel will be rock-hard and impossible to remove. To prevent this happening, a significant excess of acid is added (with rapid mixing to ensure that the pH throughout is uniform) in order that that the resultant pH is 4.0 or less. The gelation time is now sufficiently long to allow the sol to be removed from the mixing vessel before it can gel.

9 This polymerization process of silicic acid was once used in a domestic product: the so-called water-glass used for preserving hens' eggs. The silicic acid was provided as a powder and was made up as a "solution" (more correctly, a sol) in water. The eggs were then immersed in the sol and this gradually gelled, providing a matrix through which oxygen from the air could not penetrate, preserving the freshness of the eggs. As the concentrations were quite low, the gel was never hard enough to prevent eggs being removed easily when required for use. Inserting one's hand was sufficient to break the local gel structure that reformed once one's hand was removed.

The gel formed is a mixture of a solid matrix of interlinking polymeric silica chains, water, and sodium sulfate. The sodium sulfate (or other salt if another acid is used) is removed by washing and the gel is then dried to remove excess water. This procedure sounds simple but the properties of the resultant anhydrous gel are very dependent on the final pH of the gelation process, on the pH, the temperature, and the time of any aging process that follows; they also depend strongly on the temperature of the dehydration step (drying and calcination) applied after washing the material. If the aging is carried out at relatively high pH, the surface area of the resultant gel can be as low as 200 m^2g^{-1} but if the pH is low, the area can be as high as 900 m^2g^{-1}. The pore volumes and diameters are also strongly dependent on the aging, washing and calcination conditions. As a result, a large range of different materials is available commercially. It is also interesting to note that a variety of different physical forms are also available and that some of these can be prepared making use of a well-controlled gelling process; for example, very uniform spheres can be obtained if spherical drops of the sol are extruded into an inert matrix (e.g., an immiscible oil) in which the polymerization process can be completed prior to the washing, drying, and calcination steps. (See also Task 4.4.)

Task 4.4 Infrared Spectroscopy of Alumina and Silica

Look up J.B. Peri in Scopus and follow up his papers that relate to the use of spectroscopic methods in determining the surface structures of alumina and silica. Following up on the papers on alumina will lead to the paper by C. Morterra and G. Magnacca in Catal. Today 27 (1996) 497 included in Task 4.3. This in turn will lead to a monograph on IR spectroscopy in catalysis by J. Ryczkowski in Catal. Today [68 (2001) Issue 4].

While the pore structure of the gel is of importance in determining the properties of the catalysts made using the gel, the dehydration/calcination step is of importance in determining the way in which the gel interacts with metal species during the preparation of the catalyst. J.B. Peri, using infrared spectroscopy and other methods, studied the behavior of the hydroxyl species on the surface of a silica gel as a function of dehydration temperature.[10] He showed that pairs of OH groups are attached to each surface silicon atom even after drying at 600°C. Dehydroxylation at very high temperatures results in the creation of isolated OH species. Although the surface hydroxyl groups on silica are reasonably acidic, a more acidic solid is obtained by cogelling silica and alumina.[11] The resultant material, often shown as $SiO_2-Al_2O_3$, is used alone as a cracking catalyst or together with other components when an acidic support is needed. Solids with zeolitic structures are now more commonly used for these purposes. (See Section 4.3.4.)

[10] J.B. Peri, Infrared study of OH and NH_2 groups on the surface of a dry silica aerogel, J. Phys. Chem. 70 (1966) 2937; see also J.B. Peri, A.L. Hensley, The surface structure of silica gel, J. Phys. Chem. 72 (1968) 2926.

[11] J.B. Peri, Infrared study of adsorption of carbon dioxide, hydrogen chloride and other molecules on acid sites on dry silica-alumina and gamma-alumina, J. Phys. Chem. 70 (1966) 3168.

In concluding, it is worth noting two factors that are of some importance in relation to silica as a catalyst support. On the positive side, it should be recognized that silica exhibits no phase transitions on heating and so one is only limited by the possibility of sintering; with a Tamman temperature of 987K (Table 4.1), one can expect to be able to use silica as a support for react ions occurring up to about 700°C. However, on the negative side, one has to recognize that silica is not an inert solid. For example, it can react with water vapor to form a volatile hydrosilicate at higher temperatures and so it cannot be used in, or even as a component of, catalysts to be used in steam reforming or related processes. It can also react with alkali or alkaline earth metals to form the corresponding silicates and so it cannot be used in any situation where such species might be present as the silicates formed are often mobile and can cause, inter alia, severe pore blockage.

4.3.4 SiO_2–Al_2O_3 and Zeolites

SiO_2–Al_2O_3 is an amorphous mixed oxide material prepared by mixing silica and alumina sols. According to Richardson, the former is a 30% mixture of water glass and water (see preparation of silica earlier) and the latter a mixture of 4 M HCl and 0.5 M $Al_2(SO_4)_3$. These are rapidly mixed in the ratio 2:1 at a temperature of 5°C when rapid gelation occurs. After aging the gel, it is treated with 2% $Al_2(SO_4)_3$ solution to ion exchange the Na^+ ions trapped in the gel and then washed free of $SO_4^{=}$ ions; it is then dried and calcined at 550°C. It is a much more acidic material than either of its components due to weakening of surface OH bonds by the proximity of the Si^{4+} and Al^{3+} centers:

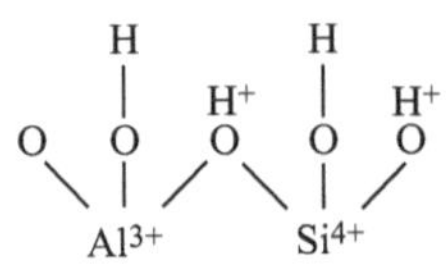

In this structure, the Si^{4+} ion replaces the tetrahedral Al^{3+} with a more electropositive center, thereby weakening the OH bond and increasing the acidity. These silica–alumina materials have been used as cracking catalysts or as acidic supports.

Zeolite materials are crystalline modifications of silica–alumina. They are prepared by hydrothermal treatment of cogels of silica and alumina in an autoclave at up to 10 atm. and at temperatures between 90 and 180°C, in the presence of OH^- ions, a suitable organic template (e.g., an organic amine or an alkyl ammonium compound), as well an appropriate seed crystals. A wide variety of different structures have been prepared and characterized, these having a large range of Si/Al ratios as well as differing pore geometries and sizes.[12]

[12] There are many possible sources of additional information on this topic, for example, "Introduction to Zeolite Science and Practice", Edited by H. van Bekkum, E.M. Flanigen, P.A. Jacobs, J.C. Jansen, Stud. Surf. Sci. Catal., Elsevier Science Publishers, Vol. 137 (2001).

Alumino-phosphate molecular sieves may also be synthesized and other oxides such as TiO_2 can also be incorporated. Zeolites are now used in many different applications, many of which are discussed in chapters by different authors in the book referred to in footnote 12. (See also Task 4.5.)

Task 4.5 Zeolitic Coatings

Van Bekkum and his colleagues [J.C. Jansen et al., Microporous Mesoporous Mater. 21 (1998) 213] have reviewed the preparation and potential uses of zeolitic coatings in catalysis. Read this review and examine some of the papers that have quoted it, paying particular attention to those that describe the preparation of zeolitic membranes.

4.3.5 Titania, TiO_2

Titania is used as a support in certain reactions, such as in the selective catalytic reduction (SCR) of NO_x (see Box 3.7 and Chapter 13: Environmental Catalysis) for which it is used as a support for vanadia. However, because of its semiconductor properties, it can also be used on its own in photocatalytic processes.[13] The most common use of TiO_2 is as a pigment for paints and it is made by one of two processes: hydrolysis of $TiOSO_4$ ("The Sulfate Route") or by the vapor phase oxidation of $TiCl_4$ ("The Chloride Route"). The sulfate process is no longer favored because of the problems of disposing of the large quantities of impure concentrated sulfuric acid formed as a byproduct.

As discussed earlier, TiO_2 may exist in several structural forms: rutile, anatase, and brookite. The anatase form is more usually used as a support and the surface area of typical materials is of the order of 80 m^2g^{-1}, relatively low compared with alumina and silica. It transforms to rutile, losing surface area, at about 550°C. This means that the applications of titania as a support are limited to reactions operating at relatively low temperature such the SCR reaction mentioned earlier. Much of the titania used as a catalyst support (and also for studies of photocatalysis by the oxide itself) in work reported in the literature has been based on a pigment grade titania, given the name P25, supplied by Degussa, now Evonic Industries http://corporate.evonik.com/en/Pages/default.aspx. However, many other suppliers of titania exist.

4.3.6 Zirconia, ZrO_2

Zirconia has recently attracted much attention as a catalyst support, partly because it is more chemically inert than the more traditional supports such as silica or alumina and also partly because it has been shown to have specific interactions with the active phase, giving differences in the activities and selectivities of the resultant catalysts. Zirconia may be

[13] See, for example, S. Malato, P. Fernandez-Ibanez, M.I. Maldonado, J. Blanco, W. Gernjak, Decontamination and disinfection of water by solar photocatalysis: recent overview and trends, Catal. Today 147 (2009), 1–59.

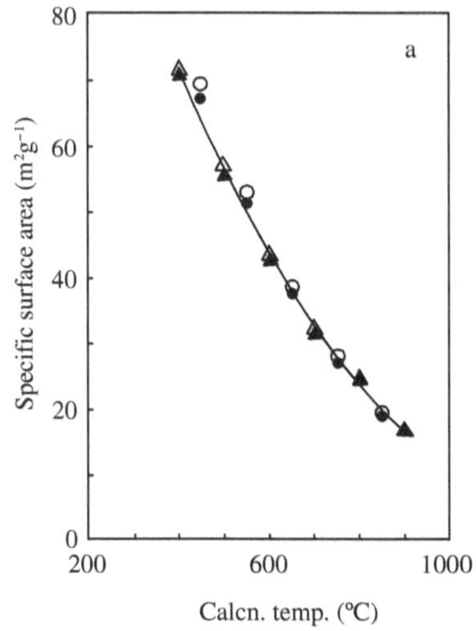

Figure 4.5
The effect of calcination temperature on the specific surface area of an undoped monoclinic zirconia sample ●, Undoped; □, MgO; △, CaO; ○, Y_2O_3; ▼, La_2O_3. *From Mercera et al., Appl. Catal. 71 (1991) 363–391. With kind permission of Elsevier.*

formed with a number of different phase structures. The unstable tetragonal phase is formed by gel precipitation but the resultant material is unstable on calcination at higher temperatures, the area dropping by as much as 97% on heating to 850°C. This sintering was found to occur by a surface transport mechanism resulting in a phase transformation to the monoclinic form, the latter being the most stable phase at normal temperatures of usage.[14] Mercera et al. showed that it is possible to prepare pure samples of the monoclinic phase by careful gel precipitation followed by careful washing in ethanol.[15] The monoclinic phase so formed was stable up to temperatures of 1170°C. However, there was still a significant reduction of surface area (from about 72 to 17 m^2g^{-1}) when the calcination temperature was increased from 450 to 900°C (see Fig. 4.5). The stability of the monoclinic phase could be improved significantly by the addition of species such as Ca, Y, or La, as shown in Fig. 4.6; as little as 20%–50% of a theoretical monolayer of one of these ions was sufficient to prevent sintering by preventing significant surface mobility of the Zr ions. Fig. 4.6 shows that the area of the zirconia, particularly with the addition of the rare earth oxides of Y and La, was now relatively unaffected as long as the calcination temperature was below about 700°C. Fig. 4.7 shows the effect of the concentration of the most effective

[14] P.D.L. Mercera, J.G. van Ommen, E.B.M. Doesburg, A.J. Burggraaf, J.R.H. Ross, Zirconia as a support for catalysts: evolution of the texture and structure on calcination in air, Appl. Catal. 57 (1990) 127–148.

[15] See, for example, P.D.L. Mercera, J.G. van Ommen, E.B.M. Doesburg, A.J. Burggraaf, J.R.H. Ross, Zirconia as a support for catalysts. Influence of additives on the thermal stability of the porous texture of monoclinic zirconia, Appl. Catal. 71 (1991) 363–391.

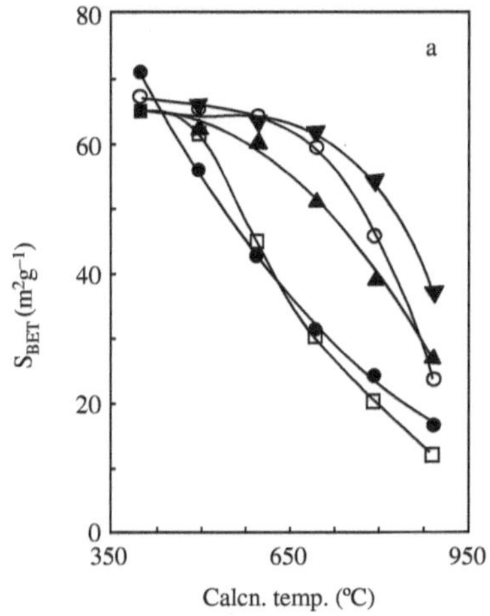

Figure 4.6
Influence of various additives (c. 5 mol%) on the thermal stability of monoclinic ZrO_2 ●, Undoped; □, MgO; △, CaO; ○, Y_2O_3; ▼, La_2O_3. *From Mercera et al., Appl. Catal. 71 (1991) 363–391. With kind permission of Elsevier.*

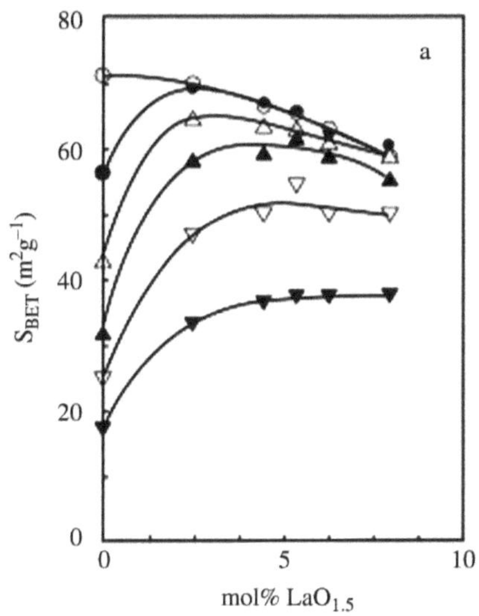

Figure 4.7
The effects of La_2O_3 concentration on the surface area of ZrO_2 as a function of calcination temperature: ○, 400; ●, 500; △, 600; ▲, 700; 800; ▼, 900°C. *From Mercera et al., Appl. Catal. 71 (1991) 363–391. With kind permission of Elsevier.*

additive, La, as a function of the La_2O_3 concentration for a series of different calcination temperatures; up to a calcination temperature of 700°C, as little as 3%La had a very significant stabilizing effect; even after calcination at 900°C, the addition of La caused a significant improvement in surface area (with values up to c. 35 m^2g^{-1} compared with about 17 m^2g^{-1} for the unpromoted material). We will return later to discuss the use of ZrO_2-supported catalysts in several different applications. (See also Task 4.6.)

Task 4.6 Zirconia as a Support

Examine the literature on the preparation, characterization, and uses of zirconia. A useful starting point is to examine the work of P.D.L. Mercera and his colleagues and the references that cite their papers.

4.3.7 Carbon

Carbon is often used as a catalyst support, particularly in very specific organic reactions, in the so-called Fine Chemical Industry. The resultant catalysts are frequently used in hydrogenation reactions that are carried out in the liquid phase (see Chapter 14) and it would appear that carbon-supported materials are particularly suited to such conditions, perhaps because they are less likely than many other supports to interact with the liquids used. There is also some indication that the pore structure is such that physical adsorption of the reactant molecules occurs in the narrow pores, bringing the reactant into closer proximity to the active site (often metal microcrystallites). The "activated" carbons used as supports are most commonly made from naturally occurring materials such as wood, coal, and peat.[16]

A consequence of the preparation routes used (normally pyrolysis), the resultant material can have surfaces containing a wide variety of different surface entities (carboxyl, keto, hydroxyl, etc.), the concentrations of each depending on the preparation and pretreatment conditions. It is therefore very unlikely that any two catalysts designated, for example, as 1 wt% Pd on C will have identical properties and so it is essential to have as much information as possible about the carbon support used.

4.3.8 "Forming" of the Support

Before the introduction of the active phase into any support, it is generally "formed" by granulation, pelleting, or extrusion, as discussed later, to give a suitably strong matrix in which to place the catalytically active material. Some typical examples of particle geometries used in commercial reactors are shown in Fig. 4.8. The objective of the forming process is to create suitably sized particles with large accessible external surface areas that have adequate strength so that they do not break down (fragment) in the reactor, particularly during the charging process; if such fragmentation occurs, it would be responsible for the build-up high back pressures or even reactor blockage.[17] The catalyst must also have long operating lives (often of the order of years of continuous operation)

[16] One of the more commonly used active carbon supports is made by the company Norrit in the Netherlands, now part of the Cabot Corporation http://www.cabotcorp.com, using peat from the peat bogs in that country. Norrit carbon was originally sold predominantly to remove the color from sugar syrup in the production of white sugar.

[17] S.P.S Andrew once described a typical industrial test for a mechanically strong catalyst: "take it into the Gents' washroom and throw it into the air; if it lands on the tiled floor without breaking into bits, it is strong enough."

Figure 4.8
Some typical catalyst geometries used in full-scale reactors.

and a good resistance to poisoning and sintering. However, first and foremost, it is essential in almost all cases that the catalyst particles have adequate pore geometries to enable the reactants to diffuse to the catalyst surface and for the products to leave the surface.

Granulation methods are used to form the spherical geometries depicted in Fig. 4.8. A technique more generally applied in the pharmaceutical industry, granulation allows the compaction of precursor powders, usually in slurry form, into well-defined spheres of the desired diameter and these are subsequently dried and calcined at the desired temperature. Granulated catalyst supports are not very common. However, granulation could be applied effectively if an egg-shell structure involving an inner and outer component was required, for example a layer with smaller particles (and hence smaller) pores on top of a more open structure or vice versa. (See also Task 4.7.)

Task 4.7 Granulation

Do a web search on granulation, paying particular attention to the types of equipment used for the process.

The pelleting process is similar to that used in the forming of pharmaceutical tablets: the appropriate powder[18] is added to a cylindrical cavity and pressure is applied above and below by pistons. The tablet is then released from the cavity and the procedure is repeated. The process is fully automated and typical tableting presses contain a large number of such cavities and pistons arranged round the circumference of a circular plate as depicted schematically in Fig. 4.9. Often, to enable the powder to flow properly, lubricants are added to the powder either before the process is started or as the powder is added to the die, these additives including surface-active agents such as stearates or even graphite. In some cases, other materials are also added to give the pellets formed extra strength. The material is then frequently "calcined" to remove any lubricant materials added during the pressing step. As a general rule, the strength of the final pellet increases as the applied pressure during pelleting is increased: necks form between the catalyst particles and these grow larger with

[18] The pelleting process can be applied either to the catalyst carrier alone or to the final formulation; in the case that only the carrier is pelleted, the active component is added to the preformed carrier by impregnation as discussed in the next section.

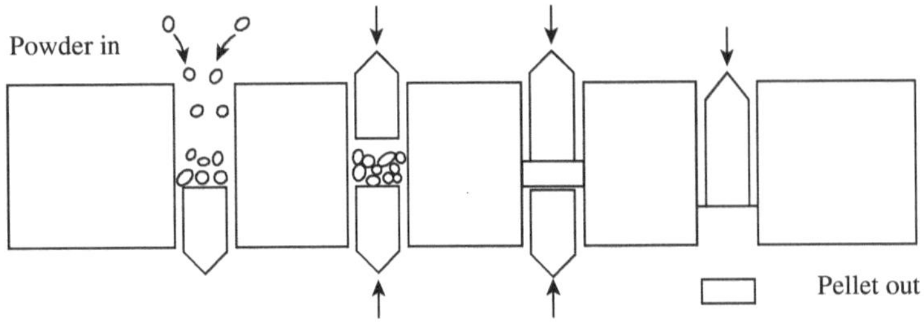

Figure 4.9
Schematic representation of the pelleting process.

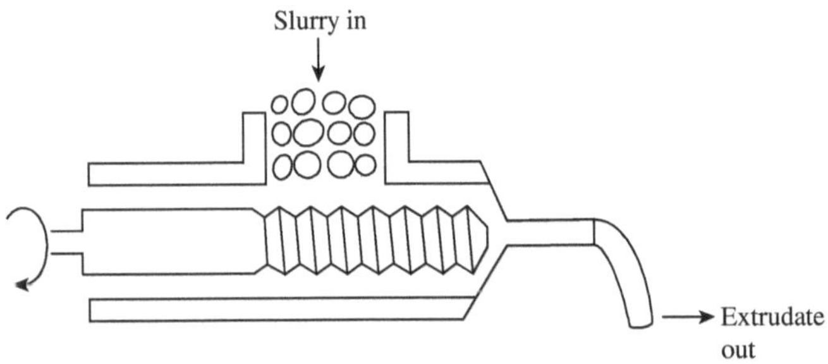

Figure 4.10
Schematic representation of the extrusion process.

increased time of pressing and applied pressure. The calcination process may also add to the strength of the pellet, especially if the precursor powder had not been calcined.

The extrusion process is similar to that encountered when squeezing toothpaste from a tube or mincing meat with a mincer (see schematic representation in Fig. 4.10): a slurry of the desired material is added to a hopper at the top and is then pressed through a nozzle of the desired shape using a screw-shaped central cylinder. Again, the extrusion properties of the slurry can be improved by the addition of surfactants, etc. The shape of the extrudate can be varied by using different dies; for example, Raschig rings and trilobes can be formed as well as ribbons and solid cylinders. Once formed, the extrudates must be dried carefully and they are then generally calcined to improve their strength.

We will now proceed to discuss the methods commonly used to prepare supported heterogeneous catalysts by impregnation or deposition/precipitation of the preformed support particles.[19] An alternative method of preparation, coprecipitation, will be described in the subsequent section (4.4.2).

[19] Catalyst preparation used to be described as an art rather than a science and it is only within the last 40 years that a more scientific approach to catalyst preparation has become more prominent. This is partly because of the initiation of the series of International Symposia on the Scientific Bases for the Preparation of Heterogeneous Catalysts which commenced in 1975 (Volume 1 of Studies in Surface Science and Catalysis) but also because of the realization that someone studying a particular catalyst or catalytic system cannot simply take it "off the shelf" but needs to have a full characterization of the specific material used.

4.4 Supported Catalysts

The active catalytic phase of a catalyst may be introduced into the support structure by either impregnation or by deposition–precipitation prior to drying and calcination. In both methods, a suitable solution is made up and introduced into the pores of the support material. For academic studies, this is often done using powders of the support but, for industrial use, the support would always be preformed as described earlier.[20]

4.4.1 Impregnated Catalysts

In the case of impregnation, the method depends on an interaction occurring between the surface of the support and the species in solution. This interaction may be as a result of ion exchange with the acidic hydrogen ions of surface hydroxyl groups or it may be more of a physical nature, especially if the concentration of impregnation species exceeds the concentration of surface hydroxyl groups. If the volume of impregnation solution introduced is equal to or less than the volume of the pores, the technique is known as "pore volume" or "dry" impregnation and the method depends more on retaining the species within the pores during drying rather than on a specific interaction; however, if the volume of solution is in excess of that required to fill the pores, the technique is known as "wet impregnation" and specific interactions play a much more important role. The advantage of dry impregnation is that the weight of added component incorporated in the catalyst can easily be controlled. However, the resultant material may not be as uniform as that prepared by wet impregnation; for example, there may be regions of the catalyst with larger pores in which there will be larger concentrations of catalytically active material than in others. As is shown schematically in Fig. 4.11, the larger pores will take up larger volumes of solution and the result is that there is likely to be larger crystallites of the precursor in these pores. In addition, if the drying process is not carried out carefully, the catalytically active material may become concentrated at the mouths of the pores or may even move onto the outer surface of the support (see later). If the support has been preformed, the distribution of pores and hence the distribution of the active material in the calcined catalyst will also be dependent on the parameters of the pelleting process.[21]

Wet impregnation is more difficult to control than is dry impregnation. If the concentration of solute ions is low, then all of the solute ions may be taken up at the mouth of the pores

[20] A preferable approach for academic studies is to impregnate preformed shapes of the type described earlier and then to powder and sieve the resultant material to give the desired particle sizes rather than just impregnating a powder of the support material. This is to avoid the build-up of back-pressure in the reactor and to optimize mixing of reactants and products, as will be discussed in Chapter 8, Mass and Heat Transfer Limitations and Other Aspects of the Use of Large Scale Catalytic Reactors.

[21] If the impregnation step is carried out on the powdered support prior to pelleting, more even distributions will be obtained as long as the particle size is quite uniform. However, changes occurring during the subsequent pelleting process will then be difficult to control.

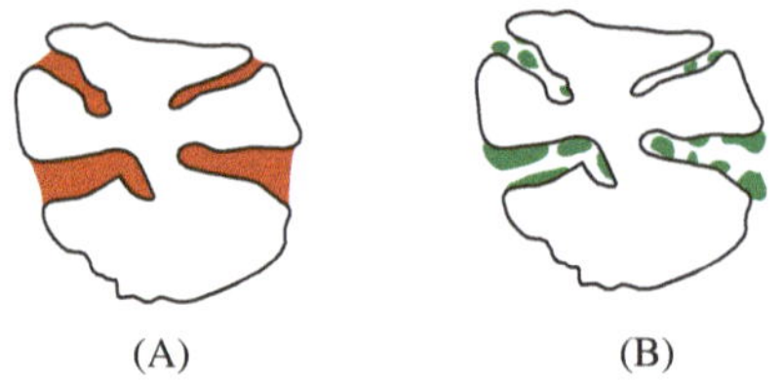

Figure 4.11
Pore filling (A) during wet impregnation and (B) the effect of subsequent drying.

(e.g., by ion exchange with acidic surface groups on the support surface near the pore mouths) and none will be found in the bulk of the support. The formation of an "egg-shell catalyst" with a high concentration of the active component at the outer surface of the pellet may of course be desirable either if this component is expensive (e.g., Pt or Rh) or if the reaction occurs predominantly near the outer surface of the catalyst pellets as a result of pore diffusion limitations (see Chapter 8). If, however, the reaction is totally kinetically controlled and all of the bulk of the catalyst is utilized (the "Effectiveness Factor" is 1.0), it is desirable that the catalytically active component is spread throughout the pellet. In that case, multiple impregnations may be necessary to ensure that the active component is present in all parts of the support and in sufficiently high concentrations.[22] (See also Tasks 4.8 and 4.9.)

Task 4.8 Chemical Reactions during Impregnation

In the above discussion of the preparation of impregnated catalysts, many aspects of the process have been omitted. For example, we have not considered the importance of the chemical step occurring when impregnation is carried out. Interaction with acidic groups on the surface can play a role (Brønsted vs Lewis sites) and the structure of the double layer forming at the support/solution interface can also determine the ease with which the reaction occurs. Paying attention to such parameters, follow some of the literature on the preparation of supported catalysts. For example, using the name of K.P. de Jong (University of Utrecht), use Scopus or its equivalent to identify some relevant papers to study. You could start with a paper from his group on the preparation of supported Pd catalysts [M.L. Toebes et al., J. Mol. Catal. 173 (1997) 281] and then read some of the papers either cited in this paper or those citing it. Other starting references could include M. Che, Chem. Rev. 97 (1997) 305; C. Parego, P. Villa, Catal. Today 34 (1997) 281; S.P.S. Andrew, Chem. Eng. Sci. 36 (1981) 1431. The paper by Parego and Villa, while giving a general description of catalyst preparation, also gives details of industrial practice; the paper by Andrew concentrates on examples of the preparation of industrial catalysts.

[22] It is possible to make catalysts in which the distribution of the active component is at a maximum somewhere between the outer surface and the interior or even that the maximum concentration is found right at the center of the pellet. Such distributions may have some practical advantages. [S.Y. Lee, R. Aris, The distribution of active ingredients in supported catalysts prepared by impregnation, Catal. Rev. Sci. Eng. 27 (1981) 207.]

Task 4.9 Monolayer Catalysts

A particularly well-studied system is the case of vanadia supported on titania for which various different preparation methods have been used. For example, the catalysts examined in a paper by Bond et al. (Brunel Univ.) and Gellings et al. (Univ. Twente) included aqueous impregnation of $(NH_4)_2[VO(C_2O_4)_2]$ and reaction with $VOCl_3$, $VO(OR)_3$ (R = iBu), and $VO(acac)_2$ in organic solvents (acac = acetonylacetonate). [Appl. Catal. 22 (1986) 361]. Using this reference as a starting point, examine the literature on vanadia/titania catalysts, paying particular attention to the effect of preparation method on the catalyst structure. Your study should include a more recent paper by I.E. Wachs and B.M. Weckhuysen [Appl. Catal. A 157 (1997) 67] that discusses the structure and reactivity of vanadium oxide species on oxide supports.

With both dry and wet impregnation, the drying process carried out after the impregnation step is very important. If the drying process is carried out too quickly, the liquid contained in the pores may be driven out of them by vapor forming deep in the pores, this resulting in deposition of the active phase on the exterior surface of the support. This is undesirable since such material is unlikely to adhere sufficiently strongly to the support and so "dusting" may occur in use: accumulation of a dust of the active component down-stream of the catalyst bed, this sometimes even resulting in blockage of the exit of the reactor. If the catalyst is to be calcined before use, the way in which this is carried out is also important. If the calcination step involves the decomposition of a salt such as a nitrate, the nitrogen oxides formed should be flushed out of the catalyst by a carrier gas to prevent the occurrence of undesired reactions. Even the elimination of any water vapor formed during the decomposition step should be carefully controlled as water vapor can encourage sintering in some oxides. Finally, if the catalyst has to be reduced before use, the reduction is best carried out slowly, often with a gradually increased temperature of reduction.[23]

4.4.2 Co-Precipitated Catalysts

4.4.2.1 Precipitation of single cations

Before considering coprecipitation, we will discuss briefly the precipitation of individual ions. The solubility of a compound, A^+B^-, is a function only of the temperature and is determined by the solubility product, K_s, a thermodynamic quantity:

$$K_s = a_A^+ \cdot a_B^- \tag{4.6}$$

[23] Reduction is most frequently carried out in the reactor where the catalyst is to be used as there are then no problems of reoxidation prior to use. However, there are cases in which a prereduced catalyst is supplied by the catalyst manufacturer. In such a case, some method has to be worked out to passivate the catalyst so that it is safe to transport and so that it can readily be reactivated in the reactor of the purchaser. Passivation is often carried out by very careful reoxidation of only the surface layer or layers by exposure of the catalyst to a low concentration of oxygen. (Often, the contamination level of oxygen in very pure nitrogen is sufficient for this purpose.) It should be noted that a prereduced catalyst is a potentially dangerous chemical: it can burn "pyrophorically" if the rate of reoxidation is sufficiently high and uncontrolled.

Table 4.2: Values of the solubility product (K_{sp}) for some common hydroxide compounds at 298K

Compound	Formula	K_{sp} (298K)
Aluminum hydroxide	$Al(OH)_3$	3×10^{-34}
Cobalt hydroxide	$Co(OH)_2$	5.92×10^{-15}
Copper (II) hydroxide	$Cu(OH)_2$	4.8×10^{-20}
Iron (II) hydroxide	$Fe(OH)_2$	4.87×10^{-17}
Iron (III) hydroxide	$Fe(OH)_3$	2.79×10^{-39}
Magnesium hydroxide	$Mg(OH)_2$	5.61×10^{-12}
Nickel hydroxide	$Ni(OH)_2$	5.48×10^{-16}
Zinc hydroxide	$Zn(OH)_2$	3×10^{-17}

where a_A^+ and a_B^- are the solubilities of the cation and anion, respectively. (For our purposes, B^- is often the hydroxide ion, OH^-). For hydroxide compounds containing multicharged cations, the expressions for the solubility product are more complex; as an example, the solubility product for aluminum hydroxide, ($Al(OH)_3$), is given by

$$K_s = a_{Al}^{3+} \cdot (a_{OH^-})^3 \tag{4.7}$$

Table 4.2 shows the values of K_s for a number of metal hydroxides commonly encountered in catalyst preparation. All of the hydroxides are very insoluble and so we can conclude that if OH^- ions are added to a solution containing one of the metal ions (cations) shown, there will be effectively 100% precipitation of the corresponding hydroxide. Further, we can make an estimate of the pH at which precipitation each of these hydroxide species will occur when the OH^- ions are added.[24] For example, if we start with a 1 M solution of aluminum nitrate, $Al(NO_3)_3$, the Al^{3+} starting concentration is 1.0 and so precipitation will start when the OH^- concentration is such that $K_s = 3 \times 10^{-34} = a_{Al}^{3+} \cdot \left(a_{OH}^-\right)^3$. In other words, $a_{OH}^- = 6.7 \times 10^{-12}$. In water, $K_w = a_H^+ \cdot a_{OH}^- = 10^{-14}$(i.e., $pH + pOH = 14$) and so this OH^- ion concentration corresponds to a pH of about 2.83.

If we now consider the case for a 1 M solution of $Ni(OH)_2$, we find that the pH at which precipitation will occur is about 6.37. In other words, if hydroxide ions are added to a mixture of Ni and Al ions, the aluminum species should precipitate at a much lower pH than should the nickel species. As we shall see in the following section (4.4.2), this is not the case.

4.4.2.2 Precipitation of several cations together: coprecipitation

When two cations are present in the precipitation solution and the pH of the solution is gradually increased by the addition of an alkali,[25] one would expect the two ions to

---

[24] In these calculations, we assume that the concentrations and activities are approximately the same. For more precise calculations, we would need to know the activity coefficients of the different species.

[25] This has been termed "the rising pH method" of coprecipitation.

precipitate separately at two different values of the pH corresponding to the individual solubility products; for example, with a solution containing nickel and aluminum ions, one would expect, as discussed earlier, that the aluminum species precipitate at a pH of less than 3 and that the nickel then precipitates at a pH of above 6. What appears to happen in practice is that the aluminum begins to precipitate at low pH but it then re-dissolves and coprecipitates together with the nickel at pH values well below 6. What is then formed is a mixed layered hydroxide compound (with a "brucite" structure) having a structure (see later) corresponding to that of minerals that are members of the so-called hydrotalcite group; in the case of the Ni–Al system the naturally occurring mineral is called "takovite." Coprecipitated Ni–Al materials were first introduced by British Gas as catalysts for the steam reforming of naphthas to produce a methane-rich gas for heating purposes. In the early work on these Ni materials, the formation of the layered structures was not recognized and it was only in a series of patents from BASF in 1970 that their importance was reported.[26] In work that we published in 1975, without knowledge of these patents, we showed that there was some form of interaction between the Ni and the Al ions in the precipitates.[27] In a later detailed study of the coprecipitation of Ni/Al catalyst precursors and the properties of the resultant materials, we showed[28] that the composition of the layered structure depends very much on the way in which the material is prepared. In particular, a much more clearly defined material could be obtained by carrying out the precipitation at constant pH: by adding a solution of the nitrates to water at the same time as adding alkali, maintaining the flows so that the pH remains constant. If the pH of precipitation is 5.0, the resultant precipitate contains nitrate ions whereas, if the pH is above 6.0, carbonate species become predominant. A relatively wide range of composition is possible but much of the work reported in the literature deals with an Ni/Al ratio of 3.0. The layered compound containing Ni and Al species in this ratio and prepared at a pH of 10.0 has the composition:

$$Ni_6Al_2(OH)_{12}CO_3 \cdot 5H_2O \tag{4.8}$$

[26] F.J. Brocker, L. Kainer, German Patent 2,024,282 (1970), to BASF AG, and UK Patent No. 1,342,020 (1971), to BASF AG.

[27] T. Beecroft, A.W. Miller, J.R.H. Ross, The use of differential scanning calorimetry in catalyst studies. The methanation of carbon monoxide over nickel/alumina catalysts, J. Catal. 40 (1975) 281.

[28] Some of the work carried out in a collaborative effort between the University of Bradford (United Kingdom) and the Technical University of Delft (The Netherlands) is summarized in the following publications and they can be consulted for further details: G. van Veen, E.C. Kruissink, E.B.M. Doesburg, J.R.H. Ross, L.L. van Reijen, Rn. Kinet. Catal. Lett. 9 (1978) 143; E.C. Kruissink, L.E. Alzamora, S. Orr, E.B.M. Doesburg, L.L. van Reijen, J.R.H. Ross, G. van Veen, in Preparation of Catalysts, II, B. Delmon, P. Grange, P. Jacobs, G. Poncelet (Eds.), (1979) 143; E.C. Kruissink, L.L. van Reijen, J.R.H. Ross, J. Chem. Soc., Faraday Trans. I, 77 (1981) 649; L.E. Alzamora, J.R.H. Ross, E.C. Kruissink, L.L. van Reijen, ibid, 77 (1981) 665.

Table 4.3: Some examples of hydrotalcite catalyst precursors prepared by coprecipitation

M^{2+}	M^{3+}	Balancing anion
Mg^{2+}	Al^{3+}	$CO_3^{=}$, OH^-, NO_3^-, etc.
Ni^{2+}	Al^{3+}	$CO_3^{=}$, $SO_4^{=}$, $V_{10}O_{28}^{6-}$, etc.
Mg^{2+}	Al^{3+}	$V_{10}O_{28}^{6-}$, $Mo_7O_{24}^{6-}$
Fe^{2+}	Al^{3+}	$CO_3^{=}$
Co^{2+}	A^{3+}	$CO_3^{=}$
Cu^{2+}	Al^{3+}	$CO_3^{=}$
Mg^{2+}	Fe^{3+}	$CO_3^{=}$
Cu^{2+}, Zn^{2+}	Al^{3+}	$CO_3^{=}$
Cu^{2+}, Co^{2+}	Al^{3+}	$CO_3^{=}$
Cu^{2+}, Zn^{2+}	Cr^{3+}	$CO_3^{=}$
Cu^{2+}, Co^{2+}	Cr^{3+}	$CO_3^{=}$

Source: *F. Cavani, F. Trifirò, A. Vaccari, Hydrotalcite-type anionic clays: preparation, properties and applications, Catal. Today 11 (1991) 173.*

If the pH is lower, then the $CO_3^{=}$ ions are replaced by two NO_3^- ions and the following composition is obtained:

$$Ni_6Al_2(OH)_{12}(NO_3)_2 \cdot 3H_2O \tag{4.9}$$

The nitrate or carbonate ions balance the excess charge introduced into the structure by the Al^{3+} ions. The structure of takovite is shown schematically in Fig. 4.9 in which the preparation and subsequent calcination and reduction of the resultant materials are also depicted. The Ni and Al ions are randomly distributed in an octahedral environment surrounded by OH^- ions in a brucite-like layered structure (brucite is $Mg(OH)_2$) and the brucite layers are stacked one above another with interlayers that are occupied by the charge balancing carbonate or nitrate ions plus the water of crystallization. When two nitrate ions are present in place of one carbonate ion, there is less space remaining for the water; the interlayer spacing is also dependent on the charge balancing anion, the being greater with the nitrate ions (0.85 nm) than with the carbonate ion (0.77 nm). It is worth noting at this stage that a wide variety of materials with hydrotalcite structures can be prepared (with cations that do not differ greatly in ionic radii from that of Mg^{2+}) and that many of these can be used as catalyst precursors, examples being the CuZnAl system. (See Table 4.3.) Further, a variety of different charge-balancing anions can be placed in the interlayer spacing.[29]

One of the greatest advantages offered by these coprecipitated materials containing the hydrotalcite structure is the ease and reproducibility of preparation: as long as the material is made using the same set of conditions and it is washed thoroughly before

[29] By precipitating a solution of Co and Al nitrates using ammonium molybdate, it was possible to make a layered structure in which the molybdate anion was located in the interlayer and the Co and Al ions were in the brucite structure. On calcination and subsequent sulfidation, this material had equivalent behavior for hydrodesulphurization as did a commercial $CoMoAl_2O_3$ catalyst.

further treatment to remove traces of undesirable alkali metal species introduced during preparation, the properties are identical from batch to batch. It should be noted that even when carbonate ions are carefully excluded during the preparation, the precipitate often contains at least some level of carbonate species as CO_2 is very easily picked up from the atmosphere. Removal of alkali ions is greatly facilitated if the precipitate is dried at c. 100°C to break down its gel structure prior to washing.

If the coprecipitated catalyst is to be used in a commercial reactor, it must be formed into pellets in the same way that the supports used in supported catalysts must be formed. For the Ni–Al materials discussed earlier, this is usually done by adding a lubricant of some sort (e.g., graphite) to the precipitate prior to calcination and pelleting the resultant mixture in a commercial tableting machine. It is then calcined (see Fig. 4.12) to remove the hydroxyl ions and carbonate species from the structure and, before use, is reduced in a flow containing hydrogen in the same way as for a supported catalyst. If the calcination is carried out at a relatively low temperature (350–500°C), the resultant material contains small particles of NiO stabilized by the presence of some Al^{3+} ions together with an alumina-rich phase containing trace quantities of Ni^{2+} ions; if the calcination is at higher temperature, the calcined material contains phases of NiO and $NiAl_2O_4$. Reduction of the material calcined at low temperature gives small Ni particles and these are stabilized by the alumina on the surface (see Fig. 4.13); the material calcined at higher temperatures contains large Ni particles that are not so well stabilized. (See also Task 4.10.)

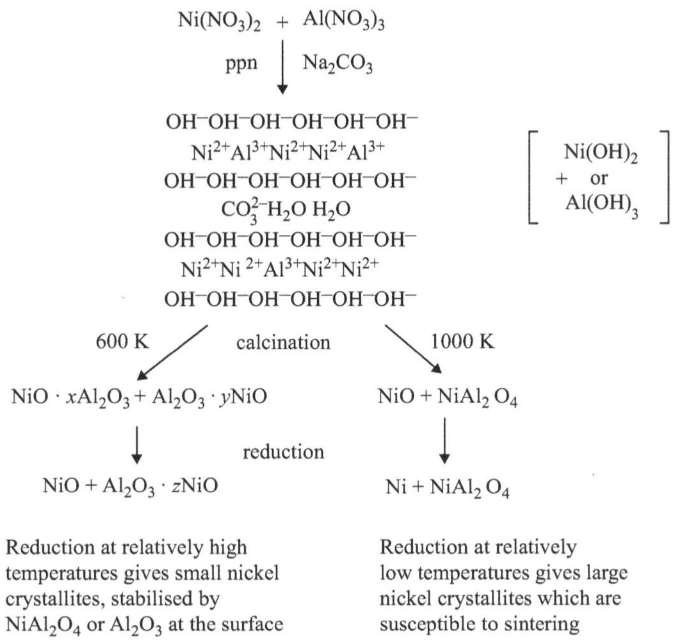

Figure 4.12

Steps in the preparation of coprecipitated nickel–alumina catalysts. *From L.E. Alzamora, J.R.H. Ross, E.C. Kruissink, L.L. van Reijen, J. Chem. Soc. Faraday Trans. I, 77 (1981) 665–681. With kind permission of Royal Society of Chemistry.*

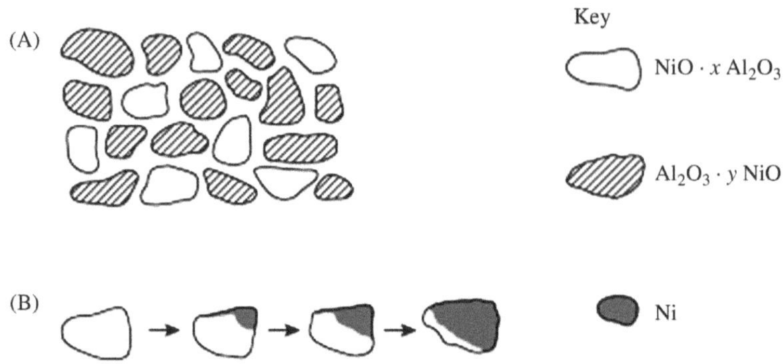

Figure 4.13

Model for the calcined Ni–Al catalyst prepared by coprecipitation and its reduction. (A) Calcined material. (B) Progression of reduction. The NiO particle size depends on the calcination temperature. The reduction rate is limited by the growth of an Ni nucleus in an NiO particle of constant size; sintering may occur after completion of the reduction by agglomeration of the Ni crystallites but is hindered by the presence of alumina on the surface of these particles. *From L.E. Alzamora, J.R.H. Ross, E.C. Kruissink, L.L. van Reijen, J. Chem. Soc., Faraday Trans. I, 77 (1981) 665–681. With kind permission of Royal Society of Chemistry.*

Task 4.10 Catalysts Derived from Hydrotalcites

Using the paper by Cavani et al. (see Table 4.3), examine the literature in relation to one or more of the uses of hydrotalcite precursors in catalysis. Various approaches to the vast literature on hydrotalcites could be used. For example, you could move forward in time (e.g., by following up on a highly cited author who has cited the Cavani review) to examine more recent catalytic work using these materials. Alternatively, you could concentrate on finding whether there has been any advance in structural knowledge or in the range of compounds that have been examined.

4.4.2.3 Precipitation and coprecipitation using urea decomposition

Urea decomposes in aqueous solution at temperatures of about 90°C to give NH_4^+ and OH^- ions. The method of preparation termed "deposition–precipitation" was first developed by Geus et al.[30] and involves adding an aqueous solution of a salt of the metal to be deposited (typically $Ni(NO_3)_2$) and urea to a support such as silica. The suspension is then heated and the urea hydrolyzes, giving OH^- ions evenly throughout the solution. The nickel species then precipitate homogeneously on the support surface. This is followed by calcination and reduction as required.[31]

[30] J.A. van Dillen, J.W. Geus, L.A.M. Hermans, J. van der Meijden, Proceedings of the 6th International Congress on Catalysis, G.C. Bond, P.B. Wells, F.C. Tompkins (Eds.), Chem. Soc. London (1977) 677.

[31] Such precipitation onto a support has also been achieved simply by adding an alkali dropwise to a suspension of the support in the salt solution with very vigorous stirring; however, this process is much more difficult to control and generally results in uneven distribution of the active components, there being a tendency to accumulate more on the external surfaces of the support rather than within the pores. Hence, deposition–precipitation is more reliable.

Urea decomposition has also been used successfully[32] to prepare coprecipitated hydrotalcite structures within the pores of a preformed support such as α-Al_2O_3. In this case, the urea solution contained not only $Ni(NO_3)_2$ but also $Al(NO_3)_3$ and $La(NO_3)_3$ and it was vacuum impregnated into the macropores of the α-Al_2O_3; the excess solution was then removed prior to heating the support. Precipitation of the layered material (in which the La was also incorporated in the brucite structure) was thus achieved; if necessary, the procedure could be repeated several times in order to increase the total Ni loading by partially decomposing the precipitate prior to each impregnation/deposition step. Ni contents of up to 8 wt% Ni were achieved. The addition of La was found to increase the Ni dispersion in the resultant catalysts and to stabilize the structure of the calcined and reduced catalysts as well as to improve the stability of the catalysts against carbon deposition during use in steam reforming. (See also Box 4.4.)

BOX 4.4 Ni/Al/La/α-Al_2O_3 Catalysts

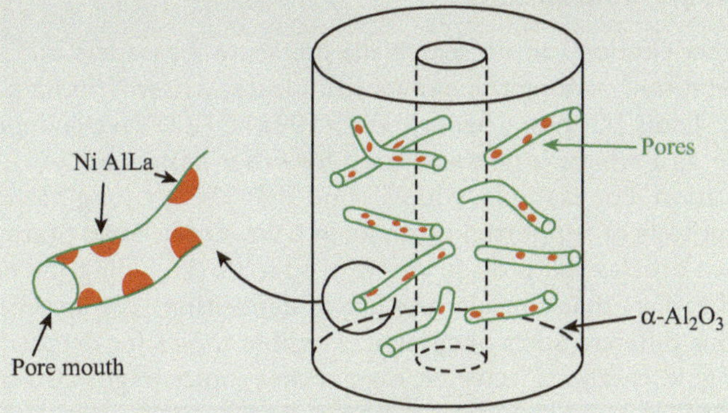

α-Al_2O_3 (corundum) is generally a dense ceramic material with low surface area (much less than 1 m^2g^{-1}) and its formation in a catalyst during use is generally considered to be very detrimental. The Ni/Al/La materials discussed in the main text are precipitated within the pores of α-Al_2O_3 extrudates that are in the form of Raschig rings, shown schematically in the figure. The pores are built into the structure of these supports by incorporating organic fibers of the desired dimensions into the aluminum hydrate precursor paste used in the extrusion process. The extrudates, in the form of hollow spaghetti-type tubes, are then sliced to the desired length and calcined at high temperatures to burn out the organic fibers and form the α-Al_2O_3. The hydrotalcite phase $Ni_6(Al + La)_2(OH)_{16}CO_3 \cdot 4H_2O$ is then incorporated in the pores by urea decomposition (see main text) and this, after decomposition and reduction, gives rise to the active promoted Ni catalyst. This catalyst has been used successfully for the production of hydrogen by steam reforming of methane and also for the steam reforming of higher hydrocarbons. (See Chapter 11: Production of Hydrogen and Syngas From Methane and Some Other Reactions of Methane).

[32] K.B. Mok, J.R.H. Ross, R.M. Sambrook, Preparation of Catalysts III, G. Poncelet, P. Grange, P.A. Jacobs (Eds.), Stud. Surf. Sci. Catal. 16 (1983) 291.

4.4.2.4 Decomposition of amino complexes

Schaper et al. have reported an alternative to homogeneous precipitation for the preparation of the hydrotalcite structures discussed earlier.[33] They prepared the amino complex of Ni by adding concentrated ammonia to a solution of $Ni(NO_3)_2$ containing concentrated nitric acid and then adding sodium hydroxide solution until the pH was 10.5. The appropriate alumina support, in the form of extrudates, was then added to the solution and heated to 90°C (while bubbling CO_2 through the solution) and the pH changed rapidly to 7.5 and then dropped further by a slower process. It was found that some of the alumina dissolved and was then reprecipitated, together with the Ni, as the hydrotalcite compound. After calcination and reduction, the resultant materials had similar properties to those made by the more conventional coprecipitation route discussed earlier. (See also Tasks 4.11 and 4.12.)

Task 4.11 Deposition–Precipitation

Use Scopus or another citation tool to search the literature for papers on "deposition–precipitation." Among the papers you locate, you will find a paper by P. Burattin, M. Che, C. Louis [J. Phys. Chem. B 102 (1998) 2722] on the deposition of Ni on silica. Use this paper as a source article to search for other papers on deposition–precipitation. For example, you will find that K.P. de Jong has cited this paper in his article on the synthesis of supported catalysts in Curr. Opin. Solid State Mater. Sci. 4 (1999) 55. These two articles will provide you with sources (including the paper by Van Dillen et al. referred earlier) that consider the use of deposition–precipitation for the preparation of various different sorts of catalyst. Possible topics for detailed study might be supported nickel or gold catalysts; however, many other topics might also be selected.

Task 4.12 Copper-Containing Catalysts for Methanol Reforming

Breen and Ross* have examined the behavior of a series of Cu–Zn–Al–Zr catalysts prepared by a sequential precipitation technique and have studied the effect of method of preparation and pretreatment on the total and Cu metal areas as well as on the activity for the steam reforming of methanol. Study this paper to learn more about the preparation and characterization methods used, especially the N_2O decomposition technique for measuring Cu surface area. You will also be able to see the effect of promoters on the activities and stabilities of the Cu catalysts. Where necessary, study some of the references given as well as carrying out a citation search on the paper to identify important papers published more recently on the subject.

*J.P. Breen, J.R.H. Ross, Methanol reforming for fuel-cell applications: development of zirconia-containing Cu-Zn-Al catalysts, Catal. Today 51 (1999) 521–533.

[33] H. Schaper, E.B.M. Doesburg, J.M.C. Quartel, L.L. van Reijen, Synthesis of methanation catalysts by deposition–precipitation, Preparation of Catalysts III, G. Poncelet, P. Grange, P.A. Jacobs (Eds.), Stud. Surf. Sci. Catal. 16 (1983) 301.

CHAPTER 5

Catalyst Characterization

Chapter Outline

5.1 The Importance of Catalyst Characterization

Having prepared a catalyst, it is essential then to characterize it. The main purpose of catalyst characterization is to be able to ascertain those aspects of the catalyst that have been prepared that are responsible for, or are related to, its catalytic behavior; an understanding of the catalyst structure can then, for example, be used to work out how the catalyst functions and how one might improve the material prepared. A secondary purpose is to be able to check the reproducibility of any preparation carried out, enabling one to show that each sample prepared under the same conditions and with the same chemical composition has the same physical and chemical (including catalytic) properties; after all, there is little point in preparing a new catalyst that performs a particular function, perhaps in a new reaction or in a very specific and novel way, if the catalyst cannot be described properly in such a way that others can reproduce the results and be sure that they have prepared the same material. Scheme 5.1 shows schematically the some of the approaches and methods used and Table 5.1 gives some details of the experimental techniques often applied. (See also Fig. 5.1.) The performance of Task 5.1 will provide additional information on the various techniques.

The techniques used for characterization range from a straightforward determination of the all-over chemical composition and phase structure of a catalyst to details of the active

Contemporary Catalysis.
DOI: https://doi.org/10.1016/B978-0-444-63474-0.00005-9

Table 5.1: Some experimental methods used for catalyst characterization

Method	Abbreviated Title	Description	Reference
Atomic absorption spectroscopy	AAS	Determination of all-over composition of catalyst sample dissolved in suitable solvent	Chapter 5, Catalyst Characterization
X-ray fluorescence	XRF	Determination of average composition of catalyst	Chapter 5, Catalyst Characterization
X-ray diffraction	XRD	Determination of phase structure and particle size	Chapter 5, Catalyst Characterization
Electron microscopy	EM	Determination of particle size and the composition of each particle	Chapter 5, Catalyst Characterization
Temperature programmed reduction	TPR	Probes any interaction between components that affect the reduction	Chapter 7, The Kinetics and Mechanisms of Catalytic Reactions
Thermogravimetry	TG	As TPR; can also be used to study adsorption and phenomena such as carbon deposition	Chapter 7, The Kinetics and Mechanisms of Catalytic Reactions
X-ray photoelectron spectroscopy	XPS	Gives the surface composition and also probes the electronic state of surface components	Chapter 1, An Introduction to Heterogeneous Catalysis—Chemistry in Two Dimensions See also http://www.xpssimplified.com
Low energy ion scattering	LEIS	Surface composition and structure	Task 4.12
Auger electron spectroscopy	AES	Electronic structure of surface components	
Secondary ion mass spectroscopy	SIMS	Gives surface composition	
Brunauer, Emmet and Teller surface area determination	BET	Gives total surface area and pore size distribution	Chapter 2, Surfaces and Adsorption
Low energy electron diffraction	LEED	Gives surface structure of single crystal samples	
Mercury porosimetry	Hg porosimetry	Gives pore size distribution	Chapter 5, Catalyst Characterization
Chemisorption	Chem	Can be used to determine metal areas or to look at acidity or basicity	Chapter 2, Surfaces and Adsorption
Infrared spectroscopy	IR	Used to examine adsorption processes, etc.	
Raman spectroscopy	Raman	Used mainly for studying surface phases of supported catalysts	
Extended X-ray absorption fine structure	EXAFS	Used to study surface phases of supported catalysts	

Note: A very useful series of descriptions of important analytical methods for surface characterization is given at http://www.mri.psu.edu/ under Materials Characterization Lab/Characterization Techniques.

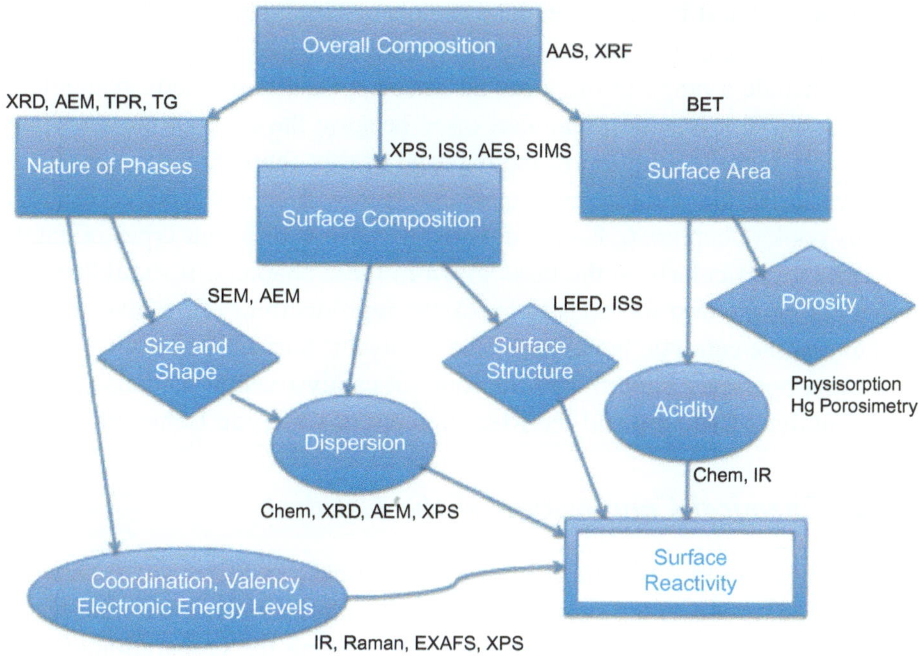

Figure 5.1

Schematic representation of some of the approaches of importance in characterizing a catalyst.

Task 5.1 Sources of the Types of Equipment Listed in Table 5.1

Use the web to identify suppliers of some of the equipment shown in Table 5.1 and add the details to the final column of the table.

---

species present on the catalyst surface (under conditions as close as possible to actual reaction conditions) and how they are arranged on the surface. In other words, one attempts to answer the question: what is the surface structure of the active catalyst? Remember that we are here dealing with what is essentially a two-dimensional structure (although it is sometimes important to understand the structure of underlying layers as well) and thus that the concentrations of the species in question are very low. As a consequence, many of the analytical techniques used routinely for studying the species taking part in a conventional homogeneous chemical process are either inapplicable or have to be modified to enable them to be used to examine the reactive surface species at the very low concentrations associated with adsorbed complexes.

The following sections (5.2–5.7) are focused predominantly on the more routine methods used in the majority of laboratories in which catalytic properties are examined, the use of which is often encountered in papers on catalytic phenomena; some of these are given alongside the various topics shown in Scheme 5.1 and are listed in more detail in Table 5.1. These lists also include a range of the more advanced specialist methods available to the catalytic chemist but in less widespread use, often because the costs of the associated equipment are very high. In addition to the material of this chapter, some specific examples of catalyst characterization and the techniques involved will be encountered in the later chapters of this book (Chapters 6, 8–15) devoted to various different types of catalysts and catalytic processes, particularly in the tasks given in these chapters. It should be stressed at this stage that one of the best methods to check the reproducibility of a given preparation technique is to test the catalytic behavior of the catalyst ("Surface Reactivity" in Scheme 5.1) under normal operating conditions. The topic of catalyst testing is considered in detail in Chapter 6, Catalytic Reactors and the Measurement of Catalytic Behavior.

5.2 Overall Chemical Composition of the Catalyst

The overall chemical composition of a catalyst is a critical parameter and it should always be given in any description of the properties of a particular sample. While the composition is often exactly the same as that which was aimed for during the preparation of the sample and can be calculated from the concentrations of the reagents used in the preparation, this is not always the case. For example, samples prepared using the impregnation method may have metal ion loadings below those calculated by assuming that all the species added were adsorbed during preparation, the difference being a result of incomplete adsorption. Similarly, not all the ionic species present in solution during a coprecipitation procedure may be precipitated or some may be redissolved during washing stages. Hence, the chemical composition should be obtained to make sure that the sample composition is as expected. The following sub-sections (5.2.1–5.2.3) outline some of the commonly used methods of analysis.

5.2.1 Wet Analytical Techniques

The chemical composition of a catalyst can be obtained by conventional analytical methods: dissolving up the sample in a suitable solvent (often an acid) and then carrying out standard wet analytical procedures (either volumetric or gravimetric) to obtain appropriate figures. However, such methods can present a variety of problems; for example, the various components of a multicomponent catalyst can cause interference with one another during analysis or one or other component may not be completely soluble under the dissolution conditions applied. Spectroscopic methods are also frequently used to determine the composition of the solution resulting from the dissolution step, the most commonly used being atomic absorption spectroscopy (AAS). Several other spectroscopic methods, for example, UV–visible spectroscopy, have also been used for analysis purposes for specific catalyst compositions, each having special advantages for a particular species, and it is advisable to study the relevant literature for any type of catalyst to find which method is most commonly used for that formulation.

5.2.2 X-Ray Fluorescence

Another commonly method applied to obtain the elemental composition of a catalyst is X-ray fluorescence (XRF), a technique that determines the elemental composition of a small sample of the material. A useful tutorial on the technique is given at: http://www.horiba.com/scientific/products/x-ray-fluorescence-analysis/tutorial/ and a more general discussion of the method is given at: http://en.wikipedia.org/wiki/X-ray_fluorescence. To carry out an analysis, a sample of the catalyst is powdered and formed into a glass with, for example, sodium borate, to give a disc with a flat surface that can be inserted into the appropriate XRF instrument. The resultant peaks are compared with those of standard mixtures of the various components prepared in the same way to enable the composition to be determined. The XRF technique cannot be applied to hydrogen, helium, and lithium ($Z \leq 3$) and is most widely applicable for elements heavier than sodium. Also, because the escape depth of secondary X-rays from a sample containing the lighter elements is relatively short, the sample must be quite uniform and the disc must be flat. (See also Task 5.2.)

Task 5.2 X-Ray Fluorescence

Carry out a web search on the XRF technique, paying particular attention to details of the type of equipment available and the methods of sample preparation used to examine catalyst samples. It should be noted that there is a very readable article on the subject available on Wikipedia.

5.2.3 Nature and Structure of the Phases Present in a Catalyst

Having determined the chemical composition of a catalyst, it is also often necessary to learn more about the phase structure. This is generally achieved using X-ray diffraction, a standard method that can be used to give the structures of the phases present (http://en.wikipedia.org/wiki/X-ray_crystallography). It is important to note that XRD requires a three-dimensional structure to be present and that it therefore gives its most valuable information when applied to catalysts consisting of a number of different components that form distinct structures or from which new structures can be formed.[1] It is also of great value in examining the structures of support materials. It should be recognized that the technique seldom supplies much useful information when applied to dilute supported phases

1 Several examples of the application of XRD were given in Chapter 4, Catalyst Preparation.

such as monolayers of oxide on a support. This is because the concentration of these phases is not sufficiently high to give coherent diffraction patterns. However, as will be discussed in Section 5.5.2, the method can also in some cases be valuable in determining particle sizes of any active phases on a support that are visible by determining the extent of so-called "line broadening." As with many of the other physical techniques, it is important to study the catalyst under conditions approaching as closely as possible the conditions used for catalysis and, as a result, in situ (or operando) methods are now commonly used. These methods are particularly valuable for examining any changes in the physical structure of a catalyst that may occur during its use. (See also Task 5.3.)

Task 5.3 X-ray Diffraction to Study the Structure of a Catalyst

An excellent example of the application of XRD to the study of catalyst structures is a review article by D.C. Puxley, I.J. Kitchener, C. Komodromos, N.D. Parkyns (The effect of preparation method upon the structure, stability and metal/support interactions in nickel/alumina catalysts, in Preparation of Catalysts III, G. Poncelet, P. Grange, P.A. Jacobs (eds.), Studies in Surface Science and Catalysis, Elsevier, 16 (1983) 237–271). Using this paper as a starting point, identify other papers which concentrate on the use of XRD, in particular those using in situ or operando methods.

5.3 Surface Composition

As we have seen, catalysis involves reaction at a surface and so surface composition is all-important in determining the behavior of a catalyst. Such characterization must be carried out under conditions that are as close as possible to those to be encountered by the catalyst in use. For example, there is little point in characterizing an oxidized metal catalyst if the material is to be used in its reduced form as the reduction process may bring about significant structural changes as well. As almost all metals are immediately oxidized, at least at the surface, upon exposure to even a very low partial pressure of oxygen, it is necessary to be able to study a catalyst with a metallic surface under conditions such that the surface is not reoxidized during the measurement procedure and this usually requires handling the sample under vacuum conditions. Similar considerations also apply to the characterization of many oxide materials since the oxidation state may be changed by variation of the surrounding atmosphere. As was discussed in Chapter 1, many new surface-sensitive analysis systems became generally commercially available in the latter part of the 20th century following the development of ultra-high vacuum techniques and of dependent methods such an X-ray photoemission spectroscopy (XPS), auger spectroscopy (AES) (see Tasks 5.4 and 5.5), secondary ion mass spectrometry (SIMS), Rutherford backscattering spectrometry (RBS), low energy ion scattering (LEIS), electron microscopy (measuring back-scattered X-rays in a technique known as energy dispersive analysis of X-rays,

EDAX) and extended X-ray absorption fine structure (EXAFS).[2,3] With these techniques, the signals measured emanate from the surface, or from near the surface, and the results obtained are related predominantly to surface rather than to bulk composition. Of these methods, LEIS is probably the most surface-sensitive as it depends on the interaction of bombarding ions with the surface atoms or ions. (See also Task 5.6.) Some of these

Task 5.4 Electron Escape Depths in XPS and Auger Spectroscopy

The degree of surface sensitivity of any spectroscopic technique involving the emission of electrons from a surface (e.g., XPS or AES) depends on the kinetic energy of the electrons and can vary from about 0.5 nm to about 10 nm, this corresponding to from several layers to many layers of atoms. According to Somorjai (*Chemistry in Two Dimensions: Surfaces,* Cornell University Press, Ithica, 1981), the optimum surface sensitivity is obtained with electrons in the range 25–200 eV. Do a web search on the topics of XPS and AES and find some examples of applications to catalytic systems. Can these techniques be applied to in situ studies?

Task 5.5 X-Ray Photoelectron Spectroscopy

The topic of the use of XPS under reaction conditions has been reviewed by Knop-Gericke and his colleagues (Advances in Catalysis, 52 (2009) 213–272). Examine the recent progress that has been made in studying catalysts under reaction conditions and the use of synchrotron radiation. Use this reference as a starting point to examine some of the most recent applications of the technique in a currently highly researched area of heterogeneous catalysis of your choice.

Task 5.6 Application of LEIS to Supported Catalysts

R.H.H. Smits and his colleagues (R.H.H. Smits, K. Seshan, J.R.H. Ross, L.C.A. van den Ootelaar, J.H.J.M. Helwegen, M.R. Anantharaman, H.H. Brongersma, A low energy ion scattering (LEIS) study of the influence of the vanadium concentration on the activity of vanadium-niobium oxide catalysts for the oxidative dehydrogenation of propane, J. Catal. 157 (1995) 584–591), have compared LEIS and XPS results for coprecipitated vanadia-niobia catalysts and have shown that LEIS is the more surface-sensitive technique. Use this reference as a starting point to examine the current status of the method, especially in relation to studying supported catalysts.

---

2 As noted by Delannay and Delmon in 1984 (F. Delannay, B. Delmon, Characterization of Heterogeneous Catalysts, F. Delannay (ed.), Marcel Dekker, Chemical Industries, No. 15, 1984, 1–28), the relative use of the various methods appears to have depended to a greater extent on the commercial availability (and cost) of the equipment rather than on the relative usefulness. Such an argument probably still applies.

3 These and other techniques are admirably described in some detail in a book by J.W. Niemantsverdriet entitled Spectroscopy in Catalysis. An Introduction, first ed., Wiley VCH, Weinheim, Germany, 1993, updated regularly since.

techniques will be discussed further in subsequent chapters in relation to particular catalytic systems. In the present context, it is sufficient to recognize that these techniques exist and that their application (often by specialists in the experimental methods concerned) can give very important information not only concerning surface composition but also about changes in electronic structure of surface species.

5.4 Surface Area and Porosity

As will be discussed in Chapter 8, the pore structure of a catalyst is frequently important parameter in determining its behavior in use since the pore diffusion is quite often a contributor to the all-over kinetic behavior of a working catalyst. The pore structure of the underlying support will also be critical in relation to the preparation of the catalyst, particularly when incipient wetness methods of preparation are used (Chapter 4). In a typical paper discussing the preparation of a new catalyst, one of the first sets of parameters given will be the total surface area as well as the pore volume and diameter. The methods used to measure total surface area and pore size distribution by measuring the physical adsorption of a gas such as nitrogen were discussed in some detail in Chapter 2.

The pore size and distribution of pores can also be determined by mercury porosimetry. This technique depends on the penetration of mercury through the pores of a solid under the influence of an applied pressure. If mercury is forced into a pore of diameter D and the contact angle of the mercury with the surface of the pore is θ, the pressure P necessary to give penetration of the pore is given by the Washburn equation:

$$D = -4\gamma\cos\theta/P$$

Hence, by measuring the volume of mercury forced into the pores as a function of the applied pressure with each increment of pressure change, the volume of the pores in each corresponding range of diameter can be determined. A good description of the techniques involved and a full derivation of this equation can be found on the website of one of the equipment managers: http://www.micromeritics.com/ and also on the website of the Materials Characterization Lab of Penn State University: http://mri.psu.edu/; the latter site also gives information on a wide variety of other materials characterization methods, many of which are applicable to catalysts. (See also Task 5.7.)

Task 5.7 Mercury Porosimetry

Do a web search on the topic of porosimetry and examine the theory behind the measurements, making sure that you understand the terms contact angle and surface tension. Attempt to find some recent publications in which the technique is applied. Can it be applied universally to all porous catalyst samples or are there restrictions as to its use? Discuss these limitations.

5.5 Size, Shape, and Dispersion

5.5.1 Electron Microscopy

The physical texture of a catalyst or its precursor(s) is often examined using scanning electron microscopy (SEM). (The associated EDAX measurements, see above, can be used to analyze the composition of selected areas of the image). Because the resolution of SEM is about 5 nm, the method does not give information on an atomic scale but can give useful data on the shape and size of micropores and, for example, changes occurring during pelleting of a catalyst sample. The much higher resolution obtainable using transmission electron microscopy (TEM) allows one to observe even individual atoms of a surface, this being particularly useful when studying small metallic or oxidic crystallites on a particular support. Careful and systematic measurements using TEM also allow estimates to be made of the particle size distributions occurring in a particular sample. Detailed information on such crystallites can also be obtained using Scanning Probe Microscopic methods such as atomic force microscopy (AFM) and scanning tunneling microscopy (STM). Modern electron microscopic methods (see Task 5.8) include developments in hardware that allow the study of particles in gaseous environments that simulate catalytic reaction conditions, as illustrated in Fig. 5.2. This involves the addition of a section in the microscope body that

Task 5.8 Electron Microscopy

Electron microscopy is a technique that has been very well established for decades and it has very frequently been used as a standard method for characterizing catalysts and to gain an estimate of the sizes and size distributions of metallic particles on a support. Such measurements have always been characterized by some uncertainty regarding the clarity of any images obtained, particularly for very small particles (so-called nanoparticles), due to aberration problems in the microscope and other factors. However, modern developments of the hardware and software associated with the most up to date electron microscopes have made it possible to study very small particles and even to identify individual atoms. The most up-to-date developments in the subject are described in an excellent review by D.S Su, B. Zhang, R. Schlögl (Chem. Rev. 115 (2015) 2818–2882) where it is shown how the resolution and clarity of electron microscope images have been improved such that it is now possible to view features as small as 0.5 Å; see, for example, Fig. 5.3. Using this or some similar review as starting point, examine some of the most recent applications of electron microscopy, for example, in examining novel supported nanoparticle Pd catalysts for formic acid decomposition or Ni catalysts for the dry reforming of methane. It could also be helpful to examine the web pages of the manufacturers and suppliers of electron microscopes as some of these give very clear diagrams illustrating the techniques involved.

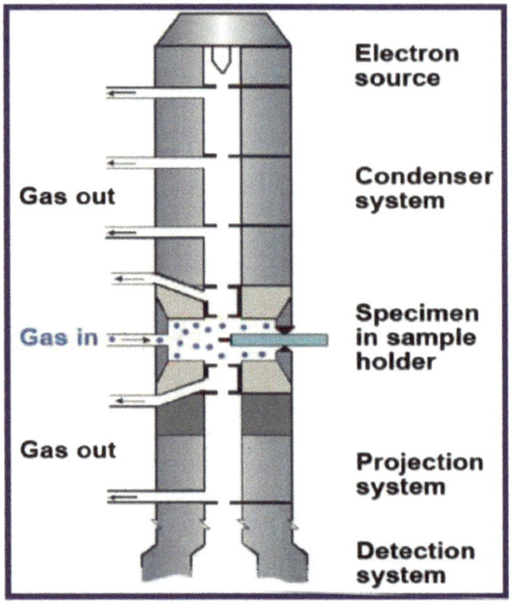

Figure 5.2
Schematic representation of an electron microscope arrangement to allow differential pumping to enable the study of solid particles in gaseous atmospheres. *Reproduced from: J.R. Jinschek, S. Helveg, Micron, 43 (2012) 1156–1168 with permission from Elsevier.*

allows differential pumping either side of the specimen chamber that can itself be maintained at a higher pressure.[4]

The resolution of electron microscopes is now so high that it is possible to examine the structures of catalysts in minute detail. Fig. 5.3 shows an example of a micrograph obtained by depositing a monolayer of iron oxide on top of a (1,1,1) single crystal of Pt and then depositing Pt on top of the iron oxide layer. The micrograph shows clearly that the iron oxide from the coating layer has diffused over the surface of the deposited Pt crystallite (Fig. 5.3.A). The enlargements of the area around the Pt particles (Fig. 5.3.B and C) show that the iron creeps up the side of the Pt particle and stops as soon as it reaches the top. This is a very clear demonstration the fact that it is now possible to image individual atoms on a surface. The example shown is a clear illustration of the phenomenon of a so-called "Strong Metal Support Interaction" (see Task 5.9), the partial or complete spreading of a support over the surface of a metal crystallite that might otherwise have been expected to expose its total metallic surface to the reacting atmosphere above it.

4 Similar approaches have been used in XPS systems to allow in situ studies under pseudo-reaction conditions.

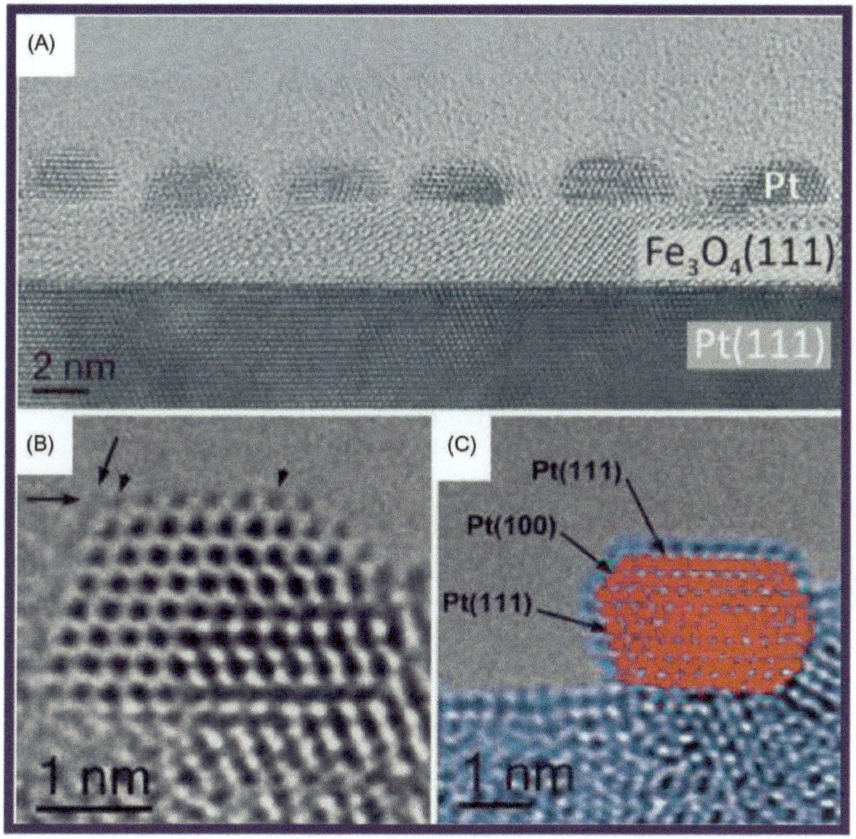

Figure 5.3

(A–C) Model catalyst for the identification of SMSI of iron oxide over Pt nanoparticles. *From: M.G. Willinger, W. Zhang, O. Bondarchuk, S. Shaikhutdinov, H.-J. Freund, R. Schlögl, A case of strong metal – support interactions: combining advanced microscopy and model systems to elucidate the atomic structure. Angew. Chem., Int. Ed. 53 (2014) 5998–6001. Reproduced by kind permission of Wiley-VCH Verlag, Weinheim.*

Task 5.9 Strong Metal Support Interaction (SMSI) Effects

The topic of strong metal-support interactions was introduced in 1978 to describe the changes that had been found to occur with particles of noble metals attached to the surface of titania (TiO_2), the result being a drastic change in hydrogen chemisorption properties. The topic has been reviewed many times, one of the most cited of these reviews being that by S.J. Tauster in Acc. Chem. Res. 20 (1987) 389–394. (Tauster shows many electron micrographs of the quality attainable at that time and these should be compared with that shown in Fig. 5.2). Using this reference, do a literature search on the topic of SMSI, paying particular attention to modern developments of the subject.

5.5.2 X-Ray Line Broadening

The technique of X-ray line broadening can be used to give information on the particle sizes present in a catalyst sample. This method uses the Scherrer equation:

$$B(2\theta) = K\lambda/L \cdot \cos\theta$$

where the peak width, $B(2\theta)$, at a particular value of 2θ (θ being the diffraction angle, λ the X-ray wavelength) is inversely proportional to the crystallite size L; the constant K is a function of the crystallite shape but is generally taken as being about 1.0 for spherical particles. X-ray line broadening gives average values of crystallite size which in most cases agree quite closely with those obtained by TEM; the great advantage of the technique is that the problems of obtaining representative results are less exacting.

5.6 Acidity or Basicity

There are many catalyst systems with which the surface acidity or basicity of the catalyst is of importance and these are commonly probed by carrying out "titrations" of the surface using bases or acids respectively. Various methods can be used to follow these titrations but the most common are simple colorimetric methods or the use of infrared (or UV) spectroscopic methods.

5.7 Catalytic Activity, Selectivity and Stability Under Operating Conditions

The most important properties of a catalyst are its activity and selectivity under operating conditions. Hence, an essential feature of characterizing a newly synthesized catalyst is the determination of the catalytic properties of the new material. It is hardly necessary to say that there is no sense in making a new catalyst if it does not exhibit good catalytic properties. Nevertheless, it is amazing to find the number of publications that do not give adequate data on the properties of the samples under investigation, particularly on the relation of the behavior of the new material for the reaction of interest relative to that of known catalysts under practically relevant conditions. Further, in making comparisons, it is essential to have a full understanding of the effects of differing testing conditions between one laboratory and another on the results obtained. Laboratory methods of testing catalysts from the point of view of characterization and comparison will therefore be the topic of Chapter 6. However, to make meaningful comparisons, it is often essential to know the kinetic expression describing the reaction in question over the new catalyst. The following chapter (Chapter 7) therefore considers the kinetics of catalytic reactions and discusses some of the more commonly encountered mechanistic approaches.

CHAPTER 6

Catalytic Reactors and the Measurement of Catalytic Behavior

Chapter Outline

6.1 Introduction

In Chapter 3, How Does a Catalyst Work?, it was shown that a catalyst speeds up a reaction by offering an alternative, more energetically favorable route for the reaction. Having seen in Chapters 4 and 5 how a catalyst can be made and characterized, we now examine an important aspect of catalyst characterization that was omitted from the latter chapter: how to examine the behavior of a catalyst for the reaction of choice. This chapter therefore answers the question: how does one measure the rate of a catalytic reaction in a laboratory and compare one catalyst with another? The problems of scale-up to a full-sized reactor will be discussed in the Chapter 8, Mass and Heat Transfer Limitations and Other Aspects of the Use of Large-Scale Catalytic Reactors.

It is assumed that the reader is conversant with methods of studying the kinetics of homogeneous reactions as well as with the concepts of order of reaction and the effect of temperature on reaction rate (the Arrhenius equation). These are well covered in many texts on Physical Chemistry and one of these should probably be perused as revision prior to

Contemporary Catalysis.
DOI: https://doi.org/10.1016/B978-0-444-63474-0.00006-0

studying this chapter. In studying the kinetics of a homogeneous liquid phase reaction such as the hydrolysis of an ester by an acid, one needs to be able to measure the concentration of at least one of the reactants or products as a function of time at a specific temperature. This is most commonly done in a vessel such as a stirred beaker immersed in a suitable thermostat (a typical constant volume reactor) from which small aliquots of the reaction mixture can be withdrawn for analysis (e.g., by titration) at regular intervals. The data are then plotted as a graph of concentration of reactant or product as a function of reaction time. In order to determine the kinetic expression governing the reaction, the data are tested in various integrated rate expressions and rate constants are determined; the activation energy for the reaction is then calculated. (See also Box 6.1.)

BOX 6.1 Kinetics of a Homogeneous Reaction

Consider the reaction in which the starting concentrations of A and B are the same:

$$A + B \rightarrow C + D$$

It is assumed that the stoichiometry of the reaction is given by the chemical equation given and so the rate of reaction can be expressed by:

$$\text{Rate} = -k_T \cdot [A]^n (= -k_T \cdot [B]^n)$$

where k_T is the rate constant at temperature T (in K) and n is the order of the reaction. If the reaction is first order, $n = 1$ and the equation may be integrated to give:

$$\ln([A]/[Ao]) = k_T \cdot t$$

where [Ao] is the initial concentration of A and t is the time of reaction. Hence, a plot of ln [A] versus t gives a straight line of gradient $-k_T$. The experiment can be repeated at various temperatures, T, and the values of k_T can be determined at each.

As the rate constant is related to the activation energy E_r by the Arrhenius equation:

$$k_T = A\exp(-E_r/RT)$$

a plot of $\ln k_T$ versus $1/T$ yields a straight line of gradient $-E_r/R$, where R is the gas constant.

A similar approach would be adopted if one were to follow the kinetics of a gas-phase reaction such as the oxidation of NO to give NO_2:

$$NO(g) + 0.5O_2(g) \rightarrow NO_2(g) \tag{6.1}$$

in which (g) indicates that the reactants and product are in the gas phase.

In this case, an analytical method has to be used which would allow the partial pressures of the reactants and products to be measured; if the stoichiometry of the reaction is followed exactly (e.g., neither N_2O nor N_2O_5 is formed), a simple pressure change measurement using a constant volume reaction system suffices, for example, a spherical Pyrex vessel carefully thermostatted in a suitable container.[1] However, many other analytical methods, for example, gas chromatography or mass spectrometry, can be used to monitor the partial pressures of the individual reactants and products.

If a fast reaction is being studied, then it is not practicable to use a constant volume system of the type described earlier as insufficient analyses can be carried out through the duration of an experiment. A wide range of techniques have therefore been developed for studying the kinetics of fast reactions, these including flow reactors using spectrophotometric measurement of the concentrations of reactant and product. In such reactors, reactants are brought together rapidly and thoroughly mixed at the entrance to a flow tube. The concentrations of reactants and products are then measured at various distances from the point of mixing, these corresponding to the time of reaction. Alternatively, in the so-called stop-flow reactor, the flow can be stopped at a given moment and the concentrations measured as a function of time at a particular position downstream of the point of mixing.

6.2 Static Reactors

By analogy with the reactors used for homogeneous systems discussed earlier, it is possible to inject the reaction mixture into a constant reaction volume containing a sample of the catalyst (Fig. 6.1) and to measure the concentrations of the reactants and products as a function of time of reaction. If such an approach is adopted, it is important to recognize an important factor in studying the rates of such reactions: the rates at which gas-phase mixing of reactants and products can take place, this in turn depending on the *mean free paths* of the constituent molecules. The mean free path, λ, of a molecule is given by:

$$\lambda = RT/\left\{\sqrt{2}\cdot N_A\cdot\sigma\cdot p\right\} \qquad (6.2)$$

where R is the gas constant, T is the temperature (K), N_A is Avogadro's number, σ is the collision cross-section of the molecule,[2] and p is the pressure. For oxygen at room temperature, the mean free path at a pressure of one atmosphere can be calculated to be 7.3×10^{-8} m = 73 nm. It is thus clear that there will be many more collisions between the constituent molecules of pure gas or of a gas mixture than there will be with the walls of a reactor of a diameter of several mm or greater. This has several very significant consequences. Most important for our present purposes is that because effective mixing of

[1] For such measurements at higher temperatures, the reactor may be immersed in a heated liquid or in a fluidized sand bath or furnace.

[2] The value of σ depends on the molecular weight of the molecule in question; for hydrogen, the value is 0.27 nm^2 while for oxygen it is 0.40 nm^2 and for benzene it is 0.88 nm^2.

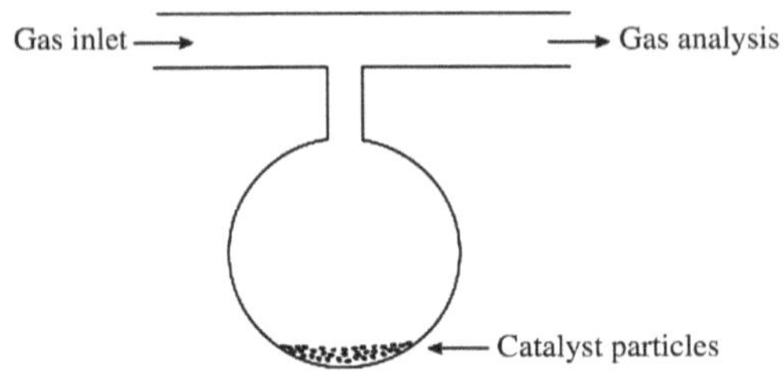

Figure 6.1
Low-pressure constant volume reactor used for the study of a catalytic reaction on a catalyst powder or granules.

two or more gases depends on the movement of the molecules past one another and that this movement will be hindered by collisions between the molecules, it will be difficult to maintain perfect mixing in a constant volume reactor; this is because the reaction takes place only in one part of the reactor—at the surface of a catalyst bed located at the base of the reactor—rather than throughout the reactor as would be the case in a homogeneous reaction. This means that if a catalytic reaction is carried out on a small sample of catalyst in such a reactor at a pressure of 1 bar and analysis is carried out using, for example, mass spectrometry, the spectrometer entrance being located some distance from the catalyst sample, the lack of mixing can lead to spurious results.[3]

A solution to the problem of mixing is to carry out the reaction at lower pressures, for example, in a preevacuated reaction volume, introducing the reactant gases via suitable valves. As a rough rule of thumb, if the mean free path approximates to the dimensions of the reactor and the numbers of collisions with the walls becomes significant relative to the number of gas-phase collisions, perfect or near-perfect mixing can be expected. In practice, this means that the upper pressure of the reactor should not exceed about 10^{-3} bar (mean free path approximately 10^{-4} m or 0.1 mm) if mixing rates are to be sufficiently high to avoid analytical problems. Box 6.2 gives an example of an investigation carried out using such a reactor system: the steam reforming of methane over a prereduced Ni/alumina catalyst. The use of a low-pressure system has two advantages here: first, the catalyst can be prereduced in situ and the reduction process can be monitored easily; and second, since the steam reforming of methane is an endothermic equilibrium reaction, the conversion approaches 100% at lower temperatures (in this case, c. 600°C) than if the reaction was to be studied at higher pressure; at higher pressures, temperatures above 800°C

[3] Similar arguments regarding the importance of the mean free path in relation to mixing of gases apply also to the situations encountered with rate determinations with catalyst pellets: in this case, limitations may be due to internal and external diffusion. We have already referred to these in relation to Fig. 3.5 and will return to the subject again later. Another way of avoiding diffusion limitations is to use microchannel reactors, as will be discussed later.

BOX 6.2 Low-Pressure Reactor for Kinetic Measurements for Steam Reforming

As an example of the use of a reactor operated at low pressure, we will consider work carried out using a low-pressure system with mass-spectrometric analysis in which the steam reforming of methane over Ni/Al_2O_3 catalysts was examined[a]:

$$CH_4 + H_2O \rightarrow CO + 3H_2 \tag{i}$$

Analysis of the reaction mixture was carried out using a small continuously evacuated mass-spectrometric "residual gas analyzer" attached to the constant volume reaction system by a capillary leak. This leak was situated some cm from the reactor. To improve gas circulation and mixing, a small centrifugal glass pump was included in the system and this allowed the system to be used as a small gas-recirculation reactor (see later). It was found that when a dose of one gas was added to a gas already in the reactor, what was effectively instantaneous mixing of the two gases could be attained. As a consequence, the reactor could be operated using a shallow bed of powdered catalyst and kinetic data could be obtained which were effectively free of problems of internal or external diffusion limitations.

It was found that the rate of reaction of methane with a 75 wt% Ni/Al_2O_3 catalyst was given by the following equation:

$$-dn/dt = 1.78 \times 10^{18} P_{CH_4}{}^{1.0} P_{H_2O}{}^{-0.5} \exp\left[\frac{-29000}{RT}\right] \tag{ii}$$

Assuming that the reaction adheres to Eq. (i) under the conditions used, for equimolar initial concentration of methane and water the rate of reaction can be rewritten as:

$$dP_{CH_4}/dt = k\, P_{CH_4}{}^{0.5} \tag{iii}$$

Integration gives:

$$P_{CH_4}{}^{0.5} = \text{const} - kt/2 \tag{iv}$$

Hence, plots of the square root of the methane pressure versus time should give straight lines. Figs. (A)–(D) show such plots calculated from data obtained at 873K. Fig. A shows the results of a number of different experiments and demonstrates the reproducibility of the results. It also shows that straight-line behavior is obtained for times of reaction greater than about 50 s. It would appear that the initial deviation from straight-line behavior is due to the fact that the initial stages of reaction occur with the catalyst under oxidizing conditions but that once reducing conditions are obtained, the expected behavior results.[b] Fig. (B) shows the effect of adding hydrogen to the feed. The major effect seen is that the addition of increasing amounts of hydrogen causes a reduction of the initial deviation from linear behavior. Finally, Figs. (C) and (D) show that both CO and CO_2 have inhibiting effects on the reaction, presumably due to competition by these molecules for the active Ni (or NiO) sites.

(*Continued*)

BOX 6.2 Low-Pressure Reactor for Kinetic Measurements for Steam Reforming (Continued)

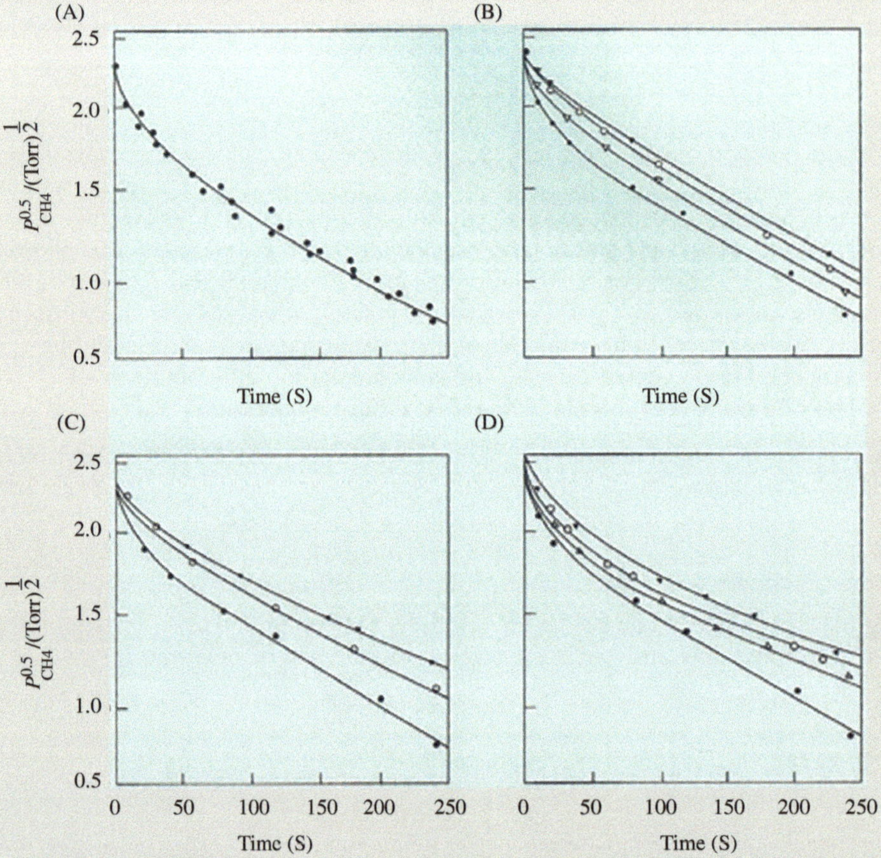

[a]J.R.H. Ross, M.C.F. Steel, J. Chem. Soc. Faraday Trans. I 69 (1973) 10.

[b]The topic of oxidizing/reducing conditions was considered in Section 3.3 of Chapter 3, How Does a Catalyst Work?.

are required to give complete conversion and the rate of reaction is less likely to be kinetically controlled.[4]

[4] Another recent example of the use of static reactors will be discussed in Chapter 15, Catalysis in Biomass Conversion, in connection with the production of hydrogen from formic acid at ambient temperatures. In this case, a formic acid solution is added to a closed stirred vessel containing a catalyst in powdered form and the total pressure (equimolar proportions of hydrogen and CO_2) above the catalyst is measured as a function of time. (See also Box 6.3 below.)

BOX 6.3 The Kinetics of Formic Acid Decomposition

There has recently been significant interest in the decomposition of formic acid to give hydrogen and CO_2:

$$HCOOH \rightarrow CO_2 + H_2$$

at ambient temperatures, a reaction that can be catalyzed by a variety of different supported Pd catalysts (see Chapter 15: Catalysis in Biomass Conversion). Much of the work is aimed at finding new catalysts with very high dispersion that are active and stable under reaction conditions used. As will be discussed further in Chapter 15, the addition of sodium formate to the reacting formic acid gives particularly good conversions at and around room temperature. The equipment used for this work is remarkably simple and an example is shown in the accompanying diagram. A stirred and thermostatted glass reaction vessel with suitable purge gas feed and a septum to allow introduction of the reaction mixture or analysis of the reaction mixture at any time is connected to a CO_2 trap and hence to a manometer that measures the buildup of hydrogen pressure with time. Prior to any experiment, the catalyst is placed in the reaction vessel and the system is purged with nitrogen before introducing the formic acid/sodium formate mixture. The rate of hydrogen buildup is then a measure of the rate of reaction.

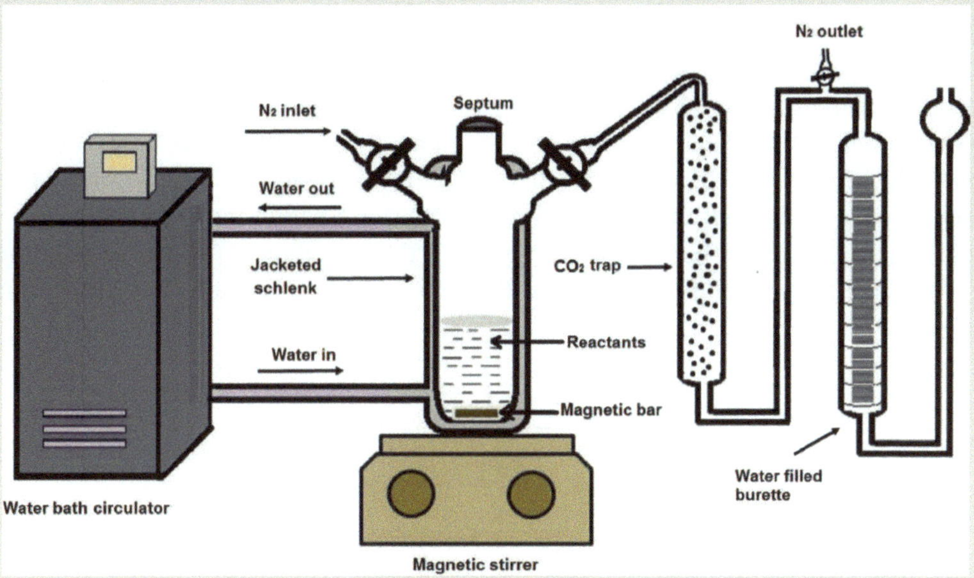

Diagram reproduced with thanks to Elsevier from S. Akbayrak, Y. Tonbul, S. Özkar, Appl. Catal., B 206 (2017) 384 (supplementary material).

6.3 Stirred and Recirculation Reactors

6.3.1 Stirred Reactors

Many of the advantages of the low-pressure constant-volume (static) reactor are also to be found in the use of stirred and recirculation reactors. Stirred reactors are more commonly used for batch reactions without any flow through the reactor and are often used for solid/liquid or solid/liquid/gas reactions in which the solid catalyst is suspended in powder form in a liquid which may be an inert or one of the reactants. The concentrations of reactant(s) and product(s) are measured as a function of time and the concentration versus time of reaction data are handled in the same way as for the constant volume reactor discussed earlier. In principle, there can also be a flow of reactants into the reactor and of products out of the reactor and in such a case the data are handled in the same way as for the stirred recirculation reactor discussed in the next section.

6.3.2 Recirculation Reactors

In recirculation reactors, the reaction mixture is recirculated externally at a rate sufficiently high to ensure that the concentrations of the reactants and products are effectively constant throughout the catalyst bed. In most cases, this is combined with a steady flow of reactant into the reactor and removal of the reaction mixture at another position relative to the bed. The difference between the rates of flow of reactant into and out of the catalyst bed is equivalent to the rate of reaction, as shown schematically in Fig. 6.2. As long as the recirculation rate is high enough compared with the rate of reaction, the concentrations of the reactants and products at the exit are the same as those within the bed. A typical recirculation reactor is the Berty Reactor (Fig. 6.3), the operation of which is well described in the Autoclave Engineering brochure (see Box 6.3) and another is the Carberry Spinning

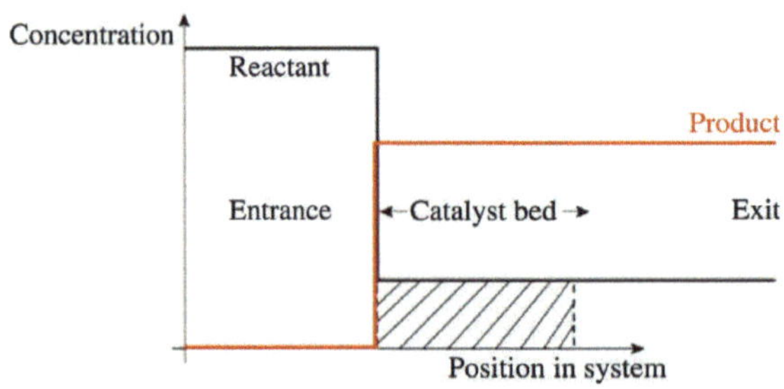

Figure 6.2
Typical concentration profiles through a recirculation reactor.

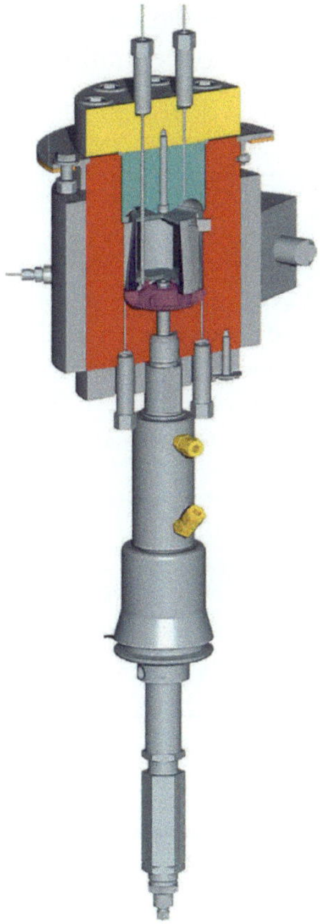

Figure 6.3
A Berty-type recirculation reactor. *Courtesy of Autoclave Engineers.*

Catalyst Basket Reactor described in the same brochure. The ideal concentration profile through a recirculation system incorporating a Berty or Carberry Reactor will be the same as shown in Fig. 6.2 for a recirculation reactor. On the left-hand (entrance) side of the reactor, the reactant concentration is constant until the point at which the flow enters the reactor. The concentration then drops immediately to the concentration that prevails throughout the reactor and this is the concentration that leaves the reactor at the exit. The product concentration increases just after the entrance to a level that remains the same throughout the reactor and also prevails at the exit. To obtain such a profile, it is necessary that the recirculation rate far exceeds the rate of reaction; in practice, this may not always be the case and so there may a slight deviation from the ideal profile shown. In practice, it is likely that there will be a second reactant as well as additional products; there will be an additional profile for each additional reactant and product but these are omitted for clarity. The recirculation reactor has the advantage over other reactor types that it allows the rate of the reaction to be determined directly from the rate of flow of reactant and the conversion; the conversion is given by the change in concentration divided by the initial concentration. (See also Box 6.4; also Task 6.1.)

BOX 6.4 Three-Phase Reactors

A typical example of a process using a three-phase stirred tank reactor is the hydrogenation of fats and oils using a powdered Ni/Al_2O_3 catalyst. The catalyst is suspended in the liquid to be hydrogenated in a stirred tank and hydrogen is introduced as either a static atmosphere or as a flow. The level of hydrogenation achieved can be determined easily by following the hydrogen consumption. Further information on the hydrogenation process can be obtained by following the Wikipedia link while data on the types of reactor used can be found by registering at the Parr website.

http://en.wikipedia.org/wiki/Hydrogenation/

http://www.parrinst.com/

Other examples of stirred reactors are found in a document available at the website of Autoclave Engineers (http://www.autoclaveengineers.com/ae_pdfs/CR_Catreact_SelGd.pdf).

Task 6.1 The Berty Reactor

Show how you would carry out an experiment in a Berty reactor to obtain kinetic data such as those represented by Eq. (ii) of Box 6.2.

6.4 Flow Reactors

By analogy with the flow reactors used to study fast homogeneous reactions (see Section 6.1), most laboratory investigations of catalytic reactions are carried out using flow reaction systems. In such reactors, the catalyst is placed in a *packed bed* that is normally held between two porous pads or frits. The reactor is usually mounted vertically to make sure that the catalyst bed is packed as uniformly as possible and the gas flow is usually directed from top to bottom to ensure that the packing of the bed is not disturbed by the flow.[5] A typical plug-flow reactor setup is shown in Fig. 6.4. The gas flow entering the reactor is carefully controlled, generally by mass-flow controllers, and the gas exiting the bed is analyzed using a suitable system such as a gas chromatograph. The packing

[5] Horizontal beds have also been used in some circumstances and flow from bottom to top of the bed can also be employed, the latter always being the case if a fluidized bed is employed (see later).

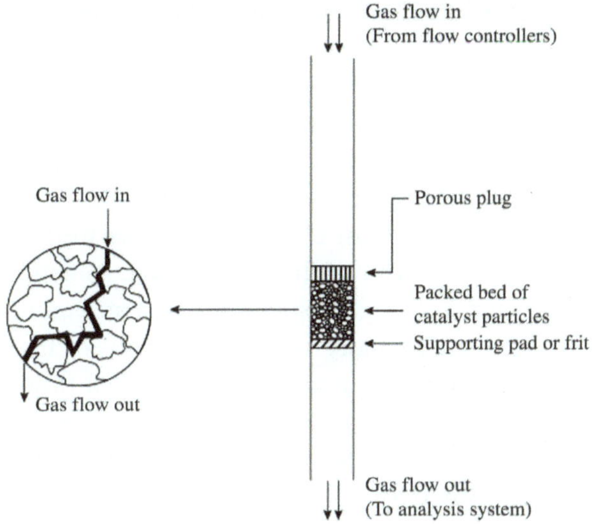

Figure 6.4
Schematic representation of a catalytic flow reactor. The inset shows the flow through a section of the bed.

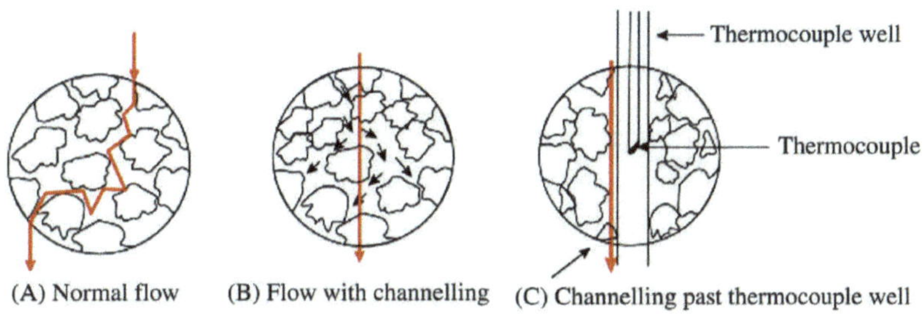

Figure 6.5
Schematic representation of channeling in catalyst bed: (A) Normal distribution as in Fig. 6.3; (B) channeling due to poor stacking of the catalyst; and (C) channeling along the side of then thermocouple well.

of the bed must be such that the gas flows through it in a tortuous fashion (see inset and Fig. 6.5A); this ensures that there is good lateral mixing and also that there are no concentration gradients at right angles to the direction of flow. To help ensure good mixing in the interparticular space, the particles of catalyst must have a diameter that is significantly smaller than the diameter of the reactor (more than 10 particles across the diameter, preferably more) and the bed depth should be more than at least three times the bed diameter to ensure that there is no back-mixing of the products with the reactants. The reactor is generally placed in a suitable furnace and a thermocouple is inserted either in direct contact with the outside of the reactor next to the bed or in a sleeve inserted in the bed itself. When a thermocouple sleeve is inserted in the bed, it is important to ensure that the flow patterns are not disturbed and that there is no chance of channeling of the reactants past the bed. Fig. 6.5 compares schematically (1) the ideal transport through a bed (as in the inset of Fig. 6.4), (2) the channeling occurring in a badly packed catalyst bed, and (3) and channeling along the wall of a thermocouple well. Another important parameter is that of thermal conductivity of the catalyst bed, something that is most important if we are dealing with a strongly exothermic or a strongly endothermic reaction: it is essential that heat is transferred rapidly between the walls of the reactor and the bed to ensure that there are no lateral temperature gradients. The ideal flow through a packed bed is then as is shown in Fig. 6.6: an even concentration front moves through the bed in such a way that there are no differences in concentration between the center and the walls, so-called plug flow. (See also Box 6.5.)

If plug flow is achieved, then the concentrations of reactants and products found at any position in the bed will be equal across the diameter of the bed and will be a function of the initial concentrations and of the time for which the reactants have been in contact with the bed, as illustrated in Fig. 6.6. The contact time, τ, will depend on the rate of flow, the cross-sectional area of the bed, and the distance from the entrance. Conventionally, the rate of flow, F, is generally expressed in cm^3(stp)/min (although other units can also be used,

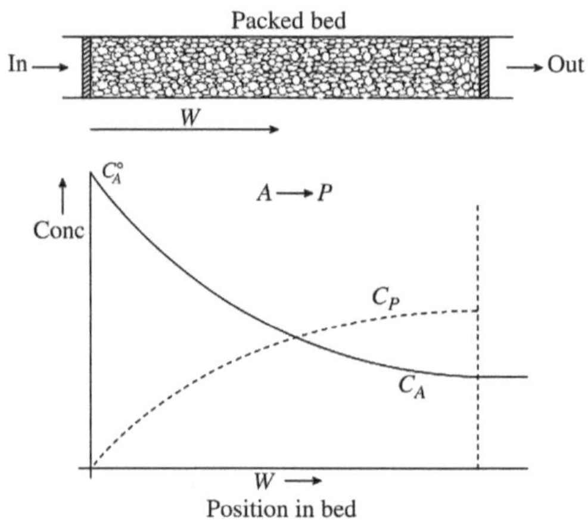

Figure 6.6
Concentration versus position in a packed bed for plug-flow conditions. The bed is shown as mounted horizontally so that the plot is equivalent to that for concentration versus time for a homogeneous reaction or a constant volume system (see Box 6.2). Note that the x-axis is shown as W, the weight of catalyst to that point in the bed; if it was expressed as W/F (F= flow rate), this would have the dimension of time. If the effective density of the bed in the latter case is 1.0 g cm^{-3}, then W/F is the time spent in the bed.

BOX 6.5 Catalyst Beds Containing Pellets or Extrudates

Fig. 6.4 depicts small irregular particles of catalyst but it is possible also to construct such a reactor using pellets or extrudates; however, because of the dimensional limitations outlined earlier, pellets and extrudates generally require larger reactors than those used in most laboratories and the use of such particle dimensions is generally only encountered in pilot-scale or full-scale reactors. Some typical reactors will be discussed in Chapter 8, Mass and Heat Transfer Limitations and Other Aspects of the Use of Large-Scale Catalytic Reactors.

for example, molecules/s or mol/h). If the total volume of the bed is V_T, then the time taken for a plug of reacting gas to pass through the bed, τ_T, will be V_T/F, assuming that there is no significant volume change in the flowing gas as a result of reaction.[6] At any position x in the middle of the bed, the time taken for the gas to reach that point, τ, will by analogy be given by V_x/F, where V_x is the volume of catalyst up to that point. The volume occupied by the catalyst will depend on the so-called packing density of the catalyst and so it is more practical to consider the weight of the catalyst from the beginning of the bed to any position of the bed (W_x) and relate it to the total weight of catalyst used (W_T) and then assume a packing density. Although it is in principle possible to measure the concentrations of reactants and products throughout the bed (e.g., using a moveable probe or by providing sampling points along the length of the bed), this is difficult to achieve. In consequence, it is more common to measure the concentrations of gases at the exit of the bed as a function of the rate of flow through the catalyst[7] (Fig. 6.7) and then construct plots of the exit concentration as a function of W/F. It is also possible to carry out a series of experiments each at different temperatures and to use these to give information on the activation energy for the reaction. (Such an approach of course assumes that the catalyst is not deactivated in any way during the measurements.)

In principle, it is possible to fit the data corresponding to plots of the sort shown in Fig. 6.7 to an integrated rate expression and hence to obtain values for the rate constant at various temperatures. In practice, this is not simple since the kinetic expressions are often not easily integrated. (See Chapter 7: The Kinetics and Mechanisms of Catalytic Reactions). Hence, it is more common to use a "differential

[6] This is a reasonable assumption if the reactants and products are diluted with a carrier gas. However, if there is no appreciable concentration of diluent, the treatment has to be modified by suitable corrections to take into account the gaseous volume change that takes place as a result of reaction.

[7] In principle, the weight of catalyst can also be varied. However, that is more difficult to achieve in a way that does not introduce errors. For example, it is difficult to make sure that the flow conditions remain the same for different bed packings and there is a danger of channeling, particularly at low bed weights. Such problems can be largely avoided by using a catalyst bed that is diluted by different quantities of a suitable inert material, making sure that the total bed weight remains constant throughout the set of experiments.

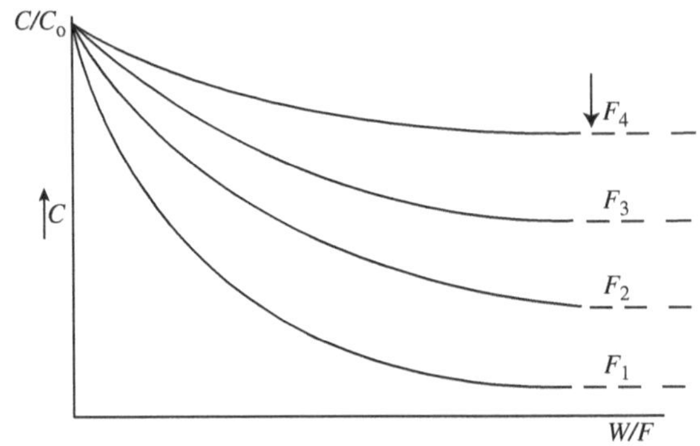

Figure 6.7
Plots of C/C_o versus W/F for different flow rates through a packed bed.

approach" by carrying out experiments to obtain the rates of the reaction under different conditions. The initial rate of the reaction shown in Fig. 6.7 can be obtained by taking a tangent to the curve as W/F tends to zero, this requiring a separate set of experiments for each value of rate. However, if the measured conversion is low (less than about 20%), the conversion approximates to the rate at the input conditions and so a single measurement suffices.[8] Hence, a set of rate data can be accumulated as a function of inlet concentrations and reaction temperature; further, by adding different amounts of product to the reaction mixture, the effects of the concentrations of the products can also be determined.

The preceding paragraphs assume that the catalyst particles in the packed bed being used are all easily accessible to the reactants and that reaction occurs homogeneously throughout each grain of the catalyst. In other words, there are no diffusion limitations. Such measurements are therefore restricted to experiments at relatively small particles in which diffusion restrictions can be neglected or to reactions with relatively low rates. In Chapter 8, Mass and Heat Transfer Limitations and Other Aspects of the Use of Large-Scale Catalytic Reactors, we will discuss the use of larger catalyst geometries (e.g., pellets) of the type used in full-scale reactors. We will discuss the problems introduced by mass transfer of reactants and products to and from the external surface of the catalyst sample (external diffusion) and/or within the pores of the catalyst pellets (internal diffusion) and we will see how these effects manifest themselves in kinetic measurements. Several other types of reactors will now be discussed.

[8] If the conversion is low, then the gradient of the chord joining the measured conversion to the origin approximates to the tangent at the origin.

6.5 Fluidized Bed Reactors

Although fluidized bed reactors are mainly used for large-scale applications (see Chapter 8), they can also be used for laboratory or bench-scale tests of a catalyst and so they will be discussed briefly in this chapter. If small catalyst particles are placed in a packed bed, severe pressure drop problems can be encountered, especially if higher gas flow rates are used. This problem can be circumvented by the use of a fluidized bed reactor. In such a system, the catalyst particles are placed in a vessel that has a porous plate at the bottom through which the reactant gases pass, as shown in Fig. 6.8.

At low gas velocities, gas bubbles will pass through the powdered catalyst bed in rather the same way that gas bubbles would pass through a liquid. However, as the gas velocity is increased, the bed becomes fluidized, increasing considerably in volume, and gas bubbles are no longer visible; in this situation, the catalyst bed is still well defined, with a separate gas phase above it. As long as the gas velocity is not increased too much, the particles remain in the reactor without any spillover to the exit tube.[9] All the many forms of fluidized bed reactors have the significant advantage over fixed bed reactors that additional catalyst can be added to the reactor using a feed system at the top of the reactor and spent catalyst can be removed from the base of the reactor. (The current item

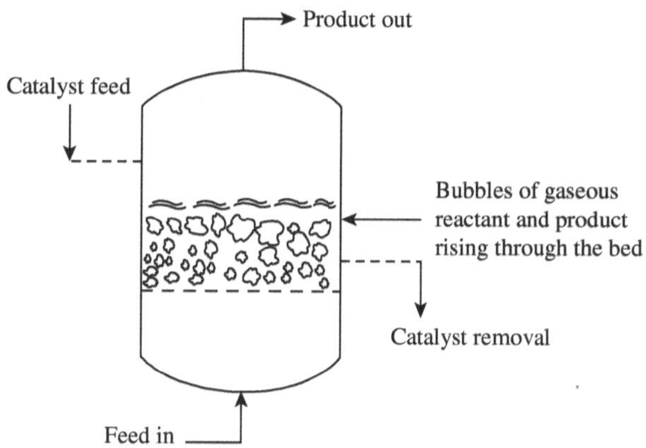

Figure 6.8
Schematic representation of a fluidized bed reactor. The reactor can be operated with or without a through flow of catalyst; the latter mode is used if gradual deactivation of the catalyst occurs.

[9] At even higher gas velocities, the gas and the solid are not separated by gravity in the reactor but in a separate cyclone as in the entrained reactor of the fluid catalytic cracking (FCC) process discussed in the next paragraph.

on fluidized bed reactors in Wikipedia gives the advantages and the disadvantages of such systems (http://en.wikipedia.org/wiki/Fluidized_bed_reactor/) but it should be stressed that this also covers the use of fluidized beds for other than catalytic reactions.) The advantages of a fluidized bed system include the elimination of heat and mass transfer limitations while the disadvantages include the fact that the reactor is much larger than a packed bed reactor and involves higher energy costs, together with the possibility of particle entrainment. (See also Task 6.2).

Task 6.2 Web-Based Animations of Fluidized Bed Reactors

A number of web sites give animated representations of fluidized bed reactors of various types. Do an appropriate search to find examples of catalytic fluidized bed operation.

It should be stressed that fluidized bed reactors are more commonly used in industrial applications than in fundamental laboratory studies as they are suitable for practical large-scale operations. However, they present difficulties in detailed modeling and so pilot-scale facilities are almost always used prior to introducing them in practice. (See also Box 6.6 and Task 6.3.)

BOX 6.6 Recirculating Solids Reactor

The principle of the entrained reactor in which small particles are recirculated after passing through a regenerator (as in the FCC reactor) has been extended by DuPont to the use of larger catalyst grains in a selective oxidation reaction, the oxidation of butane to maleic anhydride:

$$C_4H_{10} + 7/2O_2 \rightarrow C_4H_2O_3 + 4H_2O$$

In the DuPont system, the oxidation step takes place in one reactor using the oxygen of the catalyst and the partially reduced catalyst is then transported back to the second reactor where it is reoxidized using air. The advantage of this system is that the maleic anhydride is not exposed to oxygen and the production of undesired by-products such as CO_2 is thus minimized. The process requires mechanically stable catalysts that are not abraded during transportation. The energy costs associated with the transportation of such large quantities of solids are very high. Further details on the reaction are given in Chapter 14, Conversion of Hydrocarbons to Intermediates by Catalytic Hydrogenation or Oxidation.

Task 6.3 DuPont's Circulating Fluidized Bed Technology

Examine some of the literature on this process, paying particular attention to the details regarding the reactor, bearing in mind that you will be reading more about the reactor, the catalyst and the reaction in Chapter 14, Conversion of Hydrocarbons to Intermediates by Catalytic Hydrogenation or Oxidation. Start your search with the work of R.M. Contractor (DuPont's CFB technology for maleic anhydride, Chem. Eng. Sci. 54 (1999) 5627).

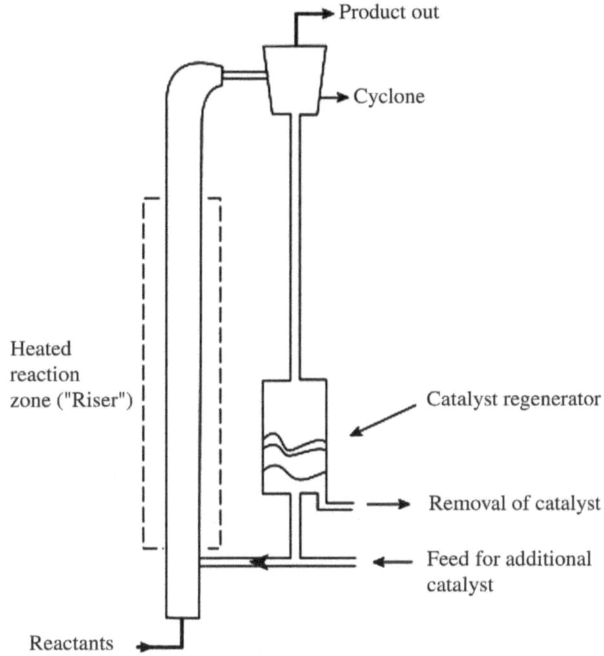

Figure 6.9
Schematic representation of a riser reactor. The catalyst particles are entrained in the flowing reactant gas and separated after reaction in a cyclone. They are then regenerated in a separate section before readmitting to the "riser."

Fig. 6.9 shows an entrained flow (or riser) reactor of the type used in the fluid catalytic cracking (FCC) process (see Chapter 10: The Production of Energy Carriers from Oil). In this case, the gas flow rate is so high that the catalyst particles become entrained in the flow and pass through the reactor at the same rate as does the gas. On leaving the vertical ("riser") part of the reactor, the gas and the catalyst are separated in a cyclone and the gaseous product is removed from the system while the catalyst is recirculated to the reactor. If required, a separate regeneration volume can be incorporated (e.g., to burn off coke formed), spent catalyst can be removed from the system and additional catalyst can be added. In some applications, additional components can be added to the catalyst stream; for example, an adsorbent can be added that will selectively remove a poison such as H_2S from the gas stream or an additional catalyst can be added to facilitate reactions parallel to the main catalytic process. One such example is the addition of a de-NO_x components to help catalyze the selective reduction in the riser section of any NO_x formed in the catalyst regenerator section. As with the more conventional fluidized bed reactors, the use of the riser reactor is generally restricted to commercial operation, although smaller pilot plants are also used to help establish the operation of such systems.

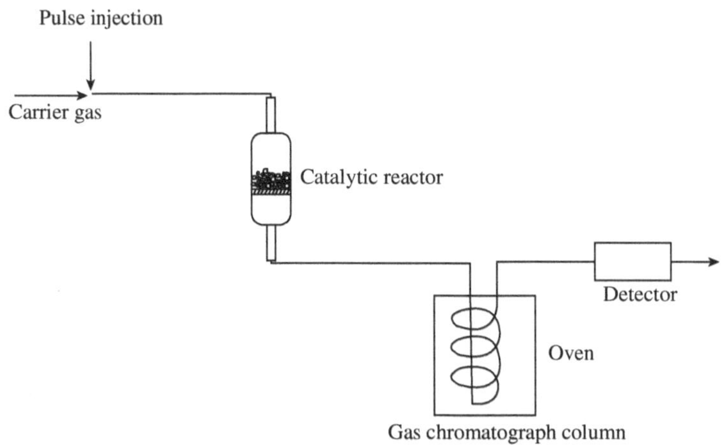

Figure 6.10
A typical pulse reactor setup.

6.6 Pulse Reactors

A pulse reactor is a system in which a small pulse of the reactant or reactants of interest is passed over a small sample of the catalyst being examined and then directly into the analysis system.[10] A typical setup is shown in Fig. 6.10; in this, the analysis is carried out using a gas chromatograph in which the column separates the mixture leaving the reactor into its separate components and the detector (after suitable calibration) gives the appropriate concentrations. Such systems have generally been used for rapid screening of a series of catalysts to find the optimum for a given reaction. However, what is possibly a more valuable application is its use in tracer experiments in which a labeled molecule is added to the reaction mixture to see where the label turns up in the products of the reaction. The technique was pioneered by Emmett and his colleagues[11] in their work on the mechanism of the Fischer Tropsch synthesis of hydrocarbons from syngas ($CO + H_2$ mixtures). The use of such systems has been reviewed by Galeski and Hightower.[12] The method has the drawback that the catalyst is not examined under steady-state conditions and so any conclusions reached apply only to the catalyst under the nonsteady state conditions used; in other words, the data often refer to a catalyst which has not been fully activated or for which the surface conditions change during the passage of the pulse. The

[10] A comprehensive review of pulse catalytic reactors is given in "Laboratory Studies of Heterogenous Catalytic processes" by E.G. Christoffel (edited by Z. Paál), Studies in Surface Science and Catalysis, 42 (1989), Chapter 4.

[11] R. J. Kokes, H. Tobin and P. H. Emmett, New microcatalytic-chromatographic technique for studying catalytic reactions, J. Am. Chem. Soc. 77 (1955) 5860.

[12] J.B. Galeski, J.W. Hightower, Microcatalytic reactors: Kinetics and mechanisms with isotopic tracers, Can. J. Chem. Eng. 48 (1970) 151.

technique is seldom used nowadays although its equivalent is sometimes used in combinatorial and high-throughput studies in which the activities of multiple small samples of potential catalysts are compared by exposing each briefly to pulses or flows of reactants.

As with pulse methods, high-throughput methods give only approximate data as the conditions used to monitor the activities and selectivities of the catalysts produced are far from ideal. Further, the approach, although saving considerable manpower, is dependent on good data-handling methods being applied. Some of the papers from the EuroCombiCat meetings (see references in Box 6.7; see also Task 6.4) are worth examining to get a feel for the level of sophistication of these methods. One of the problems of assessing the effectiveness of the methods is that although many claims have been made that new commercial catalysts have been developed as a result of work using high-throughput methods, most of the critical results are shrouded in industrial secrecy. An example of

BOX 6.7 Combinatorial and High-Throughput Catalysis

The search for new catalysts for specific reactions can be time consuming and expensive in manpower and the use of equipment and chemicals if multiple flow reactors are used to measure reaction kinetics. Hence, much effort has been devoted to the use of the so-called combinatorial and high-throughput techniques to examine multiple samples of catalysts prepared in such a way as to provide a statistical survey of potentially effective catalysts. Many different types of approach have been used, from making libraries of small samples and testing them in a series of parallel reactors simulating more conventional plug-flow reactors (see, for example, G. Grasso, J.R.H. Ross, Catalysis of Organic Reactions, in: J.R. Sowa (Ed.), Taylor and Francis, Baton Rouge (2005), p. 393) to making catalyst samples as multiple spots on a plate and then monitoring the activity of each sample in some way, for example, using pulses of reactant directed to each sample in turn and monitoring the resultant products with a snorkel device. There is a series of scientific meetings on the subject, EuroCombiCat (http://www.eurocombicat.org/) and the proceedings of two of the meetings are to be found in *Catalysis Today*: 137 (2008), Issue 1; and 159 (2011), Issue 1.

Task 6.4 High-Throughput Experimentation

A search of the web for HTE will lead to very extensive information on the subject. Concentrating on applications of the methodology to catalysis, see if you can find examples of successful development of new catalyst systems.

One of the early proponents of HTE was W.H. Weinberg of Symyx. Using his name, search the web to locate other groups working in the field. You will also find leads to the literature on the subject by using some of the key articles of the issues of *Catalysis Today* given in Box 6.7 as starting points.

a company making systems for high-throughput experimentation is HTE based in Heidelberg, a subsidiary of BASF. (See http://www.hte-company.com/ for further information on the systems available and some technical data sheets. A web search for "high-throughput experimentation" will lead to other manufacturers of such equipment.)

6.7 The Temporal Analysis of Products Reactor

The method known as "temporal analysis of products (TAP)" was developed in 1978–79 by John Gleaves (then at Monsanto) and his team. The aim of the method is to speed up a search for new catalysts for a specific process. It was introduced to the scientific community in 1984 and its use was covered by a number of patents. In essence, it is an approach that depends on the pulsing of very small amounts of the reactant gas or gases, either simultaneously or after defined time lapses, onto the surface of a catalyst using high-speed pulse valves (Figs. 6.11 and 6.12) and measuring the responses using a quadrupole mass spectrometer (QMS) which is located very close to the exit of the catalyst bed. The system can operate in a number of different modes, depending on whether a small number of pulses are added ("probe experiments," providing intrinsic kinetic information) or larger numbers of pulses are supplied in sequence ("multipulse experiments," giving information on adsorption and dissociation of different species, or "pump-probe experiments," involving

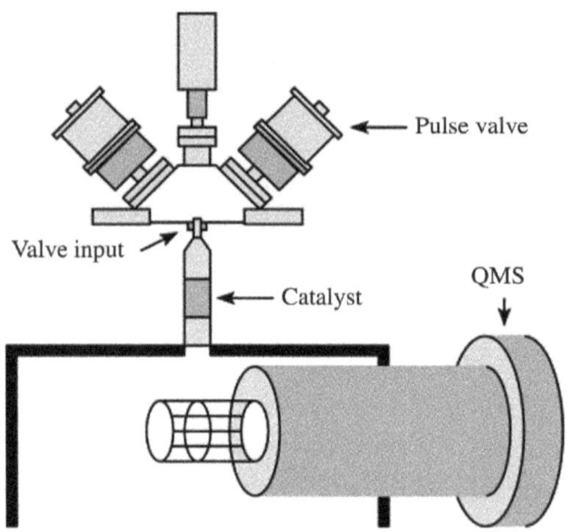

Figure 6.11
The TAP 2 reactor. *Reproduced from D. Constales, S.O. Shekhtman, G.S. Yablonsky, G.B. Marin, J.T. Gleaves, Multizone TAP-reactors theory and application IV. Ideal and nonideal boundary conditions, Chem. Eng. Sci. 61 (2006) 1878–1891, with kind permission of Elsevier.*

Figure 6.12
Photograph of the TAP 2 reactor installed at the laboratory for Chemical Technology, University of Ghent, Belgium (manufactured by Autoclave Engineers). *Reproduced courtesy of Vladimir Galvita, U. Ghent.*

Task 6.5 The TAP Reactor

Using a citation search, identify the papers by J.T. Gleaves (now at Washington University) and make use of them to identify other groups of researchers using TAP reactors. Identify papers concerned with the use of TAP to study the selective oxidation of butane to give maleic anhydride. Summarize the main conclusions of these papers in relation to a possible mechanism for the reaction.

---

admission of two or more reactants at different time intervals and giving information on the lifetime, and reactivity and selectivity, of adsorbed species). The equipment used is very expensive, depending on stainless steel ultrahigh vacuum systems (see Fig. 6.12), and so only a very limited number of such systems have been built to date. Nevertheless, experiments carried out using TAP systems have given valuable information (often confirming earlier ideas based on less definitive experiments) on a wide range of reaction systems. We will return to some results obtained with the TAP in a later chapter.[13] (See also Task 6.5.)

[13] A special issue of *Catalysis Today* (Vol. 121, issues 3–4 (2007), edited by E.V. Kondratenko and J. Perez-Ramirez, includes a very useful summary of the uses of the technique and contributions from some of the main proponents.

6.8 SSITKA

The technique known as SSITKA (steady-state isotopic transient kinetic analysis), developed originally by Happel, Bennet, and Biloen,[14] involves switching the isotopic composition of the gas flowing over a catalyst in a packed bed without any (significant) allover change in the gas-phase composition. This is done in a system in which it is possible to switch from one isotope under investigation to the other while not disturbing the other flows. Fig. 6.13 shows a typical experimental setup for an investigation of the oxidation of methane over a Pd/alumina catalyst. A flow containing $^{16}O_2$ with a trace of Ar as an inert tracer is displaced by a flow of $^{18}O_2$ and the change in the isotopic distribution of the product CO_2 is measured relative to the disappearance of the signal due to Ar. Typical results obtained at a temperature of 583K are shown in Fig. 6.14. From these and other results, it was possible to conclude that the oxygen of the catalyst was used in the

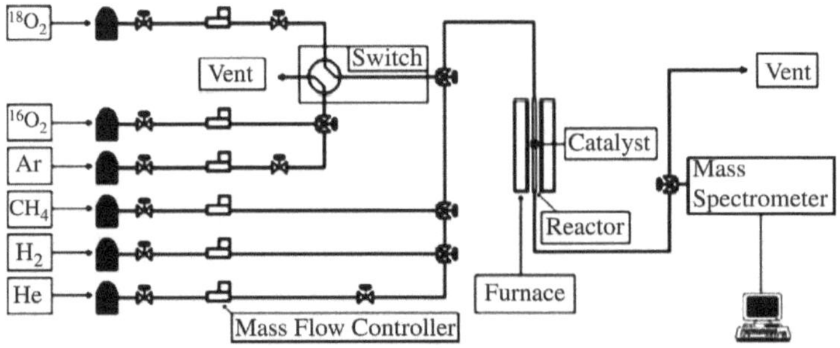

Figure 6.13

An apparatus used for SSITKA measurements on the catalytic oxidation of methane over a Pd catalyst. *Reproduced from A. Machocki, M. Rotko, B. Stasinka, Catal. Today 137 (2007) 312, with kind permission of Elsevier.*

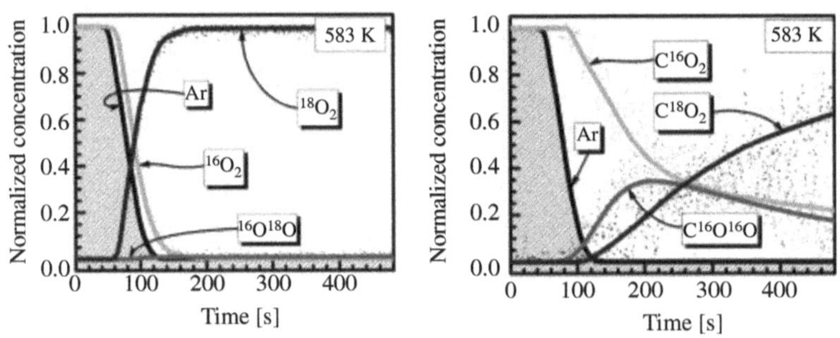

Figure 6.14

The results of switching from $16O_2$/Ar/CH_4/He to $^{18}O_2$/CH_4/He at a temperature of 583K ensuring 10% methane conversion. *Reproduced from A. Machocki, M. Rotko, B. Stasinka, Catal. Today 137 (2007) 312, with kind permission of Elsevier.*

[14] J. Happel, Chem. Eng. Sci. 33 (1978), 1567; C.O. Bennett, ACS Symp. Ser. 178 (1982) 1; J. Biloen, J. Mol. Catal. 21 (1983) 17.

reaction and that a Mars van Krevelen mechanism applied to the reaction. Full treatment of the data of SITKA experiments requires very careful mathematical analysis. Interested readers should examine the literature on the subject, using, for example, one of the references given in footnote 14 to trace some of the work on the subject. Further treatment is beyond the scope of this textbook.

6.9 In Situ/"Operando" Methods

The SSITKA technique described in the previous section is aimed at gaining information on the mechanism of a catalytic reaction under operating conditions and generally depends on curve-fitting but does not involve any direct observation inside the reactor. "In situ" and so-called Operando methods aim to identify reacting species directly under catalytic conditions.[15] A number of physical identification techniques have been used in conjunction with such measurements, these including transmission Fourier transformation infrared spectroscopy, diffuse reflectance infrared Fourier transform spectroscopy (DRIFTS), Raman spectroscopy, UV–vis spectroscopy, X-ray photoelectron spectroscopy, sum-frequency generation vibrational spectroscopy, and nuclear magnetic resonance. Although all of these methods are capable of monitoring surface species directly while carrying out a catalytic experiment, the most commonly applied method involves DRIFTS (Figs. 6.15 and 6.16; see also Box 6.8).

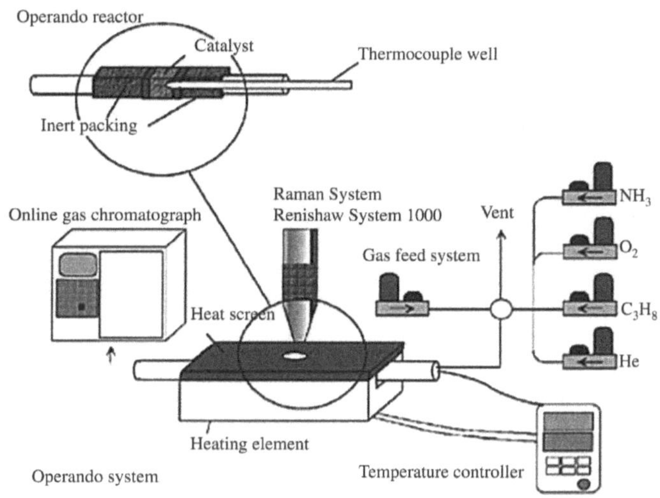

Figure 6.15
An operando Raman-GC setup. *Reproduced from M.O. Guero-Pérez, M.A. Bañares, Catal. Today 113 (2006) 48–57, with kind permission of Elsevier.*

[15] The term "in situ" implies that the measurements have been carried out using an operating catalyst but no attempt has been made to relate the measurements to actual catalytic measurements; on the other hand, "Operando" measurements are those in which there is online analysis of reaction effluent to ensure that the activity data obtained in the operando reactor are consistent with those observed in a conventional reactor.

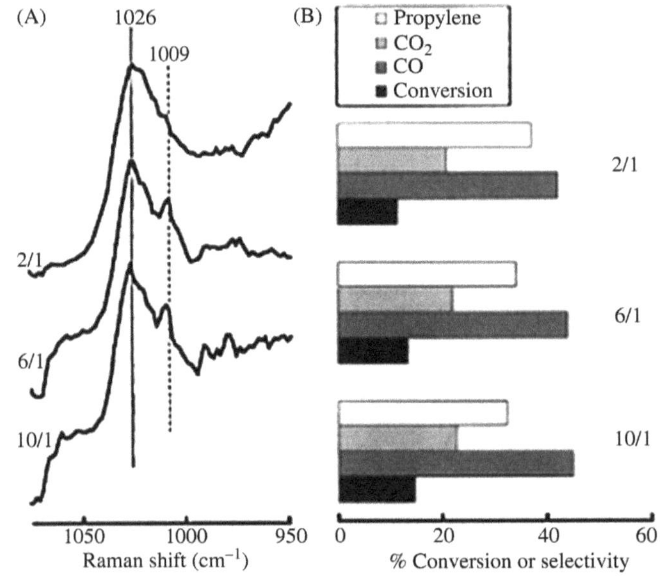

Figure 6.16
(A) Operando Raman spectra of a vanadia/alumina catalyst during the oxidative dehydrogenation of propane at 400°C as a function of C_3H_8/O_2 molar ratio. (B) Corresponding conversions of C3H8 and selectivities to CO, CO_2, and C_3H_6. *Reproduced from M.O. Guero-Pérez, M.A. Bañares, Catal. Today 113 (2006) 48–57, with kind permission of Elsevier.*

BOX 6.8 Some Meetings on Operando Spectroscopy

A series of meetings devoted to Operando spectroscopy have been held. Starting from the *Catalysis Today* issue containing the proceedings of a workshop held in Toledo, Spain in 2006 (Catal. Today 126 (Issues 1–2) 2007), identify other such meetings and examine papers relating to the use of the methodology with each of the analytical techniques mentioned in the main text.

As an example of the application of operando spectroscopy, we will take the case of the use of Raman spectroscopy[16] together with online gas chromatographic analysis of the gas phase to study the oxidative dehydrogenation of propane over VO_x/Al_2O_3 catalysts. The equipment used by Guerrero-Pérez and Bañares is illustrated schematically in Fig. 6.15 and the results of experiments carried out at different propane to oxygen ratios are shown in Fig. 6.16. The only

[16] Raman spectroscopy on its own provides information on the nature of oxide surface species; in this case, it is able to distinguish clearly between different forms of V–O bonds on the surface.

surface species visible in all the different gas compositions are V = O, although the band at 1009 cm^{-1} corresponding to isolated V = O bonds does disappear at a propane/oxygen ratio of 2.0. Parallel work using UV–vis spectroscopy showed that there was significant reduction of the catalyst during a change of C_3H_8/O_2 ratio from 10/1 to 2/1. Hence, the authors conclude that that alteration is due to reduction of surface polymeric vanadium oxide species and that only the V = O surface species are active in the oxidative dehydrogenation reaction. (See also Task 6.6.)

Task 6.6 Operando Spectroscopy

Using Scopus or another similar tool, examine the papers included in the Toledo meeting (see Box. 6.8 for the reference) and identify the most highly cited of these. Then tabulate the names of the authors, the reactions studied and the most important conclusions reached.

6.10 Microreactor Methods

There have been very significant advances in the application of microreactor techniques over the last two decades and the development of microchannel reactors has opened up new possibilities for the design of small and efficient chemical plants in which catalysis plays a pivotal role. Such developments have depended on improved engineering techniques that allow the creation of channels with very small dimensions on the surfaces of metallic plates, these plates then being stacked one on top of another to create a miniaturized reactor system. Techniques have in parallel been developed to prepare catalysts within these channels in such a way that the resultant materials have very similar properties to those of conventional catalysts. Such systems have a number of significant advantages over more conventional reactors, these including improved heat and mass transfer and ease of scale-up by simply increasing the number of plates in the assemblies. Further, because of their small volumes, these reactors can safely be used in situations in which explosive situations might be encountered.

Many potential applications of microreactors have been reported in the literature. One of the most promising developments is in the area of the production of hydrogen for mobile fuel cell applications. This is achieved by carrying out steam reforming of methane or autothermal reforming of higher hydrocarbons (diesel or petroleum). These processes, which will be discussed in more detail in later chapters, are endothermic and so heat has to be supplied to the system. Work has therefore been carried out to find systems to bring about the catalytic combustion of the anode off-gases from a fuel cell (containing hydrogen and methane) to provide the necessary heat. Fig. 6.17 shows the construction of microchannel platelets that have been developed for this purpose. These were coated with alumina (wash-coated into the channels) and then Pt/Mo or Pt/W active components were

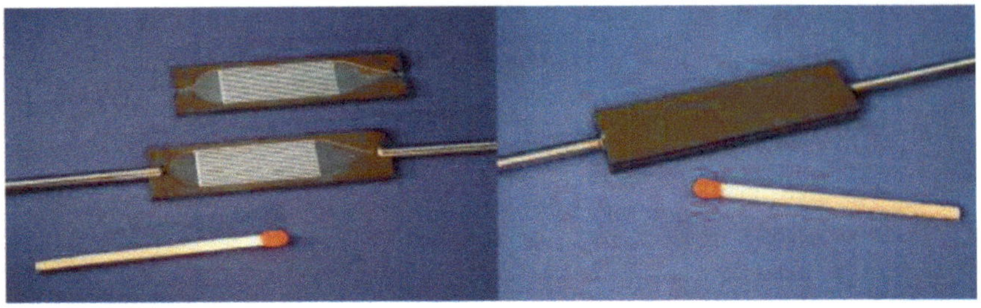

Figure 6.17

Microchannel reactor platelets used for the catalytic combustion of methane. *Reproduced from M. O'Connell, G. Kolb, R. Zapf, Y. Men, V. Hessel, Catal. Today 144 (2009), 306–311, with kind permission of Elsevier.*

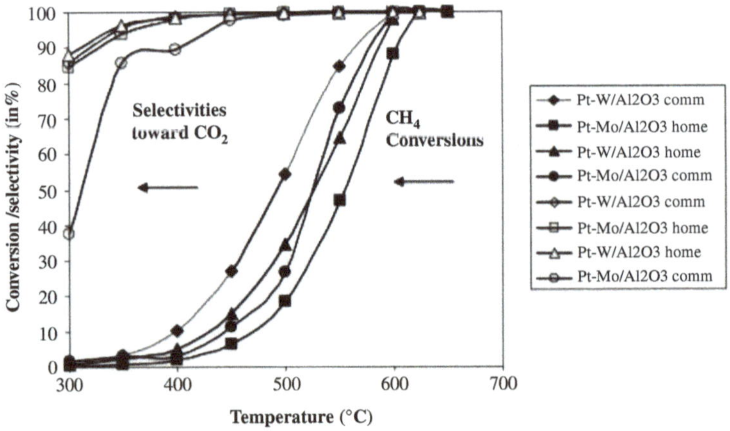

Figure 6.18

Methane conversions and selectivities for four catalysts. *For conditions, see source article. Reproduced from M. O'Connell, G. Kolb, R. Zapf, Y. Men, V. Hessel, Catal. Today 144 (2009), 306–311, with kind permission of Elsevier.*

introduced. The system was then tested for the combustion of methane; several commercial catalysts were also examined. The results of such tests are shown in Fig. 6.18.

For the purposes of this chapter, all that need to be noted from these results is that all the catalysts give 100% conversion of methane at temperatures above about 600°C and that the selectivity to CO_2 under these conditions is 100%. (See also Task 6.7.)

Task 6.7 Microreactor Technology

Carry out a citation study to find some of the papers published by V. Hassel and his colleagues and use these to identify other workers involved in the development of catalytic microreactor technology. Then list some of the reactions that have been studied by these researchers and establish whether or not commercial microreactor systems have been introduced for any of these.

6.11 Conclusions

The majority of the types of reactor discussed in this chapter are of the sort that are used to examine the catalytic behavior of catalysts on a laboratory scale, either to allow comparison between different catalysts as part of a development program or to give information on the kinetics and mechanism of the reaction. The kinetics and mechanisms of catalytic reactions will be discussed in the following chapter (Chapter 7: The Kinetics and Mechanisms of Catalytic Reactions). We will then in Chapter 8 discuss other aspects of the use of catalytic reactors: their use in full-scale processes and the problems associated with heat and mass transfer of the reactants in such reactors.

CHAPTER 7

The Kinetics and Mechanisms of Catalytic Reactions

Chapter Outline

7.1 Introduction

Kinetics and mechanism are closely interrelated and kinetic measurements are frequently used to give information on the mechanisms of reactions. Additionally, detailed kinetic equations are important in relation to the design and operation of chemical reactors. The kinetic expressions are generally quite straightforward for homogeneous systems but the situation is very frequently more complex in the case of catalytic reactions as the expressions which are determined experimentally can be much more complex than those usually associated with simple homogeneous reactions.[1] A result of this, particularly for engineering applications, is that simplified kinetic expressions ("intrinsic kinetics") are used which describe the operation of the catalyst very adequately over restricted ranges of the partial pressures of the reactants and products and also of temperature but which bear little formal relationship to the mechanism of the reaction in question. The situation is even more

[1] There are exceptions to every rule. For example, the formal kinetic expressions for chain reactions are just as complicated as those for catalytic reactions.

Contemporary Catalysis.
DOI: https://doi.org/10.1016/B978-0-444-63474-0.00007-2

complex if physical phenomena such as mass and heat transfer play a part in determining the observed rates. A discussion of the latter phenomena is included in Chapter 8.

In Chapter 3, we discussed how a catalyst works: by adsorbing the reactants on sites on the catalyst surface and by providing an alternative route by which the reaction can occur that is more energetically favorable compared to that taken in an equivalent homogeneous process. Having then discussed in Chapters 4 and 5 how to make a catalyst and to characterize it as well as to measure its catalytic properties, we examine in this chapter how the kinetics of a catalytic reaction may be described mathematically, both mechanistically and more globally. We then proceed to discuss the mechanistic aspects of a selected number of reactions that have been studied in detail.

We will start by considering in the following sections (7.2–7.5) three commonly encountered formal kinetic expressions that describe the kinetics of simple model reactions in a way that relates to the suggested mechanisms ("mechanistic kinetics"). All of these approaches use the concepts we have already encountered in relation to the Langmuir isotherm (see Chapter 2) describing the adsorption of various molecules on the surface of a uniform catalyst surface and extend these concepts to explain the rates of reaction of these adsorbed species. We will then in subsequent sections discuss some applications of these mechanisms to real catalytic systems and will give suggestions for further reading on these and other reactions.

7.2 Unimolecular Reaction of a Single Reactant A to Give Products

In Chapter 2, Surfaces and Adsorption, we described the derivation of the Langmuir adsorption isotherm Eq. (2.2) that relates the quantity of gas A adsorbed on a uniform surface at a constant temperature to the pressure of the gas P_A above the surface:

$$v_{ads}/v_m = \theta_A = b_A P_A/(1 + b_A P_A) \qquad (7.1)$$

where v_m is the monolayer capacity of the surface. The coverage θ_A is a measure of the concentration of A on the surface.

Let us now consider the rate of a reaction:

$$A_{(g)} \rightarrow P_{(g)} \qquad (7.2)$$

occurring over such a uniform surface, where the subscript (g) indicates that that species in question is in the gas phase. We then set up a scheme in which A becomes adsorbed on the surface as A_{ads}, where the subscript ads indicates that the species is adsorbed. We then consider that the rate of reaction of the surface species A_{ads} to give the gaseous product $P_{(g)}$

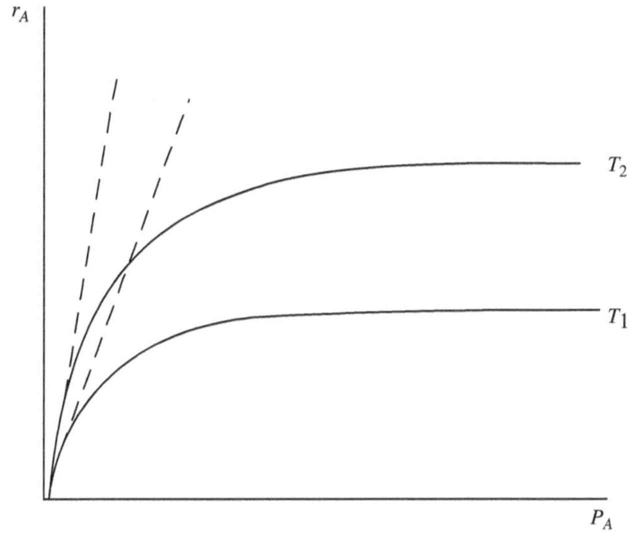

Figure 7.1
Plot of the rate of reaction 7.2 versus P_A for two temperatures, T_1 and T_2.

is rate determining and that this step is much slower than the adsorption and desorption steps so that we can consider the first step as effectively being an equilibrium:

$$A_{(g)} \rightleftarrows A_{ads} \rightarrow P_{(g)} \tag{7.3}$$

We can then write that the allover rate, r_A, of the reaction is proportional to the concentration of A_{ads}, this being given by θ_A, determined by the appropriate Langmuir isotherm:

$$r_A = -\,dc_A/dt = k_T \cdot \theta_A = k_T \cdot b_A P_A/(1 + b_A P_A) \tag{7.4}$$

where r_A is the rate of disappearance of A, k_T is the rate constant at any temperature, and b_A is the adsorption constant for the adsorption of A; b_A is essentially the equilibrium constant for the first step in Eq. (7.3). Fig. 7.1 shows a plot of this simple expression in the form of r_A versus the pressure of A, P_A, for two different temperatures, T_1 and T_2 ($T_2 > T_1$). At low pressures, each curve approximates to a *straight line* with gradient $k_T.b_A$ while, at high pressures, each curve tends to being a *straight line* parallel to the P_A axis. In other words, the low pressure data correspond to a first-order equation:

$$r_A = k_T \cdot b_A P_A \tag{7.5}$$

while the high pressure data approximate to a zero-order equation:

$$r_A = k_T \tag{7.6}$$

In between the two extremes, the full expression (7.4) applies. As we shall see later, it is possible to approximate to this expression, over very limited ranges of pressure, by an expression:

$$r_A = k'_T P_A^n \tag{7.7}$$

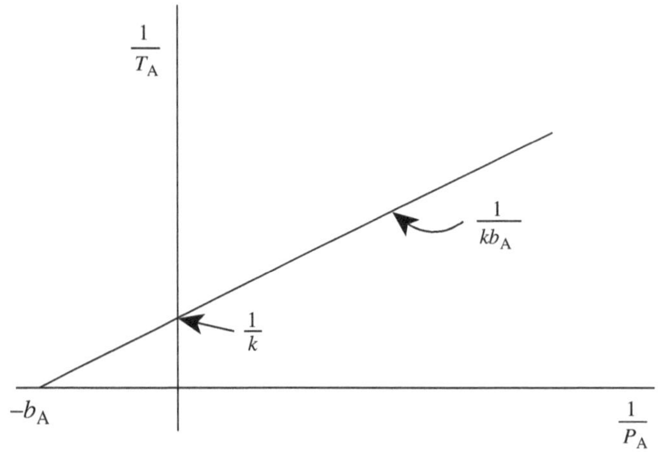

Figure 7.2
Plot of 1/rate versus 1/*T* for the data of Fig. 7.1 at a single temperature.

where n is a number between 0 and 1.[2]

Fig. 7.2 shows a so-called reciprocal plot of Eq. (7.3), in other words a plot of $1/r_A$ versus $1/P_A$, for one of the sets of data of Fig. 7.1. As shown, the values of k_T and b_A can be obtained from the intercepts on the two axes (or from one of the intercepts and the gradient). Any deviation from the *straight line* found would arise from deviations from the ideal behavior assumed in setting up the equation, for example, due to a variation in the strength of the bond between A and the surface when the coverage θ_A changes with pressure.[3] By obtaining similar plots at a variety of temperature, values for both the activation energy for the reaction, E_A, and the heat of adsorption of the reactant A, ΔH^o_{ads}, can be obtained (from plots of the values of ln k_T and ln b_A vs $1/T$, respectively).[4] (See also Box 7.1, Task 7.1).

[2] This is equivalent to making use of the Freundlich isotherm (Eq. 2.3) instead of the Langmuir isotherm to describe the adsorption equilibrium.

[3] The approach we have used here makes the simplifying assumption that the adsorbed species are in equilibrium with the gas phase and that the rates of adsorption and desorption are much faster than the rate of the reaction of the surface species to give the gaseous product, $P_{(g)}$. An alternative more rigorous approach to setting up the rate expression uses the method of the stationary state to give an equivalent expression that includes both the rate constants for adsorption and desorption as well as that for the formation of $P_{(g)}$. Then, as long as the magnitude of rate constant for the catalytic reaction is small compared to those for adsorption and desorption, the two approaches give the same result.

[4] If only two values of k_T and b_A are known, the values of E_A and ΔH_o can be obtained by solving two simultaneous equations. For example, the value of E_A is given by the equation:

$$E_A = \{2.303R \cdot T_1 \cdot T_2/(T_2 - T_1)\} \cdot \{\log_{10}k_2 - \log_{10}k_1\}$$

where R is the gas constant.

BOX 7.1 Enzyme Kinetics

An enzyme is also a catalyst and it contains active sites, albeit much more complex in structure, as with a heterogenous catalyst. An enzyme-catalyzed reaction is frequently depicted as occurring by a sequence of steps involving the enzyme E, the substrate S (i.e., the reactant), and the product P:

- formation of an enzyme–substrate complex:

$$E + S \rightarrow ES \quad \text{rate constant } k_a$$

- reverse reaction:

$$ES \rightarrow E + S \quad \text{rate constant } k'_a$$

- unimolecular formation of the products and release of the enzyme:

$$ES \rightarrow P + E \quad \text{rate constant } k_b$$

The kinetics of an enzyme-catalyzed reaction is conventionally described by the so-called Michaelis–Menton equation:

$$\text{Rate of formation of P} = k_b[S]\cdot[E_o]/([S] + K_M) \tag{B.7.1.1}$$

where K_M, the Michaelis constant, is given by

$$K_M = (k'_a + k_b)/k_a \tag{B.7.1.2}$$

and $[E]_o$ is the total concentration of enzyme, both bound and unbound.

This equation was derived by Michaelis and Menton in 1913 by assuming a rapid equilibrium in the way that we have described earlier for a heterogeneous catalyst; in the case of a rapid equilibrium, $k'_a \gg k_b$ and so $K_M = k'_a/k_a$. Hence, K_M is equivalent in all ways to b_a, the adsorption constant in the Langmuir approach.

It is also interesting to note that the method of linearizing the Michaelis–Menton equation leading to plots of the type shown in Fig. 6.2 is ascribed to Lineweaver and Burk and that these are known as Lineweaver–Burk plots. Many books have been devoted to the determination of enzyme kinetics and to the effects of inhibitors, etc. An interesting example of animated diagrams related to enzyme kinetics is to be found at:

http://www.wiley.com/college/pratt/0471393878/student/animations/enzyme_kinetics/index.html/. The diagrams shown on this site could equally well apply to a heterogeneously catalyzed reaction of the type discussed in the main text.

It is interesting to note that J.J. Carberry has pointed out in his book entitled "Chemical and Catalytic Reaction Engineering" (reprinted by McGraw Hill, NY (1976), partially available at http://books.google.com/books?id = arJLaKa4yDQC&source/), that the derivation of the rate expression for the case of enzyme kinetics described earlier was first carried out by V.C.R. Henri in 1902 (Acad. Sci. Paris, 135 (1902) 916); he also pointed out that Henri also derived

(*Continued*)

BOX 7.1 Enzyme Kinetics (Continued)

the Langmuir isotherm and the resulting rate equation. It is often complained that the current generation of young researchers is unfamiliar with the earlier literature and this is blamed on the use of electronic searches which are based on limited data. The example treated here is an excellent example of how original references can become obscured by time and over-familiarity with a specific topic, even in a time period when electronic methods were unheard of. The whole subject of communication between scientists in the early 20th century is one which deserves much more attention.

Task 7.1 Further Data from the Information given in Fig. 7.1

What would be the position of the reciprocal plot in Fig. 7.2 for the second curve of Fig. 7.1 assuming that the one shown in Fig. 7.2 is for the data at temperature T_1? What information could you obtain by using data from the two *straight lines*?

We have assumed for the unimolecular reaction (Eq. 7.3) that only the molecule A is adsorbed on the surface. However, it is possible that the product P is adsorbed in significant amounts before desorbing to the gas phase and that there is an equilibrium for the desorption step equivalent to that for the adsorption step:

$$A_{(g)} \rightleftarrows A_{ads} \rightarrow P_{ads} \rightleftarrows P_{(g)} \tag{7.8}$$

The rate expression for the reaction is now modified by the inclusion of a term for adsorbed P in the denominator:

$$r_A = -\mathrm{d}c_A/\mathrm{d}t = k_T \cdot \theta_A = k_T \cdot b_A P_A / (1 + b_A P_A + b_P P_P) \tag{7.9}$$

where P_P is the partial pressure of the product gas in the reaction mixture. We have now encountered a situation never found in simple homogeneous elementary reactions in which the concentration of the product can have an effect on the rate of the reaction. If the adsorption of *P* is relatively strong compared with that of A and $b_P P_P \gg (1 + b_A P_A)$, Eq. (7.9) tends in the limit to:

$$r_A = -\mathrm{d}c_A/\mathrm{d}t = k_T \cdot \theta_A = k_T \cdot b_A P_A / b_P P_P \tag{7.10}$$

In other words, the reaction is now first order in A but it has an order of -1 in *P*. Hence, *P* is thus now effectively a "poison" for the reaction.[5,6]

[5] In enzyme kinetics, this would be referred to as "Inhibition by products."

[6] Poisons can also be added separately but their effects on the kinetics are exactly the same, there being a term $b_X P_X$ in the denominator for each poison species X. If X is very strongly adsorbed, the reaction can be totally inhibited.

7.3 Bimolecular Reactions—Langmuir–Hinshelwood Kinetics

Sir Cyril Hinshelwood (http://nobelprize.org/nobel_prizes/chemistry/laureates/1956/hinshelwood-bio.html/) was a British chemist best known for his work on the kinetics of chain reactions. In his book entitled "The Kinetics of Chemical Change," first published in 1926, Hinshelwood expanded an earlier idea of Langmuir from 1921 to come up with what are now known as Langmuir–Hinshelwood kinetics[7] to explain the reaction between two molecules over the surface of a heterogeneous catalyst. Let us consider first the catalyzed reaction:

$$A_{(g)} + B_{(g)} \rightarrow C_{(g)} \tag{7.11}$$

The coverage of the surface sites by A_{ads} and B_{ads} are given by the isotherms:

$$\theta_A = b_A P_A / (1 + b_A P_A + b_B P_B) \tag{7.12}$$

and

$$\theta_B = b_B P_B / (1 + b_A P_A + b_B P_B) \tag{7.13}$$

Assuming that the rate-determining step for the reaction between $A_{(g)}$ and $B_{(g)}$ is given by the rate of the surface reaction of the two species A_{ads} and B_{ads} whose concentrations can be expressed by θ_A and θ_B, the rate of reaction can be written as:

$$r_A = -\,\mathrm{d}c_A/\mathrm{d}t = k_T \cdot \theta_A \cdot \theta_B = k_T \cdot b_A P_A \cdot b_B P_B / (1 + b_A P_A + b_B P_B)^2 \tag{7.14}$$

Fig. 7.3 shows a plot of the rate of the hypothetical reaction (7.11) as a function of the pressure of A, P_A, the value of the pressure of B, P_B, being kept constant. In the case of this biomolecular reaction, the rate goes through a maximum. At this point, $\theta_A = \theta_B$ and so the ratio of b_A to b_B can be established (since $b_A P_A = b_B P_B$ at that point). The figure also shows the effect of using larger values of P_B: the curves move to the right with a decrease in rate at lower values of P_A and an increase at higher values of P_A. At lower values of P_A, the rate is proportional to P_A while at higher values, the rate is inversely proportional to P_B.

---

[7] In some American literature, the names Hougen and Watson are often also encountered associated with this approach. [O.A. Hougen and K.M. Watson, Chemical Process Principles, Vol. III, Wiley, New York (1943); the treatment is outlined in detail by Carberry (Chemical and Catalytic Reaction Engineering, Clarendon, Oxford (1940); reprinted by Wiley, NY (1976); see Box 7.1]. According to Carberry, the approach, although similar in form to the Langmuir–Hinshelwood theory, is more explicit in terms of its inclusion of the effects of catalyst sites, of diffusion and of catalyst deactivation. We will, however, restrict ourselves here to the Langmuir–Hinshelwood approach as it is perfectly adequate for the present purposes.

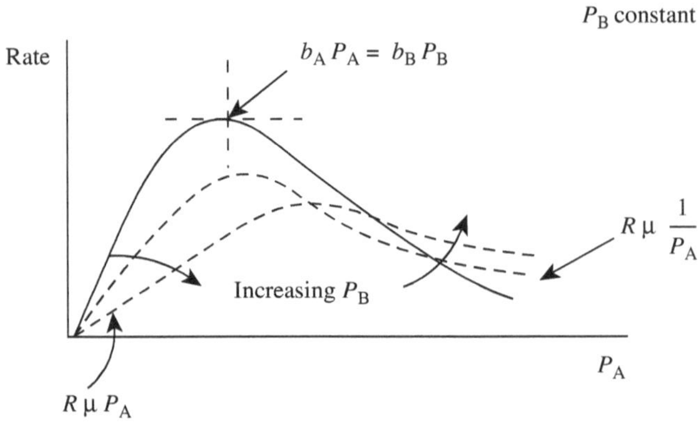

Figure 7.3
Plots of the Langmuir–Hinshelwood model for a bimolecular reaction of A and B over a catalyst as a function of P_A with P_B held constant, showing the effect of increasing the value of P_B.

If neither A nor B is strongly adsorbed, $1 \gg b_A P_A + b_B P_B$ and so Eq. (7.14) reduces to:

$$r_A = -\,dc_A/dt = +\,dC_C/dt = k_T \cdot b_A P_A \cdot b_B P_B = k_T' P_A \cdot P_B \tag{7.15}$$

In other words, the rate equation reduces to a second-order rate expression.

When one of the reactants (e.g., A) is strongly adsorbed and the other (B) is weakly adsorbed, then the equation reduces to:

$$r_A = -\,dc_A/dt = +\,dC_C/dt = k_T \cdot b_A P_A / b_B P_B = k_T'' P_A / P_B \tag{7.16}$$

In this case, the reactant B is effectively a poison for the reaction as an increase in its partial pressure causes a decrease in the rate of the reaction.

The situation is even more complicated if the products of the reaction are adsorbed to any appreciable extent or if there is a strongly adsorbed poison present in the reaction mixture. However, we will not consider such cases here as they result in even more complex expressions that are difficult to analyze in a satisfactory manner. Indeed, it is often possible to come up with several different mechanistic explanations for a complex set of kinetic data and it is only with additional supporting evidence of the sort provided by some of the methods described in Chapter 6 that it is possible to decide on the most likely mechanism. Before moving on to discuss what happens when one or other, or both, of the reactants dissociates on the surface before reaction, we will first discuss two other kinetic expressions which have some importance in relation to catalytic mechanisms.

BOX 7.2 The Effect of Temperature on a the Rate of a Catalytic Reaction

For a homogeneous (noncatalyzed) first-order gas-phase reaction for which

$-dC_A/dt = kC_A$, an increase in temperature only has an effect on the value of the rate constant k and application of the Arrhenius equation gives rise directly to the activation energy for the rate-determining step in the reaction. However, when we are dealing with a heterogeneously catalyzed reaction such as those shown in Eqs. (7.3 and 7.8), the corresponding rate equations contain in addition to rate constants adsorption constants which also have temperature dependencies. Let us consider the case of Eq. (7.4):

$$r_A = -dc_A/dt = k_T \cdot \theta_A = k_T \cdot b_A P_A/(1 + b_A P_A) \qquad (7.4)$$

At low values of P_A, the effective rate constant is $k_T.b_A$ while at high values of P_A it is k_T. As b_A is effectively an equilibrium constant with an enthalpy change of adsorption of ΔH°_a, its temperature dependence is given by the van't Hoff isochore:

$$d(\ln b_A)/dt = \Delta H^\circ_a/RT^2$$

or, in its integrated form:

$$b_A = C \exp\left(-\Delta H^\circ_a/RT\right)$$

Hence, at low P_A, the apparent activation energy E_{app} for the reaction is given by $E_a + \Delta H^\circ_a$ while at higher values of P_A, the apparent activation energy is the same as E_a, the "true" activation energy. A plot of log r_A versus $1/T$ for such a reaction at a constant value of P_A is shown in Task 7.2. How would you determine the values of E_a and ΔH°_a from such a plot?

Task 7.2 True and Apparent Activation Energies for Eq. (7.4) (see Box 7.2)

The following figure shows an Arrhenius plot for a unimolecular catalyzed reaction. Explain the shape of the plot.

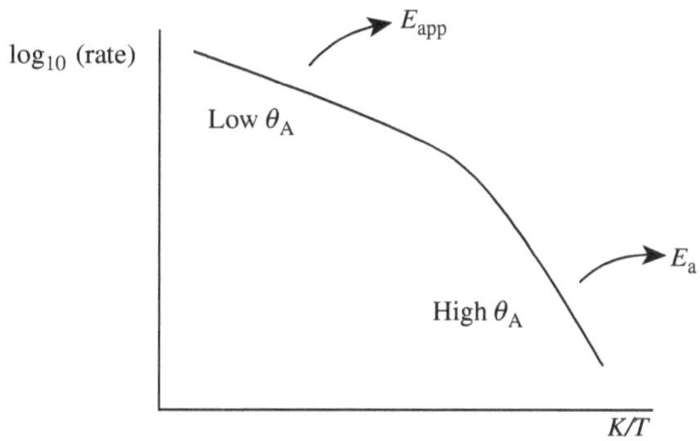

As drawn, the axes of this plot do not show any numerical values. If the reaction rate in the low temperature region increases by a factor of two when the temperature is raised from

50°C to 70°C, calculate the true activation energy, E_a, and the enthalpy of adsorption of A on the surface, ΔH_a^o.

Hint: You will have to make an estimate of the gradients of the *lines* shown at low and high temperatures.[a]

[a]As we will see later in this chapter, the gradient of the type of plot shown in Task 7.2 may also be affected by diffusion limitations and readers should therefore be aware that a change of gradient of an Arrhenius plot is not necessarily an indication of the existence of a Langmuir–Hinshelwood mechanism.

7.4 Bimolecular Reactions—Eley–Rideal Kinetics

The Eley–Rideal mechanism describes the situation which occurs in a bimolecular reaction when one of the reactants interacts directly from the gas phase with an adsorbed layer of the other reactant:

$$A_{(g)} + B_{(ads)} \rightarrow P_{(g)} \tag{7.17}$$

The mechanism was first postulated by Eley and Rideal in 1939[8] in relation to work on hydrogen reactions on metals but is more usually nowadays applied to more complicated situations. In essence, the mechanism is based on the idea that one reacting species is adsorbed on the surface at effectively full coverage and that the other reacts with it from the gas phase (or perhaps as a physically adsorbed species) to form the product:

$$D_{2(g)} + H_{ads} \rightarrow HD_{(g)} + D_{(ads)} \tag{7.18}$$

The kinetic expression is set up by using a Langmuir expression for the concentration of $B_{(ads)}$ and using P_A to express the concentration of A:

$$\text{Rate} = -\,dP_A/dt = k_T \cdot P_A \cdot \theta_B = k_T \cdot P_A \cdot b_B P_B/(1 + b_B P_B) \tag{7.19}$$

It is interesting to note that an equivalent equation could be set up if it was assumed that A and B adsorbed on different sorts of sites so that two separate Langmuir isotherms must be used to express the rate:

$$\text{Rate} = -\,dP_A/dt = k_T \cdot \theta_A \theta_B = k_T \cdot \{b_A P_A/(1 + b_A P_A)\} \cdot \{b_B P_B/(1 + b_B P_B)\} \tag{7.20}$$

This reduces to Eq. (7.18) if $b_A P_A \ll 1$, in other words if A is weakly adsorbed.

8 D.D. Eley was at the time a doctoral student working with Sir Eric Rideal in Cambridge on ortho-para hydrogen exchange over metal surfaces. It has also been named the Langmuir–Rideal mechanism as it has been suggested that Langmuit postulated the mechanism as early as 1921.

7.5 The Mars–Van Krevelen Mechanism

The third commonly encountered mechanistic approach of importance in catalysis is the Mars–Van Krevelen Mechanism. This mechanism is generally used to describe the kinetics of reactions such as selective oxidation of hydrocarbons (HCs) although it was first developed to explain the kinetics of the oxidation of SO_2 over a molten V/K salt supported on a porous silica support.[9] It has more recently been applied to many other types of reaction such as hydrodesulfurization and NO_x removal.[10]

The mechanism is based on the idea that adsorption of one molecule occurs on top of another molecule which had previously been adsorbed.[11] We will now derive the appropriate equation for a very general case, the allover reaction:

$$A_{(g)} + B_{(g)} \rightarrow C(g) \tag{7.21}$$

We assume that the reaction takes place according to the following steps, where $\square$ represents an empty site on the surface of the catalyst:

$$\begin{array}{ll} A_{(g)} + B_{ads} \rightarrow A - B_{ads} & \text{(a) Adsorption of A on top of adsorbed B} \\ AB_{ads} \rightarrow C_{ads} & \text{(b) Reaction of adsorbed complex} \\ C_{ads} \rightarrow C_{(g)} + \square & \text{(c) Desorption of product C} \\ B_{(g)} + \square \rightarrow B_{ads} & \text{(d) Adsorption of B} \end{array} \tag{7.22}$$

The adsorption of B (step d) is shown last since the empty site is created by the reaction rather than being present on the fresh catalyst.[12] In setting up a kinetic relationship, steps (a) and (b) are combined, assuming that the product is formed as soon A interacts with an adsorbed B. It is also assumed that the desorption step (c) is rapid compared either with this combined reaction or with step (b). We then assume that the reaction effectively occurs in two steps: the adsorption of B with rate constant k_2 (step (d)); and a second (a combination of all the other steps), the formation of an empty site as well as the products by the reaction:

$$A_{(g)} + B_{ads} \xrightarrow{k_1} C_{(g)} + \square \quad \text{(e)}$$

If the coverage of the surface by adsorbed B species is given by θ_B and that of empty sites is given by $\theta_\square$ (where $\theta_B + \theta_\square = 1$), the rates of these two steps are given by:

$$r_1 = k_1 \cdot P_A \cdot \theta_B \text{ and } r_2 = k_2 \cdot P_B \cdot \theta_\square = k_2 \cdot P_B \cdot (1 - \theta_B) \tag{7.23}$$

[9] P. Mars, D.W. van Krevelen, Chem. Eng. Sci. Spec. Suppl. 3 (1954) 41.

[10] An excellent review of more recent applications of the mechanism is given in an article by C. Doornkamp and V. Ponec in J. Mol. Catal. A: Chem. 162 (2000) 19–32.

[11] The original derivation was for the adsorption of SO_2 on top of a preadsorbed oxygen atom.

[12] In practice, $B_{(g)}$ is often oxygen and so we then also have to consider the dissociation of the O_2 molecule.

from which, with $r_1 = r_2$ in a steady state, we obtain

$$\theta_B = k_2 P_B/(k_1 P_A + k_2 P_B) \quad (7.24)$$

Hence the rate of reaction is given by[13]:

$$\text{Rate, } r = k_1 \cdot P_A \cdot \theta_B = k_1 k_2 P_A P_B/(k_1 P_A + k_2 P_B) \quad (7.25)$$

which can be rewritten in a much simpler form as[14]:

$$1/r = 1/k_1 P_A + 1/k_2 P_B \quad (7.26)$$

There are two limiting cases:

If $k_1 \ll k_2$, then

$$r = k_1 P_A \quad (7.27)$$

and if $k_2 \ll k_1$, then

$$r = k_2 P_B \quad (7.28)$$

The Mars–Van Krevelen equation is most frequently encountered in relation to selective oxidation reactions. The selective oxidation of an HC is treated in Box 7.3. (See also Task 7.3).

BOX 7.3 The Mars–Van Krevelen Mechanism Applied to the Selective Oxidation of a Hydrocarbon

Let us assume that the reaction occurs by two steps:

$$C_xH_y + O_{ads} \xrightarrow{k_1} \text{Oxidation products} + \square \quad \textit{(Reduction)} \quad (1)$$

and

$$0.5O_2 + \square \xrightarrow{k_2} O_{ads} \quad \textit{(Oxidation)} \quad (2)$$

The concentration of the active oxygen on the catalyst surface is determined, as in the derivation in the main text, by the relative rates of reactions (1) and (2). We assume here for

(Continued)

[13] It is interesting to note that the form of the Mars–Van Krevelen equation (7.25) is very similar to that of the Eley–Rideal equation for the reaction of A(g) from the gas phase with a surface containing both A_{ads} and B_{ads} if one assumes that the adsorption of both molecules is strong (the 1 in the denominator can be ignored) but that A does not react with B on the surface to an appreciable extent.

[14] This is equivalent to the equation giving the total resistance for two electrical resistances in series. It can be extended further by adding an equivalent reciprocal term for the rate of diffusion of gaseous molecules to the surface (external diffusion). We will discuss external diffusion limitations in this chapter. It is interesting to note that an equivalent term for internal diffusion cannot simply be added since the expressions allowing for internal diffusion include terms for the chemical kinetics.

BOX 7.3 The Mars–Van Krevelen Mechanism Applied to the Selective Oxidation of a Hydrocarbon (Continued)

simplicity* that one oxygen atom is needed to produce the (selective) oxidation products from the HC and that there is no parallel nonselective oxidation process to H_2O and CO_2.

Following the same arguments as for the general case in the main text (but now with the rate of step (2) expressed as $k_2 P_{O2}{}^{1/2}.(1-\theta_O)$, where θ_O is the coverage of the surface by oxygen species), we obtain:

$$\text{Rate} = r = k_1 k_2[\text{HC}]\cdot P_{O_2}^{1/2} / \left\{ k_1[\text{HC}] + k_2 \cdot P_{O_2}^{1/2} \right\} = 1/\left\{ 1/k_1[\text{HC}] + 1/k_2 \cdot P_{O_2}^{1/2} \right\} \quad (3)$$

or

$$1/r = 1/k_1[\text{HC}] + 1/k_2 \cdot P_{O_2}^{1/2} \quad (4)$$

There are two limiting cases as before:

$$r_{red} = k_1 \cdot P_{HC} \quad (5)$$

and

$$r_{ox} = k_2 \cdot P_{O_2}^{1/2} \quad (6)$$

Taking as an example the oxidation of an HC over a V_2O_5-containing catalyst for which the activity is due to the oxidation/reduction cycle:

$$V^{5+} + e \rightleftarrows V^{4+} \quad (7)$$

Eq. (4) would apply to the situation in which the active sites were full of oxygen (V^{5+}) and Eq. (5) would apply if the sites were fully reduced (V^{4+}).

*The conventional derivation has recently been criticized by Vannice (Catal. Today 123 (2007) 18–22) who argues that, although the redox model may apply, the mathematical treatment is acceptable only for the adsorption of molecular oxygen on a single site. He argues that the redox model is better described by a Langmuir–Hinshelwood or Hougen–Watson approach. The approach above works as long as step (2) is not an elementary step but encompasses adsorption of oxygen followed by dissociation of the oxygen to allow it to be adsorbed on two distinct sites. We discuss kinetic expressions allowing for dissociation of the adsorbed species in the following sections.

Task 7.3 Mars–Van Krevelen and Selective Oxidation Catalysis

Using the name Barbara Grzbowska (-Swierkosz) as a starting point, do a citation search for articles related to the Mars–Van Krevelen mechanism as applied to the selective oxidation of HCs over vanadia catalysts. Examine some of these articles in relation to their discussion of the mechanism of the reactions considered.

7.6 Some Practical Examples of Mechanistic Kinetic Approaches

7.6.1 The Langmuir Equation and the Pressure Function

We will now focus our attention on some examples of kinetic expressions that have been set up for reactions of practical importance. In the examples of the three main sorts of mechanism given earlier, we have considered only abstract molecules, A, B, HC, etc. We will now consider how the equations are modified if dissociative adsorption of one or more of the molecules taking part in the reaction occurs. Our first example concerns the kinetics of the synthesis of ammonia.

It is at first necessary to have a brief recap and to introduce the concept of "pressure function." In Chapter 2, Surfaces and Adsorption, we discussed the formal equations that can be set up using the Langmuir approach for the dissociative adsorption of molecules such as hydrogen and methane (as CH_3 groups) and obtained for these two molecules the following isotherms:

$$\theta_H = c_H P_{H_2}^{0.5}/(1 + c_H P_{H_2}^{0.5}) \tag{2.5}$$

which for intermediate coverages simplifies to:

$$\theta_H = c_H^n P_{H_2}^n \tag{2.6}$$

and

$$\theta_{CH_3} = kP_{CH_4}/P_{H_2}^{0.5}/(1 + kP_{CH_4}/P_{H_2}^{0.5}) \tag{2.8}$$

which again simplifies to

$$\theta_{CH_3} = k'(P_{CH_4}/P_{H_2}^{0.5})^n \tag{2.9}$$

where $0 < n < 1$.

Similar power expressions, the so-called pressure functions, can be written (Box 7.4) for many other adsorbed species resulting from dissociative adsorption. If we are concerned with a reaction involving dissociated species such as $CH_{3(ads)}$ or $H_{(ads)}$, then it is possible with such expressions to set up kinetic expressions which can be considered to have some basis in mechanism, even though the resultant expressions could also be considered to be semiempirical rate expressions. We will now deal with several examples of detailed kinetic studies reported in the literature in which such relationships are used to construct mechanistic kinetic rate laws. The reader is encouraged to follow up on some of the examples given in the following text and to search the literature for other examples.

7.6.2 Ammonia Synthesis

We have discussed the importance of the catalytic synthesis of ammonia by the Haber–Bosch Process in some detail in Chapter 2, Surfaces and Adsorption. We showed there that both the

BOX 7.4 Pressure Functions

If we wish to set up a Langmuir adsorption expression for a species such as $C_2H_{5(ads)}$ resulting from the adsorption of ethane on a uniform surface, we may consider the following equilibrium occurs:

$$C_2H_{6(g)} \rightleftarrows C_2H_{5(g)} + 1/2H_{2(g)} \quad (1)$$

and then write:

$$K_{C_2H_5} = P_{C_2H_5}P_{H_2}^{1/2}/P_{C_2H_6} \quad (2)$$

Hence, we can write:

$$P_{C_2H_5} = K_{C_2H_5} \cdot P_{C_2H_6}/P_{H_2}^{1/2} \quad (3)$$

This is known as a "pressure function" and expression (3) can be inserted in a Langmuir adsorption isotherm to give the appropriate isotherm for the adsorption of C_2H_5 species*:

$$\begin{aligned}\theta_{C_2H_5} &= b_{C_2H_5} \cdot P_{C_2H_5}/(1 + b_{C_2H_5} \cdot P_{C_2H_5}) \\ &= b_{C_2H_5}K_{C_2H_5} \cdot P_{C_2H_6}/P_{H_2}^{1/2}/(1 + b_{C_2H_5}K_{C_2H_5} \cdot P_{C_2H_6}/P_{H_2}^{1/2}) \quad (4) \\ &= c\, P_{C_2H_6}/P_{H_2}^{1/2}/(1 + c\, P_{C_2H_6}/P_{H_2}^{1/2})\end{aligned}$$

where the constant c incorporates both the equilibrium constant and the adsorption constant. For intermediate coverages, Eq. (4) can be simplified further to give a power expression:

$$\theta_{C_2H_5} = c' P^{n}_{C_2H_6}/P^{n/2}_{H_2}$$

This treatment assumes that only C_2H_5 species are adsorbed on the surface to any appreciable extent and that all the reactions up and to the adsorption of the C_2H_5 species are in equilibrium. Similar expressions can be set up for any such species as long as similar assumptions can be made. We will see in the following text that such an approach has been made in the case of the kinetics of ammonia synthesis.

*C. Kemball, Thermodynamic factors in adsorption and catalysis. Equilibria in the adsorbed phase, Disc. Faraday Soc. 41 (1966) 190.

recognition that this reaction could be carried out and the research carried out on the catalysts and plants to be used commercially were extremely important landmarks in the modern development of catalysis as we know it. Ammonia synthesis occurs by the reaction:

$$N_{2(g)} + 3H_{2(g)} \rightleftarrows 2NH_{3(g)} \quad (7.29)$$

As we will see in Chapter 8, Mass and Heat Transfer Limitations and Other Aspects of the Use of Large Scale Catalytic Reactors, the reaction is severely limited by thermodynamic considerations and so the reverse reaction occurs to a significant extent as soon as a reasonable conversion is reached. Many papers have been written on the subject of the kinetics of ammonia synthesis and many consider the effects of the reverse reaction as well as the forward reaction. Here we will consider only the forward reaction in any detail; the interested reader should consult one of the more advanced textbooks available[15] to find more details of fuller treatments, including a discussion of the idea of stoichiometric number as applied to multistep reactions. (See also Box 7.5 and Tasks 7.4 and 7.5).

BOX 7.5 Ammonia Decomposition

It is much easier to study ammonia decomposition than ammonia synthesis since the decomposition reaction, when carried out at reasonably high temperatures, can be considered to be essentially irreversible. Further, it need not be studied under high pressure and so the interpretation of kinetic measurements is much more straightforward. Unless there is a large change in reaction mechanism with change in temperature, the kinetic parameters determined for the decomposition reaction should be directly related to those for the forward (synthesis) reaction. This feature is particularly useful in making comparisons between various different catalyst compositions.

It is interesting to note that there has recently been a renewed interest in the catalytic decomposition of ammonia as a source of clean hydrogen for fuel cell applications.

Task 7.4 Kinetics of Ammonia Decomposition

Use the approach outlined earlier to derive a general power rate kinetic expression for the decomposition of ammonia over a metal catalyst.

Task 7.5 Literature Search on Ammonia Synthesis and Decomposition

Examine the recent literature on the synthesis and decomposition of ammonia. Start with a citation search for the article: J.A. Dumesic, A.A. Trevino, J. Catal. 116 (1989) 119. Using this reference, find more recent papers and reviews on the subject.

We now consider the derivation of an expression for the forward reaction to explain the kinetics of the reaction over an iron catalyst at very low conversions. This expression was originally proposed by Tempkin and Pyzhev (Acta Physicochim. USSR, 12 (1940) 3270) on the basis of the Elovich equation (which allows for a variation of the energetics of adsorption with coverage) but the approach was modified by Ozaki, Taylor and Boudart

[15] See, for example, J.M. Thomas, W.J. Thomas, Principles and Practice of Heterogeneous Catalysis, second ed., Wiley-VCH, Weinheim, 2014; C.H. Bartholomew, R.J. Farrauto, Fundamentals of Industrial Catalytic Processes, second ed., Wiley Interscience, New Jersey, 2006; R.J. Farrauto, L. Dorazio, C.H. Bartholomew, Introduction to Catalysis and Industrial Catalytic Processes, Wiley, 2016.

(Proc. R. Soc., Sect. A 258 (1960) 47) by using a Langmuir approach with the assumption of a uniform surface. In common with the earlier approach of Temkin and Pyzhev, they assume that the rate-determining step in the synthesis reaction is the rate of the dissociative adsorption of nitrogen on a nitrogen-covered surface:

$$r = k_a P_{N_2}(1-\theta_N)^2 \tag{7.30}$$

The value of θ_N is given by a Langmuir isotherm of the form:

$$\theta_N = bP_N^*/\left(1 + bP_N^*\right) \tag{7.31}$$

where $P_N{}^*$ is the effective partial pressure of nitrogen atoms (sometimes known as the "virtual pressure" of the nitrogen atoms[16]) that would be in equilibrium with the adsorbed nitrogen atoms. As this virtual pressure cannot be measured directly, it is calculated assuming that the nitrogen atoms are present in equilibrium according to:

$$N^*_{(g)} + 1.5H_{2(g)} \rightarrow NH_{3(g)} \tag{7.32}$$

for which the equilibrium constant is given by:

$$K_N^* = P_{NH_3}/P_N^* \cdot P_{H_2}^{1.5} \tag{7.33}$$

Inserting Eq. (6.32) in Eq. (6.30) gives:

$$\theta_N = K \cdot P_{NH_3}/P_{H_2}^{1.5}/\{1 + K \cdot P_{NH_3}/P_{H_2}^{1.5}\} \tag{7.34}$$

where $K = b/K_N{}^*$

Substituting Eq. (7.34) in Eq. (7.30), after slight manipulation (check!), gives:

$$r = k' P_{N_2}/\{1 + K \cdot P_{NH_3}/P_{H_2}^{1.5}\}^2 \tag{7.35}$$

This equation, which can be approximated by:

$$r = k_a P_{N_2} \cdot \{P_{H_2}^{1.5}/P_{NH_3}^2\}^n \tag{7.36}$$

where $0 < n < 1$, gives a reasonable description of the kinetics at low conversions of nitrogen to ammonia. At higher conversions, the rate of the reverse reaction becomes appreciable and so, as noted earlier, a term has to be added for the reverse reaction. The full Tempkin–Pyzhev equation that accounts for the rates of the forward and reverse reactions is:

$$r = k' P_{N_2}\{P_{H_2}^3/P_{NH_3}^2\}^m - k''\{P_{NH_3}^2/P_{H_2}^3\}^{(1-m)} \tag{7.37}$$

[16] The virtual pressure is equivalent to the pressure function referred to in Box 7.4.

It should be recognized that when $r = 0$, the reaction is at equilibrium and then

$$k'P_{N_2}\{P^3_{H_2}/P^2_{NH_3}\}^m = k''\{P^2_{NH_3}/P^3_{H_2}\}^{(1-m)} \tag{7.38}$$

or

$$k'/k'' = P^2_{NH_3}/(P_{N_2}\cdot P^3_{H_2}) = K_{eq} \tag{7.39}$$

7.6.3 Ethane Hydrogenolysis

Much work has been reported in the literature on the mechanism of the conversion of HCs over metal catalysts, both unsupported (e.g., metallic films) and supported. This work is related to a variety of different types of process: hydrogenation[17], hydrogenolysis, isomerization, and cracking/hydrocracking. Common to all these reactions is the adsorption of the HC on a clean metallic surface, this involving as a first step dissociation of a C–H bond and adsorption of the HC species, together with hydrogen atoms, on the metal surface. We will now discuss the model set up by J. H. Sinfelt for the hydrogenolyis of ethane to give methane. His analysis is based on the following steps occurring on the metal surface:

$$\begin{aligned}
&C_2H_{6(g)} \rightleftarrows C_2H_{5(ads)} + H_{(ads)} && \text{(i)}\\
&C_2H_{5(ads)} + H_{(ads)} \rightleftarrows C_2H_{x(ads)} + \alpha H_{2(g)} && \text{(ii)}\\
&C_2H_{x(ads)} \rightarrow \text{adsorbed } C_1 \text{ fragments} && \text{(iii)}\\
&\text{adsorbed } C_1 \text{ fragments} + H_{(ads)} \rightarrow CH_{4(g)} && \text{(iv)}
\end{aligned} \tag{7.40}$$

The ethane molecule adsorbs by the initial dissociation of a C–H bond and the $C_2H_{5(ads)}$ species formed then dehydrogenates rapidly and reversibly to give the species $C_2H_{x(ads)}$ which is therefore considered to be in equilibrium with the gaseous ethane. The quantity α is equal to $(6 - x)/2$. Sinfelt's analysis assumes that the $C_2H_{x(ads)}$ species is the only one present on the surface in any appreciable concentration and that it reacts to give adsorbed C1 fragments (step iii), these then giving $CH_{4(g)}$ by rapid interaction with adsorbed hydrogen species (step iv).

The coverage of the surface by $C_2H_{x(ads)}$, θ_x, is then given by Eq. (7.41):

$$\theta_x = (K\cdot P_E/P^{\alpha}_{H_2})/(1 + K\cdot P_E/P^{\alpha}_{H_2}) \tag{7.41}$$

Simplifying this expression for intermediate coverages (as we have done in several examples mentioned earlier):

$$\theta_x = (K\cdot P_E/P^{\alpha}_{H_2})^n \tag{7.42}$$

where, as before, $0 < n < 1$.

[17] The subject of hydrogenation is dealt with in Chapter 14, Conversion of Hydrocarbons to Intermediates by Catalytic Hydrogenation or Oxidation.

Table 7.1: Kinetic of ethane hydrogenolysis. Analysis of hydrogen pressure dependencies for different metals

				Exponents in Rate Expression		
				Experimental		Theoretical
Catalyst	Temp. (°C)	x	α	Ethane (n)	Hydrogen (m)	Hydrogen ($-n\alpha$)
Co	219	4	1	1.0	− 0.8	− 1.0
Ni	177	2	2	1.0	− 2.4	− 2.0
Ru	188	2	2	0.8	− 1.3	− 1.6
Rh	214	0	3	0.8	− 2.2	− 2.4
Pd	354	0	3	0.9	− 2.5	− 2.7
Os	152	2	2	0.6	− 1.2	− 1.2
Ir	210	2	2	0.7	− 1.6	− 1.4
Pt	357	0	3	0.9	− 2.5	− 2.7

Reproduced with permission from J. Catal., 3 (1969) 175.

The rate of hydrogenolysis to methane is then given by that of step (7.40.iii):

$$r = k_{iii} \cdot \theta_x = k_{iii} \cdot K^n \cdot (P_E/P_{H_2}^{\alpha})^n = k \cdot P_E^n P_{H_2}^{(-n\alpha)} \tag{7.43}$$

Sinfelt had previously reported[18] kinetic measurements for the hydrogenolysis of ethane over a range of different metals supported on a silica support. He fitted those results to Eq. (7.43) and obtained the values for the exponent for the hydrogen pressure given in the final column of Table 7.1. The agreement between the experimental and theoretical exponents (final two columns) is very close, indicating that the model works very well for the conditions studied.[19]

As noted earlier, the quantity x denotes the number of hydrogen atoms associated with the C_2 species on the surface before dissociation occurs. It can be seen that Co gives the highest value, corresponding to C_2H_4 adsorbed on the surface, while Pd and Pt give complete removal of hydrogen in the adsorbed species. (See also Task 7.6).

Task 7.6 Other Papers using Sinfelt's Approach to Kinetic Equations

Do a citation search on the paper by Sinfelt given in footnote 19 and find other work on the application of this kinetic approach. A paper by Rostrup Nielsen and Alstrop should have appeared in your search, this discussing the relationship between hydrogenolysis and steam reforming of HCs. Follow up on the work from the Topsøe laboratories on the effects of sulfur and gold on the steam reforming reaction.

[18] J.H. Sinfelt, Catal. Rev. 3 (1969) 175.

[19] In the paper cited above, Sinfelt goes on to discuss the effect of carrying out the hydrogenolysis reaction at higher temperature and shows that the dependence of the rate on hydrogen partial pressure is temperature dependent at these temperatures.

7.6.4 Methanation

The subject of the methanation of CO was the subject of much attention in the period around 1980 and it was one of the top 15 "research fronts" in the first 6 months of 1982.[20] This was largely because there was a shortage of natural gas, particularly in the Northern states of the United States, and there was a renewed interest in the possibility of producing natural gas from syngas made by coal gasification.[21] Many papers on the subject were published in that period, discussing topics varying from the fundamentals of the mechanism of the reaction to plant operation. We will return to one aspect of the operation of full-scale reactors for this reaction in Chapter 8. However, here we will concentrate on the kinetic aspects of the reaction.

In 1975, Vannice published two papers[22] in which he described the measurement of the kinetics of the methanation reaction over a series of Group A metals and fitted his rate data to a kinetic expression:

$$\text{Rate} = k \cdot P_{H_2}^{x} \cdot P_{CO}^{y} \tag{7.44}$$

An equivalent theoretical equation was derived on the assumption that the rate-determining step in the reaction was the interaction of adsorbed hydrogen atoms with CHOH species adsorbed on the surface:

$$CHOH_{(ads)} + y/2H_{2(ads)} \rightarrow CH_{y(ads)} + H_2O_{(g)} \tag{7.45}$$

Ollis and Vannice later published[23] a more satisfactory treatment based on the rate-determining step shown in Eq. (7.45) in which they derived a Langmuir expression for $CHOH_{(ads)}$:

$$\theta_{CHOH} = K \cdot P_{CO} \cdot P_{H_2}/(1 + K \cdot P_{CO} \cdot P_{H_2}) \tag{7.46}$$

which was approximated by:

$$\theta_{CHOH} \approx (K \cdot P_{CO} \cdot P_{H_2})^{n} \tag{7.47}$$

with $0 < n < 1$ as in the previous derivations. Their resulting theoretical equation has the form:

$$\text{Rate} = k\{P_{CO} \cdot P_{H_2} \cdot (P_{H_2}/P_{CO})^{y/2}\}^{1/2} \tag{7.48}$$

---

[20] E. Garfield, Current Contents, Phys. Chem. Sci. 22 (1982) 5.

[21] For a review of methanation and steam reforming, the reader is referred to the article: "Metal Catalysed Methanation and Steam Reforming," by J.R.H. Ross, in "Catalysis," 7 (1985), 1–45, Special Periodical Reports, Royal Society of Chemistry. (Available for download from www.rsc.org/ebooks/archive/free/.../CL9780851865843-00001.pdf/)

[22] M.A. Vannice, The catalytic synthesis of hydrocarbons from H_2/CO mixtures over the Group VIII metals, Parts 1 and 2, J. Catal. 37 (1975) 449–461 and 462–473.

[23] D.F. Ollis, M.A. Vannice, J. Catal. 38 (1975) 514.

in which k incorporates several constants and y is the number of hydrogen atoms involved in the rate-determining step (Eq. 7.45).

Later work on the methanation reaction showed quite clearly that the reaction proceeds not through $CHOH_{(ads)}$ species but through adsorbed carbon atoms derived from the dissociative adsorption of CO. The present author derived an expression for the kinetics of the methanation reaction making the assumption that the following equilibria occur[24]:

$$\begin{array}{ll} CO_{(g)} \rightleftarrows CO_{(ads)} & \text{(i)} \\ CO_{(ads)} \rightleftarrows C_{(ads)} + O_{(ads)} & \text{(ii)} \\ H_{2(g)} \rightleftarrows 2H_{(ads)} & \text{(iii)} \\ 2H_{(ads)} + O_{(ads)} \rightleftarrows H_2O_{(g)} & \text{(iv)} \end{array} \tag{7.49}$$

with the following reaction as rate determining:

$$C_{(ads)} + yH_{(ads)} \rightarrow CH_{y(ads)} \quad \text{(v)}$$

The species $CH_{y(ads)}$ then combines rapidly with more hydrogen species and methane is desorbed.

The resultant rate expression is:

$$\text{Rate} = \frac{k \cdot P_{CO} P_{H_2}^{(1+y/2)} / P_{H_2O}}{(1 + b_C P_{CO} P_{H_2} / P_{H_2O})^{1+y}} \tag{7.50}$$

Multiplying through by $P_{CO}P_{H2}/P_{H2O}$, one obtains:

$$\text{Rate } \alpha \ \frac{\{P_{CO} P_{H_2} / P_{H_2O}\}^{(1+y)} \cdot P_{H_2O}^{y}}{(1 + b_C P_{CO} P_{H_2} / P_{H_2O})^{1+y} P_{CO}^{y} P_{H_2}^{y/2}} \tag{7.51}$$

Making the same approximation as earlier, the resulting equation is:

$$\text{Rate } \alpha \ P_{CO}^{\{n+y(n-1)\}} P_{H_2}^{\{n+y(n-1/2)\}} P_{H_2O}^{-\{n+y(1-n)\}}$$

The most significant difference between this expression and that obtained by Ollis and Vannice is the inclusion for a term for the water partial pressure. Unfortunately, no work in which the effect of water vapor on the kinetics of the methanation reaction has been published; some work carried out at the time using a microbalance system with gas analysis of the exit gases gave strong support for this kinetic expression[25] but it was never possible to obtain sufficiently good control of the water partial pressures to obtain unequivocal results. (See also Task 7.7).

[24] J.R.H. Ross, A modified kinetic expression for the methanation of carbon monoxide over Group VIII metal catalysts, J. Catal. 71 (1981) 205.

[25] B. Höhlein, R. Menzer, J.R.H. Ross, J. Sarkar, unpublished results.

Task 7.7 Kinetic of the Methanation Reaction over an Ni/SiO_2 Catalyst

R.Z.C. van Meerten and his colleagues (Appl. Catal. 3 (1982) 29) have carried out a painstaking examination of the kinetics of methanation of CO over a nickel/silica catalyst. Study the results of this paper and discuss the model put forward. Note how the authors allow for gradual deactivation of the catalyst in their kinetic measurements.

Then carry out a citation search based on this paper and find some more recent papers citing the work, paying particular attention to several papers having J. Rostrup Nielsen as one of the authors.

7.6.5 Formic Acid Decomposition

The decomposition of formic acid has already been mentioned briefly in previous chapters (Chapters 1 and 6) as a reaction of current interest as it provides a route for the production of pure hydrogen for use, for example, as a fuel in fuel cells for mobile applications. Some of the recent work on the development of active and selective catalysts for this reaction will be described in Chapter 15, Catalysis in Biomass Conversion and so the contents of this section will be confined to a discussion of the kinetics of the reaction over various catalysts in the presence of formate anions.

Recent work from the author's laboratory[26] to be described in more detail in Chapter 15, has shown that when a Pd/C catalyst is doped with relatively high concentrations of K_2CO_3, the performance of the catalyst is improved very substantially compared with the undoped material, conversions of almost 100% being achieved at temperatures as low as 80°C, some 50°C below the temperature needed for the undoped catalyst. This change in behavior was shown to be due to the formation of an aqueous solution of potassium formate with formic acid within the pores of the catalyst, this solution then exhibiting the behavior of a buffer solution. The activation energy for the decomposition of the formic acid was now significantly different to that for the undoped catalyst and the rate of reaction changed from being first order in formic acid to being zero order. This is consistent with a mechanism in which the rate-determining step in the reaction is now the decomposition of a formate ion at the Pd surface, the concentration of these ions being maintained constant by the immediate replacement of the reacted ion by the dissociation of another molecule of formic acid, as follows:

$$COOH^- \rightarrow CO_2(g) + H^-_{ads} \tag{7.52}$$

$$HCOOH \rightleftarrows H^+ + COOH^- \tag{7.53}$$

[26] L. Jia, D.A. Bulushev, J.R.H. Ross, Catal. Today 259 (2016) 453–459.

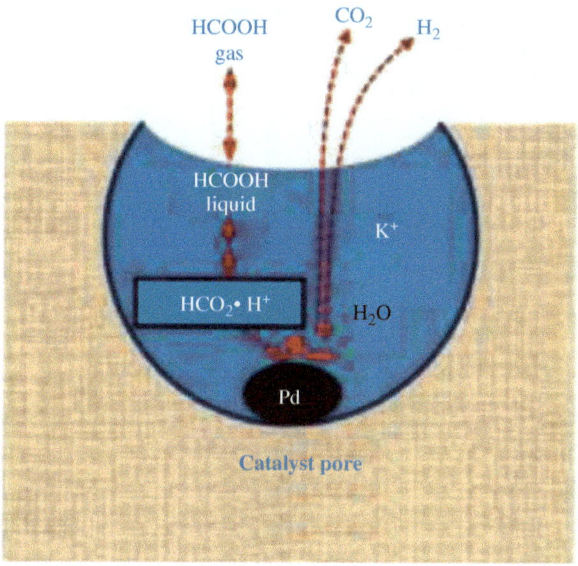

Figure 7.4
Schematic representation of the reaction of formic acid in a pore of a potassium-doped Pd/C catalyst with the formation of a formate buffer solution within the pore. *Reproduced from L. Jia, D. A. Bulushev, S. Beloshapkin, J.R.H. Ross, Hydrogen production from formic acid vapour over a Pd/C catalyst promoted by potassium salts: evidence for the participation of buffer-like solution in the pores of the catalyst, Appl. Catal. B 160–161 (2014) 35–43, with kind permission of Elsevier.*

The catalytic cycle is completed by the reaction of the H^+ ion with the H^-_{ads} species, probably with electron transfer and the intermediate formation of adsorbed hydrogen atoms on the Pd surface:

$$H^+ + H^-_{ads} \rightarrow H_{2(g)} \tag{7.54}$$

This sequence is depicted schematically in Fig. 7.4

If this mechanism is correct, one can predict that it should also apply in other catalytic situations in which formate ions are present with formic acid. There has recently been a great deal of work on the reaction of aqueous formic acid solutions containing sodium formate on supported Pd catalysts at temperatures close to room temperature[27] but rather surprisingly little attention has been given to the role of the formate ions in the reaction. However, Li et al. have recently recognized[28] that the formate mechanism probably applies to their work on the effect of Pd particle size on the rate of formic acid decomposition in sodium formate solutions. Fig. 7.5 shows some of their results that demonstrate clearly that

[27] A typical reaction system used for studying this reaction is shown in Box. 6.3.
[28] J. Li, W. Chen, H. Zhao, X. Zheng, L. Wu, H. Pan, J. Zhu, Y. Chen, J. Lu, (Hefei), J. Catal. 352 (2017) 371; and supplementary material.

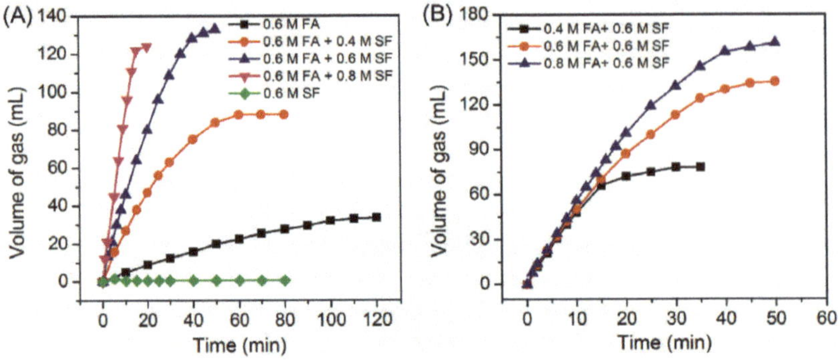

Figure 7.5

Kinetic data for the decomposition of formic acid (FA) in sodium formate (SF) solutions at a temperature of 25°C over a Pd/C catalyst with 2.1 nm diameter Pd particles showing that the rate of reaction is dependent on the formate ion concentration (A) and independent of the formic acid concentration (B). *Data from the paper by Li et al. cited in footnote 28 and reproduced with kind permission of Elsevier.*

the reaction is first order in formate ion concentration and zero order in formic acid concentration. It is interesting to note that all the curves shown for $HCOOH/HCOO^-$ are approximately linear at lower conversions, the linear behavior corresponding to zero-order dependency on the formic acid conversion; the nonlinear portion of each curve at higher conversions will correspond to when there is no longer any buffer behavior. It is also clear that the curve shown for the reaction of formic acid in the absence of Na^+ ions is characteristic of first-order behavior. Li et al. found that sodium formate on its own gave no reaction (Fig. 7.5.a) as might be expected from the mechanism outlined earlier.[29]

Another convincing example of zero-order kinetic behavior is shown in Fig. 7.6 for the reaction of a 1:1 mixture of formic acid and sodium formate at various temperatures from 25°C to 50°C over a Pd/MSC-30 catalyst. The plots for each temperature are almost exactly straight lines as would be expected for a zero-order reaction. It is also clear from the inserted plot of Fig. 7.6 that the data behave according to the Arrhenius equation. However, the activation energy obtained, 38.6 kJ mol^{-1}, is very low and may indicate that in this case that diffusion limitations apply. The topic of diffusion limitations and their effect on reaction kinetics will be treated in Chapter 8. As noted earlier, the subject of the decomposition of formic acid will be discussed further in Chapter 15. (See also Task 7.8).

[29] It is worth noting that some other papers published at about the same time reported that sodium formate alone reacted to give hydrogen. However, this is probably a function of the temperature of reaction as it has been shown [H. Kramer, M. Levy, A. Warshawsky, Int. J. Hydrogen Energy 20 (1995) 229] that formate ions can be converted to bicarbonate ions at 70°C.

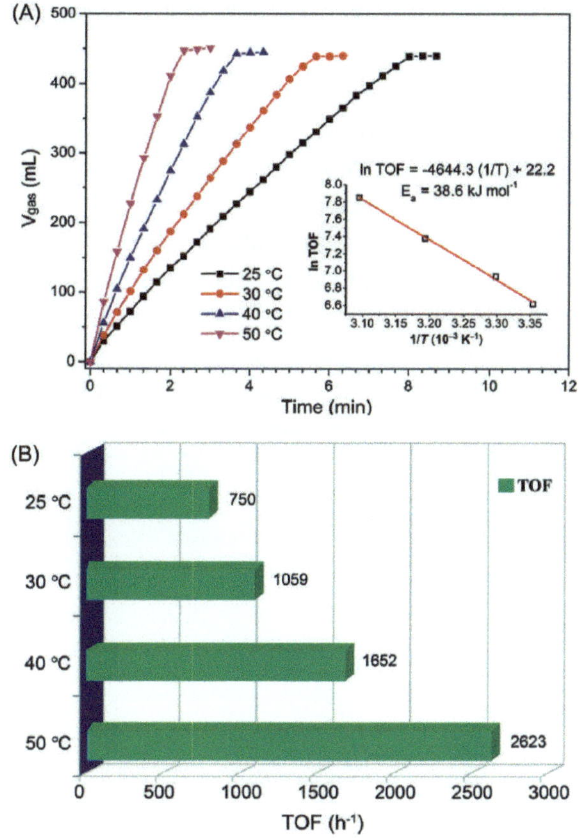

Figure 7.6
(A) Volume of generated gas (CO_2 + H_2) versus time at different temperatures with a formic acid/sodium formate mixture (1:1) over a Pd/MSC-30 catalyst prepared with NaOH. (B) Corresponding turnover frequencies at different temperatures. *Data reproduced from Q.-L. Zhu et al., Chem. Sci. 5 (2014) 195, with kind permission of the Royal Society of Chemistry.*

Task 7.8 Literature Search on the Decomposition of Formic Acid using Solutions with Sodium Formate

Using the reference given in footnote 28, examine the literature on the decomposition of formic acid from solutions of formic acid and sodium formate. You will need to use a citation search, working both backward to find earlier papers on the subject as well forward to identify newer ones. Pay particular attention to any papers discussing the mechanism of the reaction and see whether you can find any others which examine the order of reaction. (There is at least one paper that suggests that the reaction is first order with respect to formic acid.) There is some evidence that other metals than Pd may be active for the reaction. What are these? Is it possible to carry out the reaction using a metal formate other than sodium formate?

7.7 Conclusions

This chapter has considered the establishment of formal kinetic expressions to relate the rate of a catalytic process to the concentrations of reactants (and sometimes also the products). These expressions are often simplified to more empirical equations containing powers of the partial pressures of the reactants. In principle, either type of equation of the appropriate form could be used to describe that rate of reaction in any practical situation. However, an important caveat must be added: it is possible that the rate of a reaction could be governed by diffusion limitations. In Chapter 8, Mass and Heat Transfer Limitations and Other Aspects of the Use of Large Scale Catalytic Reactors, we will first discuss the topic of diffusion limitation before proceeding to discuss some practical situations in which heat transfer and also thermodynamic aspects become important.

CHAPTER 8

Mass and Heat Transfer Limitations and Other Aspects of the Use of Large-Scale Catalytic Reactors

Chapter Outline

8.1 Introduction

In Chapter 6, Catalytic Reactors and the Measurement of Catalytic Behavior, we considered the types of reactors used in laboratory studies to investigate catalytic reactions. In such studies, the work is usually done with small catalyst particles and the reaction conditions are chosen to avoid diffusion problems. As the reactors become larger, larger catalyst geometries are used (see later) and it becomes impossible to ignore problems introduced as a result of mass and heat transfer problems. A detailed description of the design and operation of industrial catalytic reactors is beyond the scope of this book and so we will restrict ourselves largely to the limitations placed on kinetic measurements in larger reactors by mass transfer restrictions and to a brief discussion of some of the more important types of industrial reactors used, particularly in relation to reactions operating close to thermodynamic equilibrium. We will start with a discussion of the problems associated with external and internal diffusion. We will then examine a number of examples of the reactors used for catalytic processes, particularly ones in which heat and mass

Contemporary Catalysis.
DOI: https://doi.org/10.1016/B978-0-444-63474-0.00008-4

transfer are of importance. Subsequent chapters (Chapters 9–15) will then show how some of the topics that have been covered in the preceding chapters are applied in practice.

8.2 The Importance of Mass Transfer in Catalysis

Having made a suitable catalyst formulation and having formed the catalyst into a robust shape, we need to know how mass transfer limitations might affect the performance of the catalyst in a reactor geometry of the type used in practice. As we discussed in Chapter 4, Catalyst Preparation (see Fig. 4.1), a reactant molecule fed to a reactor must first diffuse toward the external surface of the catalyst pellet and then enter a pore, diffusing through it to the active sites. Once it reacts at these sites, in most cases with another reactant that has also diffused to these sites, the product molecule or molecules must follow the reverse path. Outside the catalyst pellet, the reactants and products diffuse according to the normal laws governing gaseous diffusion and, in the current context, we describe this as "external diffusion." Once within the pores, as long as these are relatively narrow, the molecules move according to the so-called Knudsen diffusion—in other words, the collisions with the walls are as important as collisions with one another—and we then refer to "internal diffusion." These steps can be summarized as follows:

- Diffusion of reactants to the external surface of a catalyst particle; the main collisions are between gaseous molecules (reactants and products plus any inert species).
- Reactants enter pores and diffuse within them to the active sites; main collisions are now with the walls of the pores.
- Reaction occurs at active sites.
- Products diffuse away from the active sites and out of the pores; again, the main collisions are with the walls of the pores.
- Products leave the pores and diffuse away from the catalyst particle; the main collisions are now between products and other gaseous species.

Both the external and internal diffusion phenomena are of importance in practical applications of catalysis and we will now consider each, in turn, paying particular attention to those aspects that are of greatest importance to those more interested in the more fundamental aspects of the subject.[1]

[1] Some of the material in the following sections is derived from a set of lecture notes used by the author when he worked at the University of Twente (NL). These had evolved over the years and also contained contributions by H. Bosch, P.J. Gellings, P. Mars, J.G. van Ommen, and J. Sonnemans as well as by the author.

8.2.1 External Diffusion

An important parameter that must be considered in the design of any catalytic flow reactor, whether of small or large scale, is the linear flow rate of the reactant and product gases through the reactor. When a gas flows past the boundary of a particle in such a reactor, a static layer of gas molecules builds up alongside the particle and there is gradual transition between this layer and the free-flowing gas phase. This is shown schematically in Fig. 8.1 for a spherical particle. (The situation will be somewhat different for irregularly shaped particles but there will still be a static layer surrounded by the free-flowing gas.) The thickness of the boundary layer will depend on a number of factors, these including the pressure and the composition of the gas flowing over the surface. However, the most important parameter as far as the present discussion is concerned is the gas velocity: the thickness of the boundary layer, and hence the rate of diffusion through it, is inversely proportional to the gas velocity. In other words, if the gas velocity is high enough, the rate of diffusion through the boundary layer at the surface of a catalyst particle at which reaction is occurring will also be high enough to ensure that the concentrations of the reactants and products at the surface of the catalyst will be the same as in the body of the gas flowing past the particle; under these conditions, "external diffusion" will have no effect on the measured rate of any reaction. However, if the gas flow rate is too low, the boundary layer thickness will be significant and the rate of diffusion through it will contribute to the observed rate of reaction. The rate of diffusion through such a layer is given by:

$$R_{\mathrm{ext}} = k_{\mathrm{m}}(C_{\mathrm{g}} - C_{\mathrm{s}}) \tag{8.1}$$

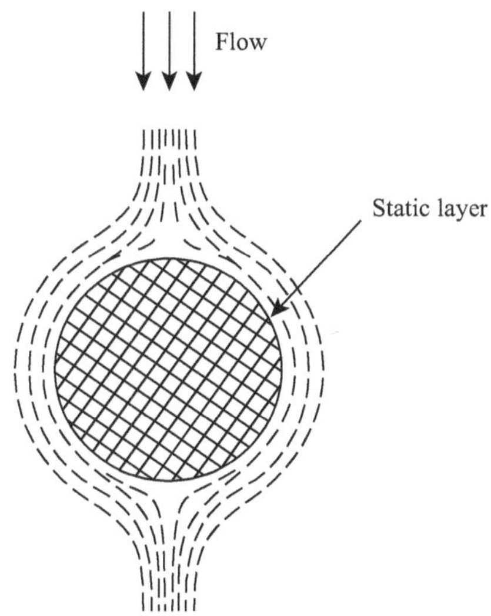

Figure 8.1
The boundary layer round a spherical particle in a constant flow of gas.

where C_g is the concentration of the gas in question in the bulk of the gas, C_s is the concentration of that gas just above the surface of the catalyst, and k_m is the mass transfer coefficient. As we will see later, the rate of the surface reaction is then given by some function of C_s. The value of k_m is determined by the physical properties of the gas mixture and from the aerodynamic flow conditions and depends on the dimensionless Schmidt (Sc), Reynolds (Re), and Sherwood (Sh) numbers. These numbers are known for most practical situations and so k_m can be worked out. Box 8.1 gives some practical examples of reactions in which external diffusion is important. (See also Box 8.2 and Task 8.1.)

BOX 8.1 Ammonia Oxidation and Formaldehyde Production

In the case of catalytic reactions in gas–solid systems, external diffusion generally only plays a significant role in reactions occurring at very high temperatures such as the oxidation of ammonia (10% ammonia in air) over a small number of woven nets (normally three) composed of Pt/Rh alloy wires of approximately 0.15 mm diameter. The reaction occurs at about 800°C with a gas velocity of about 5 m s^{-1} through the nets. The conversion obtained is about 98% and the rate is completely controlled by the rate of impact of the reactant molecules with the surface of the wires. The following figure shows schematically the setup used for ammonia oxidation.

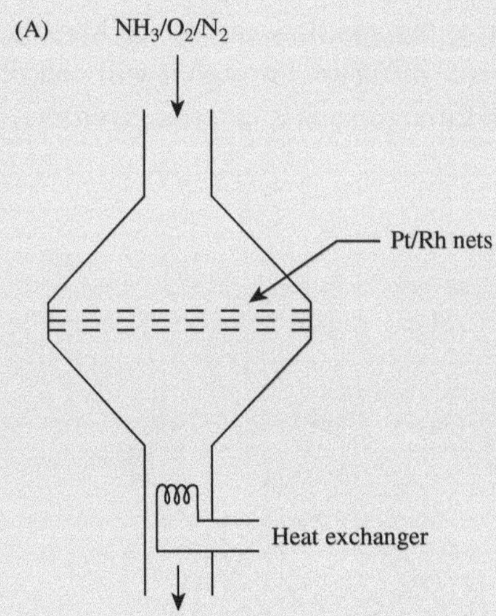

As the reaction is very exothermic, the system includes a heat exchanger downstream (in the case of flow from top to bottom) from the catalyst gauzes to recover waste heat. In practice, there is a second set of gauzes below the operating gauzes, these being kept at a slightly lower temperature in order to capture any Pt that evaporates from the catalytic gauzes.

A similar situation is encountered in the system used for one of the two commercial processes used for the oxidation of methanol to give formaldehyde:

$$CH_3OH + \frac{1}{2}O_2 \rightarrow HCHO + H_2O$$

(Continued)

BOX 8.1 Ammonia Oxidation and Formaldehyde Production (Continued)

Some dehydrogenation also occurs in parallel:

$$CH_3OH \rightarrow HCHO + H_2$$

In this case, the catalyst comprises of a bed of crystalline silver approximately 1 cm thick sitting on a suitable mesh. The reaction, which is very exothermic, is self-sustaining and operates between 500°C and 900°C, giving a conversion of methanol of up to 99% with a selectivity to formaldehyde of up to 90%. (What would the other products be?) The catalyst lifetime can be up to about 8 months, depending on the purity of the feed, but the deactivated catalyst can be readily regenerated by a simple electrolysis process.

Some further details can be found in: L. Lefferts, PhD Thesis, U. Twente, 1987.

BOX 8.2 Production of Margarine.

A completely different example of mass transfer limitation is found in the three-phase system used in the hydrogenation of fats and oils over nickel catalysts. This reaction takes place at about 180°C and the situation encountered is shown schematically in the figure.

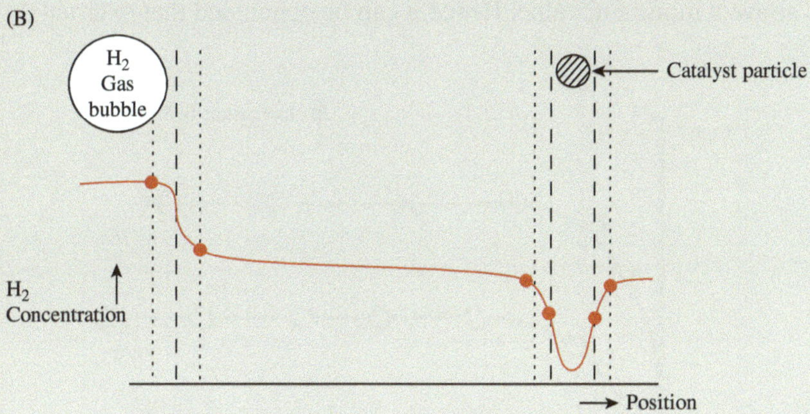

The catalyst particle, suspended in the liquid that is to be hydrogenated, is shown on the right-hand side and a bubble (of hydrogen gas) is shown on the left. Barriers to the transfer of hydrogen exist at the boundary between the gas and the liquid, through the liquid itself, outside the catalyst particle and finally within the pores of the catalyst. (The last step, diffusion within the pores of the catalyst, will be discussed further in the following text.) The first step, transfer from the gas to the liquid, can be improved by rapid stirring of the liquid, thus increasing the number of gas bubbles and thus increasing the gas–liquid interface area; the creation of more, smaller bubbles also decreases the distance for diffusion within the liquid. If the rate of agitation is sufficiently high, the external diffusion in the liquid outside the catalyst particle can also be reduced.

Task 8.1 Methanol Oxidation

By searching using the term "formaldehyde production," find out what you can about the details of the two industrial processes used commercially for the conversion of methanol to formaldehyde.

In practice, when setting up a set of experiments with a given catalyst system, one should check whether or not external diffusion limitations occur under the chosen conditions. This entails carrying out a series of catalytic conversion measurements in which the linear gas velocity is changed without otherwise changing the other conditions of the reaction being studied; in other words, the residence time (or *W*/*F*), the temperature and the total bed dimensions must all remain constant. Hence, in practice, the value of *W* is decreased in sequence by adding an appropriate diluent (e.g., silicon carbide, SiC, or α-alumina) to keep the bed dimensions the same while varying F at the same time in such a way that the value of *W*/*F* remains constant. A typical plot of the type of results to be expected is shown schematically in Fig. 8.2; this shows the conversion as a function of flow rate for two different constant values of *W*/*F*. (It could also represent results for two different temperatures.) It is clear that the conversion is unaffected by the value of the flow rate (and hence the linear flow velocity) as long as this is above a minimum value. Hence it can be concluded that external diffusion is

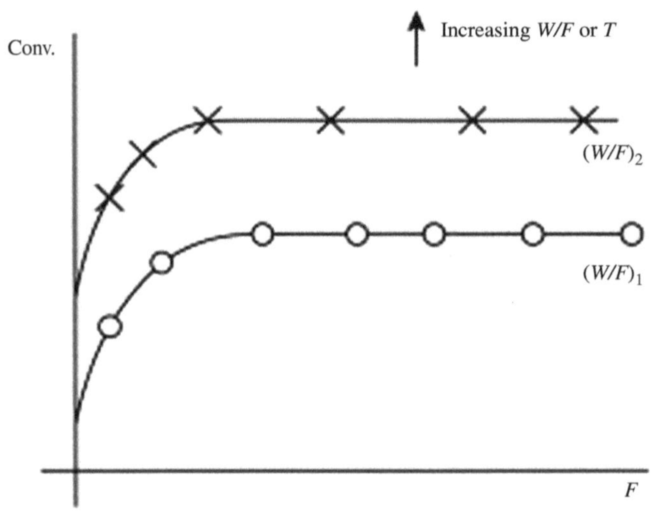

Figure 8.2
Plots of the conversion versus flow rate F for two different constant values of *W*/*F*. As W is varied, a diluent is added so that the bed dimensions are unaltered. Similar results would have been obtained if the experiments had been carried out at two different temperatures.

unimportant under these conditions. In practice, such measurements are performed only with difficulty as there are physical limits to the range of flow rates and catalyst weights that can be used in a particular reaction system. (See also Box 8.3.)

BOX 8.3 Another Test for External Diffusion Limitations

It is also possible to test for the occurrence of external diffusion limitations by examining the effect of varying the reactor geometry. This is most easily done by varying the diameter of the reactor tube but keeping the weight of catalyst used (W) and the flow rate (F) the same. Doubling the diameter decreases the linear velocity by a factor of four (as the linear velocity is inversely proportional to the reactor cross-sectional area). If the conversion does not change, it can be concluded that external diffusion limitations do not occur.[a]

[a]It is surprising how few papers state explicitly that external (or even internal) diffusion limitations do not apply to the results given. It is to be hoped that the required tests have been carried out while establishing the experimental methods to be used.

8.2.2 Internal Diffusion

In Chapter 4, Catalyst Preparation, we indicated the importance of the diffusion of reactants and products into and out of the pores of a catalyst (see Fig. 4.1). If this pore diffusion is slow (i.e., we have "internal diffusion limitations"), it can have an effect on the apparent rate of the reaction. If the rate of a catalytic reaction

$$\mathrm{A} + \mathrm{B} + \cdots + \rightarrow \mathrm{P} + \mathrm{Q} \cdots \tag{8.2}$$

that occurs without any diffusion limitations is given by:

$$\text{Rate} = k_\mathrm{T} \cdot \text{fn}(C_\mathrm{A}, C_\mathrm{B}, \ldots, C_\mathrm{P}, C_\mathrm{Q}, \ldots), \tag{8.3}$$

where fn(C_A, C_B,. . ., C_P, C_Q,. . .) is a function involving the concentrations of the reactants and products (see Chapter 7: The Kinetics and Mechanisms of Catalytic Reactions), then the reaction rate constant k_T is dependent only on the temperature of the reaction. The dependence of k_T on temperature is given by:

$$k_\mathrm{T} = A \cdot \exp\{-E_\mathrm{a}/RT\} \tag{8.4}$$

where A is the Arrhenius constant. A plot of ln k_T versus $1/T$ will give a straight line of gradient $-E_\mathrm{a}/R$ as shown by the *solid line* in Fig. 8.3. However, if diffusion limitations within the pores of the catalyst particle have an effect on the measured rate of reaction, the plot of lnk_T versus $1/T$ deviates from the ideal line as depicted by the other lines given in Fig. 8.3. The reasons for these variations from normal Arrhenius behavior will now be discussed.

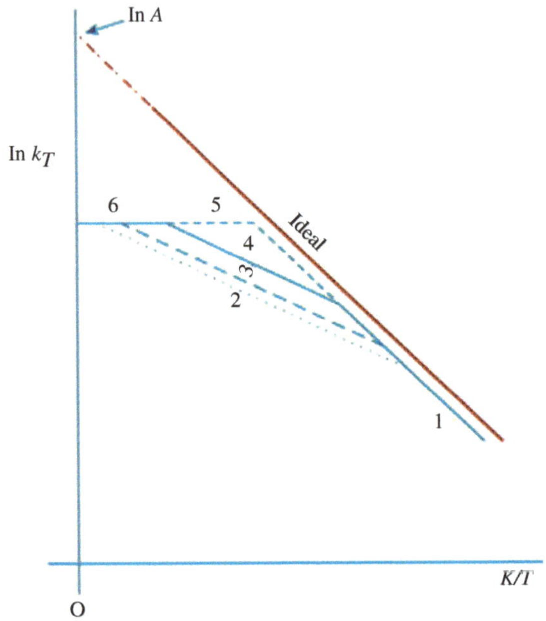

Figure 8.3
Arrhenius plots showing ideal behavior (*solid line*) and deviations due to internal and external diffusion limitations (lines 1–6); see main text for explanation.

Any deviation from the ideal behavior can be expressed by the so-called effectiveness factor, η, defined by Eq. (8.5):

$$\eta = \frac{\text{observed reaction rate}}{\text{rate without internal gradients}} = \frac{r_{obs}(C,T)}{r_{chem}(C_s,T_s)} \tag{8.5}$$

where $r_{obs}(C,T)$ is the observed reaction rate (averaged over the whole catalyst particle) at any concentration and temperature and r_{chem} (C_s,T) is the reaction rate at the particle surface at the surface concentration and temperature: in other words, the rate that would occur if no diffusion limitations occurred.

The effectiveness factor is a function of the Thiele modulus, ϕ, a dimensionless number that relates the catalytic activity to the rates of internal mass transfer[2] (Box 8.4). The relationship between η and ϕ is illustrated for a first-order reaction in Fig. 8.4. (The figure is given for a plate-shaped catalyst; the shape of the curve is slightly different for

[2] A deeper treatment of this subject is outside the scope of this book. For further details, the reader should consult a standard text on reactor engineering such as S.H. Fogler, Elements of Chemical Reaction Engineering, Prentice-Hall, London (1986); see also J.A. Moulijn, M. Makkee, A. van Diepen, Chemical Process Technology, Wiley, Chichester, New York, Weinheim, Brisbane, Singapore, and Toronto (2001) for an excellent treatment of reactor design.

BOX 8.4 Thiele Modulus, ϕ, and Effectiveness Factor, η

The value of the Thiele modulus, ϕ_{gen}, for an *n*th-order reaction is related to k_v, the intrinsic reaction rate constant per unit volume, *n*, the order of reaction, D_{eff}, the effective diffusion rate of the molecules, and C_s, the concentration at the surface of the catalyst particle (pellet, slab, grain, etc.) by the equation:

$$\phi_{gen} = \frac{V}{SA} \cdot \sqrt{\frac{k_v}{D_{eff}} \cdot \frac{n+1}{2} \cdot C_s^{n-1}} \qquad (1)$$

The quantity *V*/*SA* is the volume/external area ratio of the particle in question. For a first-order reaction with a particle of slab geometry (length *L*, with effective diffusivity D_{eff}), it can be shown that

$$\eta = \frac{\tanh\phi}{\phi} \qquad (2)$$

where

$$\phi = L\sqrt{\frac{k}{D_{eff}}} \qquad (3)$$

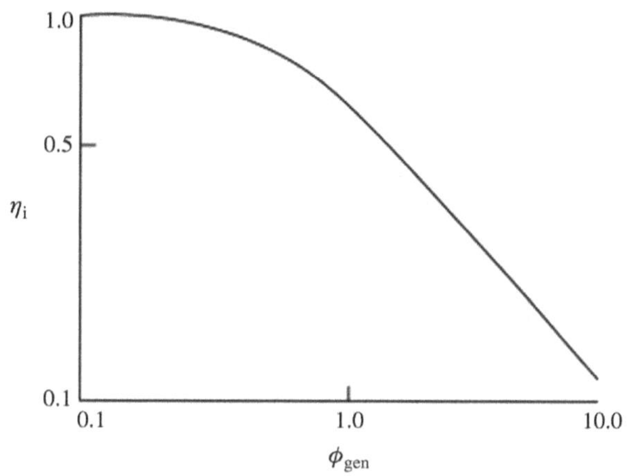

Figure 8.4
The effectiveness factor, η, as a function of the Thiele modulus, ϕ_{gen} for a first-order reaction over a catalyst in the form of a plate.

cylindrical and spherical particle dimensions.) It can be seen that the effectiveness factor tends to unity at low values of ϕ (i.e., at high values of D_{eff}; the value of tanh ϕ tends to ϕ as ϕ tends to zero) and so all of the catalyst sites contribute to the measured rates under such conditions. This corresponds to the ideal behavior (line 1) shown in the Arrhenius plot of Fig. 8.3. However, when the temperature increases, η_i decreases and the Arrhenius plot deviates from the ideal straight line. The temperature at which this deviation occurs depends on many factors, including the catalyst porosity, and the resultant plots are shown as lines 2, 3, and 4. (See Box 8.5 for a further treatment for a first-order reaction.) Line 5 shows a situation in which no effect of internal diffusion appears; instead, the line follows the ideal one until the rate reaches a limit (line 6) corresponding to the rate of mass transfer to the external of the catalyst particles. The same limiting situation corresponding to line 6 applies to the other cases (2.—4) as well.

BOX 8.5 Limiting Values of η for a First-Order Reaction

For high values of ϕ, tanh ϕ tends to unity and Eq. (2) of Box 8.4 reduces to:

$$\eta = 1/\phi \tag{1}$$

Hence, the observed reaction rate is now given by:

$$r_{obs} = \eta \cdot r_{chem} = r_{chem}/\phi \tag{2}$$

From Eq. (3) of Box 8.4,

$$\phi \infty \sqrt{k}$$

and so $r_{obs} \infty r_{chem}/\sqrt{k}$

For a first-order reaction, $r_{chem} = k \cdot (C)$

Hence,

$$r_{obs} \infty k/\sqrt{k} \cdot (C) \infty \sqrt{k} \cdot (C) \tag{3}$$

As $k = A \exp(-E/RT)$,

we obtain

$$r_{obs} = \sqrt{A} \cdot (C) \cdot \exp(-E/2RT) \tag{4}$$

In other words, for a first-order reaction, the gradient of the Arrhenius plot is one half of that for a reaction without diffusion limitations and the apparent activation energy for the diffusion-limited case is given by $E_{app} = 0.5\ E_{true}$. (Lower broken lines in Fig. 8.3.)

Fig. 8.5 shows schematically the situation that exists in a spherical catalyst particle containing narrow pores at several different reaction temperatures. For a low temperature (low ϕ), the concentration profile (line 1) is constant throughout the particle and the effectiveness factor is 1.0. However, as the temperature is increased, the concentration at the center of the sphere decreases (line 2) and at yet higher temperatures, the concentration at the center of the sphere can drop to zero (line 3, high ϕ).

An equivalent result is shown more quantitatively in Fig. 8.6 which shows, for an ideal cylindrical pore of length $2L$ open at both ends (e.g., a pore traversing a catalyst pellet), the

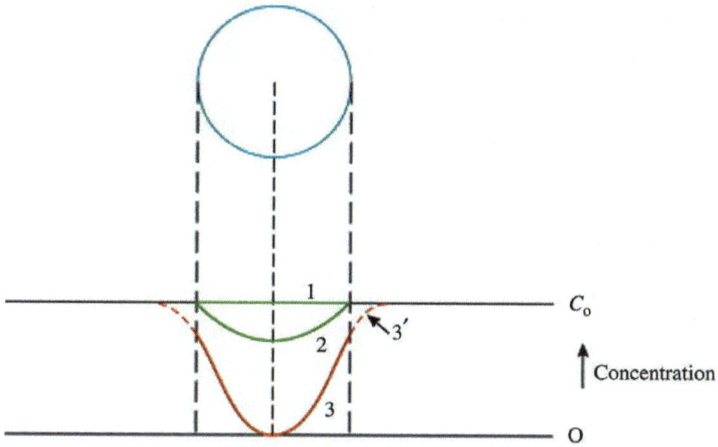

Figure 8.5
Concentration profiles of a reacting species through a catalyst particle as a function of reaction conditions; 1, low temperature, large pores; 2, higher temperature, smaller pores; 3, high-temperature, small pores.

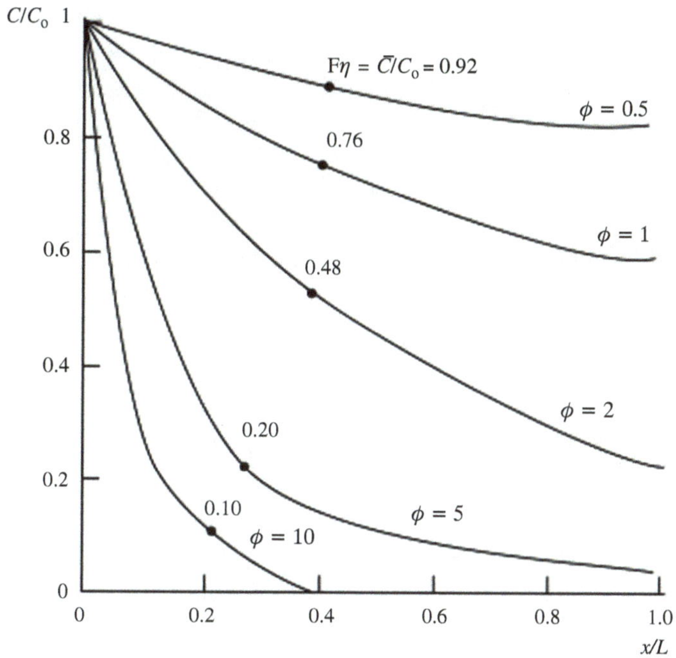

Figure 8.6
Concentration profile in a catalyst particle as a function of the Thiele modulus ϕ. The average internal concentration is also shown for each curve as a value of C/C_o.

profile of the concentration of a reactant (plotted as C/C_o) versus the distance into the pore (plotted as X/L) for different values of the Thiele modulus ϕ. (For a value of $\phi < 0.3$, not shown, the line would be parallel to the x/L axis.) The values of C/C_o shown are the average concentrations of the reactant divided by the concentration at the pore mouth, in other words, the effectiveness factors, η. It can be seen that concentrations at the center of the pore are in all cases significantly lower than the concentration at the pore mouth and that the effectiveness factors can drop below 10% at high values of ϕ.

As discussed earlier, the Thiele modulus, ϕ, a dimensionless quantity, is a measure of the rate of the catalytic reaction relative to the rate of diffusion of the reactants in the pores of the catalyst (internal mass transfer); low values correspond to situations with high rates of internal mass transfer and high values to situations where internal mass transfer determines the effective rate of reaction. In order to ensure that mass transfer limitation is at a minimum and to allow full use of high catalytic activity, it is therefore necessary to have active surface areas that are as accessible as possible. This can be achieved by using particle geometries that aid accessibility of the reactants to the active surfaces by having wide, short pores. One way to ensure that the pores are short is to use very small particles, for example, in fluidized beds (see later). Another solution, as discussed previously, is to use support shapes designed to give external areas that are as high as possible while having relatively short pores; examples of suitable geometries—spherical particles, Raschig rings, hollow extrudates or trilobes—were shown in Chapter 4, Catalyst Preparation, and Fig. 4.8. It is also possible to use geometries in which the catalyst is applied in relatively thin layers to an inert matrix, as in a monolith material or in microchanneled reactors in which the microchannels are coated with catalytically active components; see Chapter 6.

8.2.3 *Effects of Diffusion Limitations on Selectivity*

In addition to having a hand in determining the activity of a catalyst, the effectiveness factor, η, can play an important role in relation to its selectivity. Consider a *sequential reaction*:

$$A \rightarrow B \rightarrow C \tag{8.6}$$

in which B is the desired product and C is an undesired by-product. (An example would be a selective oxidation reaction in which the selective product can be further oxidized to give CO_2 and H_2O.) The active sites for the reaction are located evenly through the pores of the catalyst and both steps in the reaction can occur on these sites; A and B will compete for the sites and so further reaction of B to give C only occurs if the concentration of A is low. For simplicity, we assume that the pores are all cylindrical and open-ended. When η is unity (Fig. 8.5, line 1), the concentration of reactant remains the same throughout each pore and so all of the active sites present in the pores of the catalyst contribute equally to the conversion of the molecule A to give B; hence, the selectivity to B is optimal. However, if η is low, for example at higher temperatures, the concentration of the reactant A at the center of the pore drops to zero (Fig. 8.5, line 3) and the intermediate product molecule B can diffuse further into the pores to react further at empty sites. Hence, the selectivity of the catalyst now falls below its optimal value. To prevent such a situation occurring, it is possible to make so-called *egg-shell catalysts* in which the active phase is concentrated near the pore mouths (see Box 8.6). In such a geometry, even if B diffuses into the center of the catalyst, there are no sites at which the production of C can occur. The use of egg-shell catalysts favors higher selectivities in sequential reactions and enables the use of higher reaction temperatures than would be possible using a catalyst made in the normal way.

BOX 8.6 Egg-Shell Catalysts

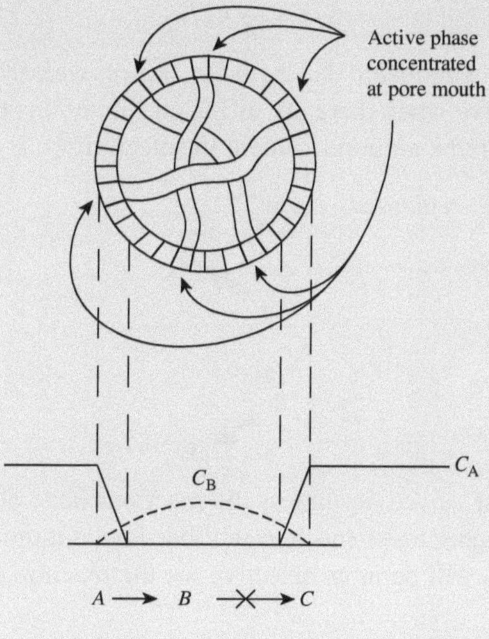

An egg-shell catalyst is one that is made in such a way that the active phase is only present near the surface of the support particle (sphere or cylinder) and the center of the support is inactive. If it can catalyze both steps in a sequential reaction such as

$$A \rightarrow B \rightarrow C$$

and it is desired to maximize the yield of B, the selectivity to B will be much improved compared with using a particle with active phase distributed throughout as there are no sites within the rest of the egg-shell catalyst particle at which B can react further. As indicated in the main text, a similar effect is possible using small particles with large pores, possibly in a fluidized bed configuration.

If a catalyst can bring about two different *competing reactions*:

$$\mathrm{A} \rightarrow B;\ \text{rate constant } k_\mathrm{A} \tag{8.7}$$

$$\mathrm{X} \rightarrow Y;\ \text{rate constant } k_\mathrm{X} \tag{8.8}$$

Without diffusion limitations, the selectivity is given by:

$$S = k_\mathrm{A}/k_\mathrm{X} \tag{8.9}$$

However, with diffusion limitations:

$$S' = \sqrt{k_\mathrm{A}D_\mathrm{A}/k_\mathrm{X}D_\mathrm{X}} \tag{8.10}$$

where D_A and D_X are the respective diffusion coefficients. As the diffusion coefficients of A and X are unlikely to be very different,

$$S/S' \propto \sqrt{k_\mathrm{A}/k_\mathrm{X}} \tag{8.11}$$

For a selective reaction, it is probable that $k_\mathrm{A} \gg k_\mathrm{X}$. Hence, we can conclude that the catalyst will be less selective when there are diffusion limitations than when there are none. Only if $k_\mathrm{X} > k_\mathrm{A}$ would there be an improvement in selectivity.

If the there are two *parallel reactions*:

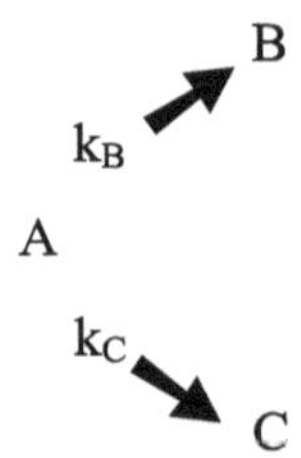

(8.12)

diffusion will not affect the selectivity unless the reactions have different orders; if the orders are different, the higher order reaction will be slowed more than the lower order reaction and so the system will be more selective for the reaction of lower order.

8.2.4 Effects of Diffusion Limitations on Catalyst Poisoning

If a reaction proceeding under diffusion limitations is poisoned by some component present in the reaction mixture by the selective adsorption of that component on the active sites, the sites at the pore mouth will become selectively poisoned (Fig. 8.7), leaving the interior of the catalyst pores unpoisoned. A small amount of poison will therefore have a major effect on the catalyst activity (A. Wheeler, Adv. Catal. 3 (1951) 349).

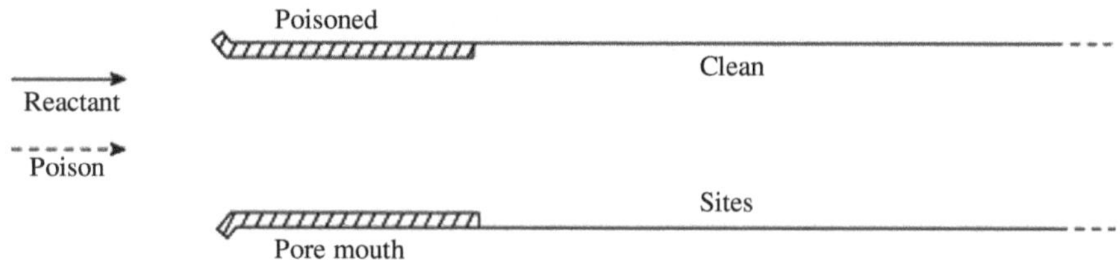

Figure 8.7
Effect of diffusion limitations on catalyst poisoning.

8.3 Heat Transfer in Catalysis

Just as mass transfer (external and internal diffusion) plays an important role in practical applications of catalysis, so does heat transfer. With few exceptions, catalytic processes are either exothermic or endothermic and so the way in which heat is either removed (exothermic reaction) or supplied (endothermic reaction) is very important. Further, as many of these processes are thermodynamically reversible, the removal or supply of heat can have a significant effect on the equilibrium position. Before discussing the design of some typical process plants, we will first consider the importance of temperature control on catalyzed reactions.

8.3.1 Sequential Reactions

We will first examine the effect of temperature on the course of a hypothetical sequential reaction (shown earlier in Eq. 8.6) that occurs in two steps:

$$A \xrightarrow{k_A} B, \qquad \text{Activation energy}, E_A \tag{8.6A}$$

and

$$B \xrightarrow{k_B} C, \qquad \text{Activation energy}, E_B \tag{8.6B}$$

We first answer the question: at what temperature does one need to operate to obtain the maximum yield of the product B? Fig. 8.8 gives two possible Arrhenius plots for these two reactions, one showing the case for $E_A > E_B$ and the other that for $E_B > E_A$. In the first case (Fig. 8.8A), the rate constant for the formation of B from A will be much greater than that of C from B as long as the temperature is sufficiently high, that is, to the left-hand side of the point of intersection. This means that the rate of the reaction from A to B will be much higher than the rate to form C, particularly when the concentration of B is low. On the other hand, for the second case (Fig. 8.8B), the rate constant and hence the rate of the

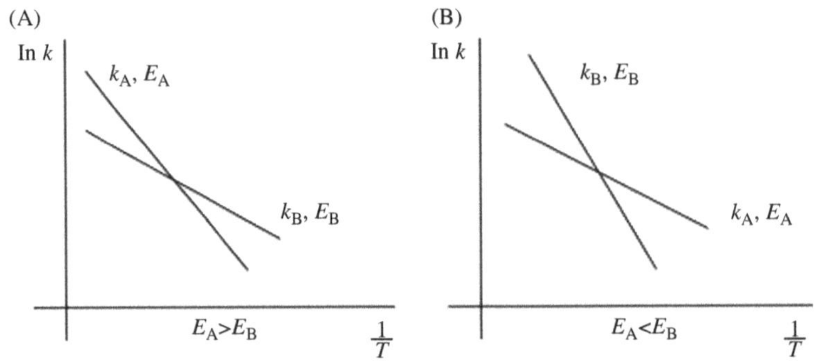

Figure 8.8
Arrhenius plots for two (consecutive) reactions: A → B and B → C with (A) $E_A > E_B$ and (B) $E_B > E_A$.

reaction of A to give B will only be significantly higher than that of B to give C if the temperature is low or if the concentration of B is low. Thus, to maintain a good selectivity to B and hence optimize the yield, the temperature must remain well below that corresponding to the point of intersection in Fig. 8.8B. The drawback of such an approach in practice is that the rate of conversion of A will be relatively low. Hence, in order to decrease the time taken to obtain B, a different approach is taken: optimizing the time at any temperature in order to optimize the yield of B without the formation of any significant amount of C. This can be achieved by starting the reaction at high temperature and then going in steps to lower temperatures, as shown very schematically in Fig. 8.9. In this figure, which plots the concentrations versus the time in the reactor (contact time), results are shown for three decreasing temperatures, $T_1 > T_2 > T_3$. During operation at temperature T_1, k_B is much greater than k_A but, because the concentration of B is still quite small, the rate of formation of C is also small. As soon as an appreciable concentration of B has formed, C also begins to form and the rate of that reaction increases as [B] increases further. At that point, the temperature is decreased, this lowering both rates. Because E_B is greater than E_A, the rate of formation of C relative to that of B will be lowered and so reaction to the preferred product B predominates. The same applies when the temperature is decreased once more.[3] Hence, remembering that the time of reaction is directly related to the catalyst bed length, the situation depicted in Fig. 8.9 corresponds to three sequential reactors, each at a different temperature descending in the order T_1 to T_2 to T_3. The situation that we have just dealt with might correspond to a selective oxidation reaction in which B is the selective

3 If the whole process had been carried out at the lowest temperature, the selectivity to B would have been higher than that achieved in the step-wise process. However, the contact time required to achieve that result would have been much greater.

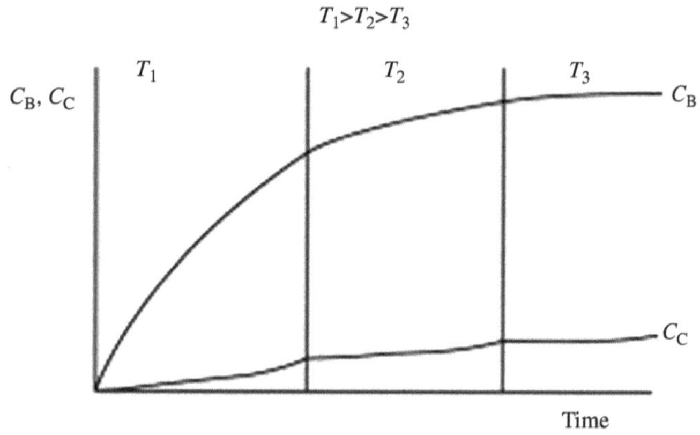

Figure 8.9
Plots of C_B and C_C at three decreasing temperatures as a function of contact time for case (B) of Fig. 8.8. This example assumes isothermal operation.

oxidation product and C is the nonselective product, CO_2. In that case, there would of course be additional reactants and products but the arguments are the same.

8.3.2 Reversible Reactions

The arguments put forward in the last section for the case of sequential reactions apply equally well to reversible reactions. Let us consider the reaction:

$$A \underset{k_B}{\overset{k_A}{\rightleftarrows}} B \tag{8.13}$$

This is entirely equivalent to the case of Eq. (8.6) in which the second reaction is the reverse of the first. Case (A) above (see Fig. 8.9) then corresponds to an endothermic reaction for which

$$\Delta H^o = E_A - E_B > 0$$

while case (B) corresponds to an exothermic reaction

$$\Delta H^o = E_A - E_B < 0$$

In order to get the optimal yield in case (A), as in the treatment above for sequential reactions, the endothermic forward reaction to be carried out at as high a temperature as possible in order to get optimum yields. An example of such behavior is the steam reforming of methane, a reaction that we discuss in Chapter 11, Production of Hydrogen and Syngas From Methane and Some Other Reactions of Methane. In that case, the temperature of reaction is about 900°C and this gives conversions of about 99%. The upper

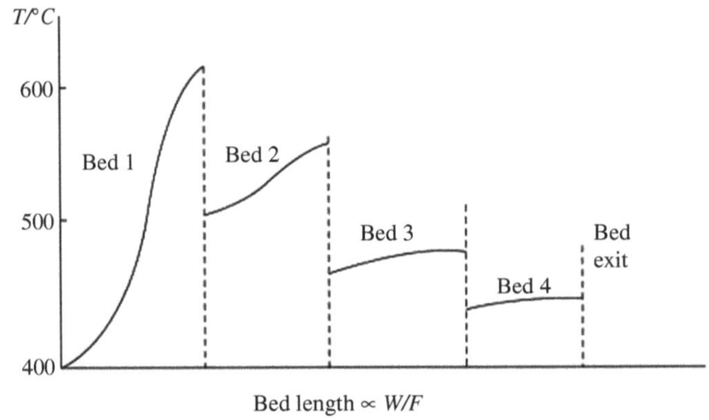

Figure 8.10
Temperature profiles in a series of adiabatic reactor beds for a reversible exothermic reaction. To achieve this profile, there has to be cooling between each bed. *Figure taken from the lecture notes referred to in footnote 1.*

temperature in this case is limited by the availability of suitable reactor construction materials. As discussed further in Chapter 11, the energy required to drive the reaction is provided by the combustion of a proportion of the natural gas feed to the plant. The reaction is carried out in long tubular reactors filled with suitable catalyst shapes, normally Raschig rings, and the tubes are heated externally along their full length by a furnace in which the natural gas is combusted.

In the case of an exothermic reaction, case (B), a different approach has to be followed. Fig. 8.10 shows the temperature profile as a function of bed position (*or* time in the reactor *or W/F*, as long as all the catalyst beds have the same composition) through a series of four adiabatic catalyst beds for a typical exothermic reaction.[4] The entrance to the first bed is at a temperature of 400°C, high enough for the catalyst to have sufficient activity to give the required rates of reaction. Immediately the reaction begins at the entrance of the bed, heat is given out and the temperature of the bed increases; as long as no heat is lost from the walls,[5] the temperature rise at any position in the bed is proportional to the conversion level at that point. The bed temperature then rises significantly until (see further in the following text) the conversion is approaching the equilibrium value at the highest temperature in that bed. As in the example of the consecutive reactions dealt with above, the reaction mixture,

4 Adiabatic operation assumes that there is no loss of heat from the reactor and that any heat generated is used to heat the catalyst in the immediate vicinity of the reaction. The example shown in Figs. 8.10 and 8.11 corresponds to the oxidation of SO_2 to SO_3 but it could equally well apply to reactions such as the methanation of CO or methanol synthesis from syngas. The catalyst for the SO_2 oxidation reaction is a molten salt of K_2S_2O containing dissolved V_2O_5 retained in the pores of a porous silica support.

5 This is a somewhat idealized picture as there will inevitably be loss of heat from the reactor. This approach also assumes that the heat capacity of the bed does not change appreciably with temperature.

including the products, is then passed to a second bed, the entrance of which is at a lower temperature. The allover rate of the forward reaction is now much lower than at the same temperature in Bed 1 since the rate of the reverse reaction has now become significant; nevertheless, Bed 2 brings about further reaction, the bed length being designed such that the conversion at the exit of this bed approaches, but does not reach, the thermodynamic equilibrium position. The same applies in two further beds.

Fig. 8.11 shows the data of Fig. 8.10 for the four beds replotted as the values of the reactor temperature as a function of the conversion. The gradient of each line is the same, being determined by the heat capacity of the beds, each assumed to have the same construction and composition. (As discussed earlier, it is assumed that the reactor system operates adiabatically. There will be significant deviations should there be nonideal operation.) Also shown in Fig. 8.11 are two curves, one corresponding to the equilibrium position for any temperature ("equilibrium line") and the another corresponding to the temperature giving the highest rate of reaction at any conversion ("optimal temperature line"). As shown in Box 8.7, the optimal temperature line can be calculated from a detailed knowledge of the kinetics of the reaction for the reversible reaction of A to give B.

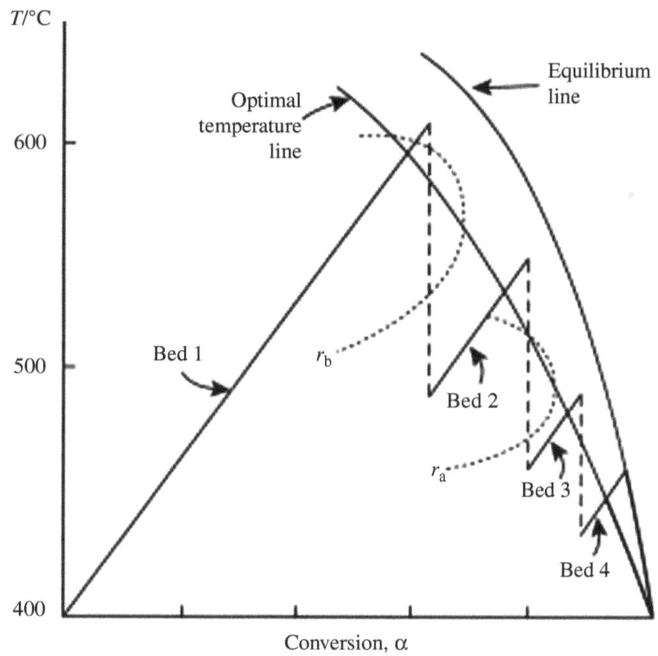

Figure 8.11
The operating lines for a series of connected reactors for a reversible exothermic reaction. The *solid curves* represent the equilibrium conversions and the optimal temperature line. The *dotted curves* r_a and r_b represent lines of constant reaction velocity (see Box 8.7).

BOX 8.7 Derivation of the Equation Governing the Optimal Temperature Line for a Reversible Reaction $A \rightleftharpoons B$

The rate of the forward reaction is given by:

$$r = r_A - r_B = k_A[A] - k_B[B] = k_A^o \cdot e^{-E_A/RT} \cdot [A] - k_B^o \cdot e^{-E_B/RT} \cdot [B]$$

The maximum rate of reaction occurs when $dr/dT = 0$. Hence, if we differentiate the last expression and set it equal to zero:

$$dr/dT = 1/RT^2\{k_A \cdot E_A \cdot [A] - k_B \cdot E_B \cdot [B]\} = 1/RT^2\{r_A \cdot E_A - r_B \cdot E_B\} = 0$$

it can be seen that:

$$r_A \cdot E_A = r_B \cdot E_B$$

and so

$$r_A/r_B = E_B/E_A$$

The ratio r_A/r_B can then be expanded as follows:

$$r_A/r_B = k_A^o/k_B^o \cdot e^{-(E_A - E_B/RT)} \cdot [A]/[B]$$

$$= k_A^o/k_B^o \cdot e^{-(E_A - E_B/RT)} \cdot \{(1-\alpha)/\alpha\}$$

At the "optimal temperature line" (Fig. 8.12), r_A and r_B are equal and so $r_A/r_B = 1$. Hence, the expression for r_A/r_B can be rearranged to give:

$$\alpha/(1-\alpha) = k_A^o/k_B^o \cdot e^{-(E_A - E_B/RT)}$$

and so the optimum conversion at any temperature can be calculated.

Similar expressions can be worked out for more complex kinetics.

*For completeness, the two *dotted lines* shown in Fig. 8.12 (labeled r_a and r_b) represent two different constant rates of reaction; the tangents to these lines are parallel to the T axis at the optimal temperature line.

8.3.3 Reactors for Exothermic Reversible Reactions

Just as for the sequential reaction A → B → C, the optimal yields for a reversible reaction are often achieved in practice in a sequence of reactors in which the reaction is started at as high a temperature as possible and then cooled down before admission to a second reactor. Fig. 8.12 shows schematically the construction of such a reactor in which interbed cooling is achieved by the introduction of a coolant between the beds. For simplicity, the jacket surrounding the whole of the reactor is omitted from the drawing. In the geometry shown, frequently applied for the production of ammonia but also to the production of methanol

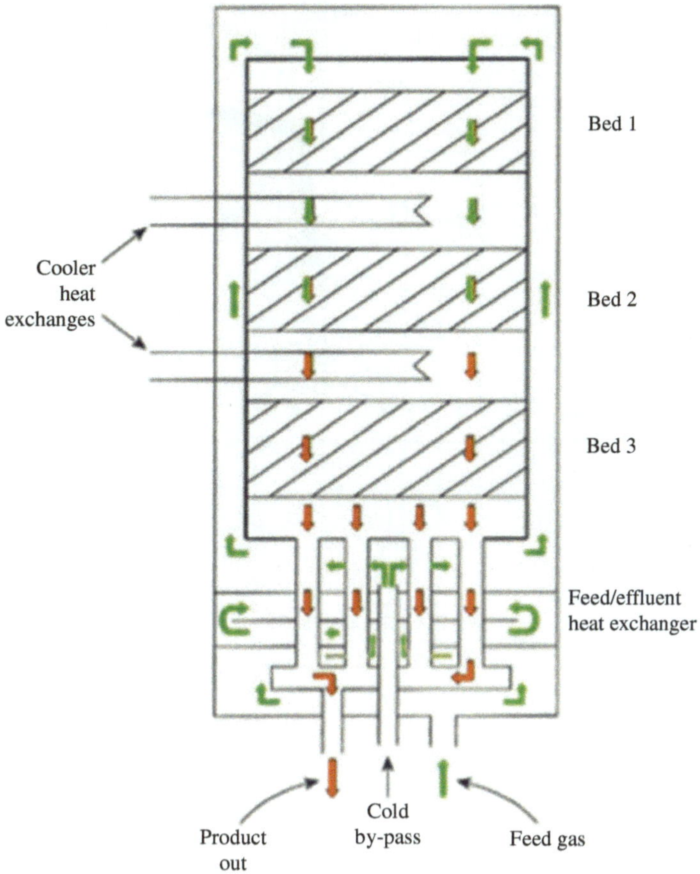

Figure 8.12
Schematic representation of an adiabatic reactor used for the synthesis of ammonia.

from syngas or to that of sulfuric acid by the oxidation of SO_2 to SO_3—all exothermic reversible reactions—the feed is first passed through the "feed-effluent heat exchanger" in which it is heated by the exiting product gas. It then passes the outside shells of each of the three reactors, absorbing some of the heat generated in each of these reactors before passing to the entrance to Bed 1. As the reaction proceeds, heat is evolved and the bed heats up. The exit gas from Bed 1 is then cooled by heat exchanger pipes before entering Bed 2, and so forth. The progress of the reaction in each bed will be given by plots similar in form to those shown in Figs. 8.10 and 8.11.

One of the problems with the geometry shown in Fig. 8.12 is that each reactor has to be packed with catalyst separately and the removal and replenishment of catalyst charges can be quite complicated and time-consuming. As a consequence, single bed reactors have been developed in which part of the reactant gas is introduced into the bed at various positions as

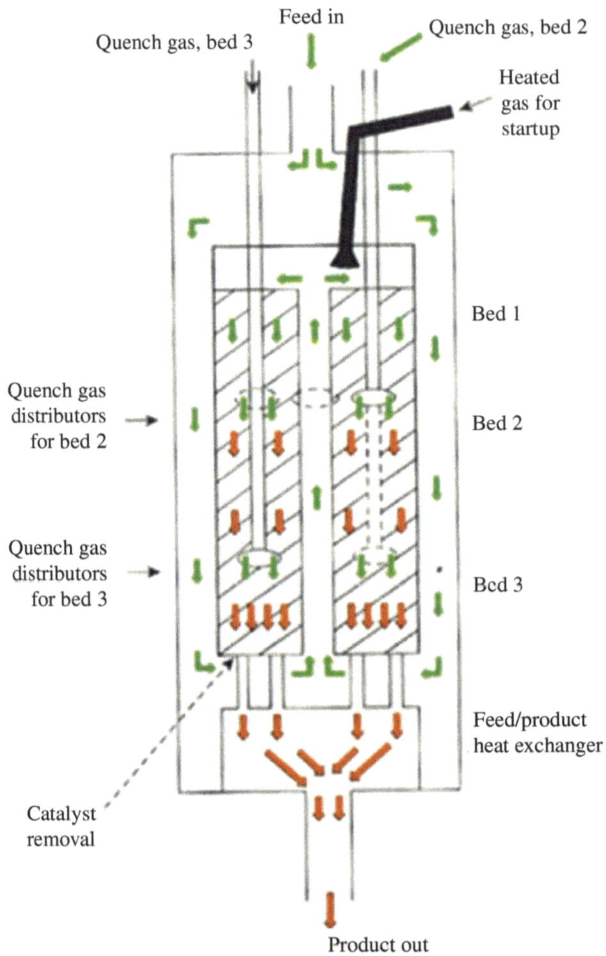

Figure 8.13
A quench reactor developed by ICI (now Johnson Matthey) for ammonia synthesis.

a quench gas through nozzles suitably placed in the bed, this giving similar temperature profiles to those discussed earlier; each addition of quench gas gives a decrease in temperature as in the reactor with interbed cooling discussed earlier. Fig. 8.13 shows such a quench reactor developed by ICI (now Johnson Matthey) for ammonia synthesis in which the quench gas is introduced through nozzles placed at two positions in a single bed in such a way that there are again effectively three distinct beds; because of the simplified construction, the catalyst recharging process can be carried out relatively easily by removal from the base of the bed and addition of a new charge from the top.

A rather different approach is shown in Fig. 8.14. This diagram depicts a "Radial Flow Reactor" developed by Haldor Topsøe for use in ammonia synthesis. Two separate beds are used but the flow is now arranged to pass radially through the bed, this having the advantage that diffusion

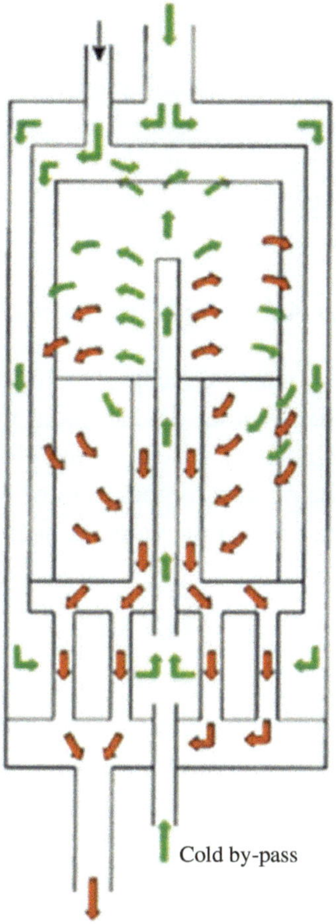

Figure 8.14
A Haldor Topsøe Radial Flow Reactor for ammonia synthesis.

lengths through the catalyst beds are kept to a minimum, thus reducing pressure drops through the beds. Such an approach can be used to allow the use of smaller catalyst particles in order to decrease internal mass transfer limitations. After passing through the usual sequence of heat exchangers, the feed gas passes up through the center of the second bed and diffuses through the wall of the inner tube of the first bed. The product gas then travels down the outer wall of the second bed through which it diffuses. The product leaving the second bed then heats the feed gas entering the first bed. A quench gas can be added at the base of the center feed tube in order to provide temperature control. It is claimed by Haldor Topsøe that this geometry gives very significant gains in the available catalyst surface area per unit reactor volume.

All catalytic processes require careful heat management and these examples have been given to illustrate the complexity of the design of the plants involved. As far as the catalytic chemist is concerned, the challenges of such types of application are many. The catalysts used must have the desired activity and selectivity, they must have suitable geometries to enable maximal use of the active sites (in other words, minimizing diffusion limitations as discussed earlier), and they must have mechanical and thermal strength to withstand use under extreme temperatures in a practical reactor. An example of a process in which heat management is of paramount importance and in which the demands on the catalysts used are extreme is discussed in the following concluding section.

8.3.4 The Chemical Heat Pipe: "The Adam and Eve Process"

The so-called the Adam and Eve Process was a concept developed in Germany in the 1970s at the Kernforschungsanlage (KFA) Jülich. Under development in the KFA at that time was a high-temperature nuclear reactor in which the spherical fuel elements were cooled by helium gas, the He exiting the reactor being at a temperature of 800°C or above. The EVA reactor made use of this high-temperature He to provide the heat necessary for the steam reforming of methane to produce syngas:

$$CH_4 + H_2O \rightarrow CO + 3H_2 \qquad (8.14)$$

Whereas in the normal steam reforming reactor the reforming tubes are placed in a gas-fired furnace and the tube walls are heated directly by radiant heat, the EVA reactor used a countercurrent flow system in which there was direct heat transfer through the reactor walls from the helium to the reactor. As helium has exceptionally good heat transfer properties, the efficiencies of this reactor were very high.[6,7]

Having carried out the reforming reaction at the site of the nuclear reactor, the syngas produced was to be transported by pipeline to some distance (50–100 km) from the nuclear plant to a region where the energy stored in the syngas was required. The energy was then to be reclaimed by carrying out the reverse "methanation" reaction in the Adam reactor:

$$CO + 3H_2 \rightarrow CH_4 + H_2O \qquad (8.15)$$

After condensing out the water formed, the methane would then be transported back to the nuclear reactor using a parallel pipeline and the cycle was completed. A schematic representation of this process is given in Fig. 8.15 and a more detailed diagram of the reactor setup is shown in Fig. 8.16.

[6] Some preliminary work with this system is described in a paper by B. Höhlein, R. Menzer, J. Range, Appl. Catal. 1 (1981) 130.

[7] The Adam and Eve concept was tested in two pilot plants of differing scale. Because of the difficulty of coupling the system to the nuclear reactor, both these pilot plants used an electrically heated He stream. Although both plants operated well, the concept was never taken to practical application as the high-temperature nuclear reactor development work was abandoned.

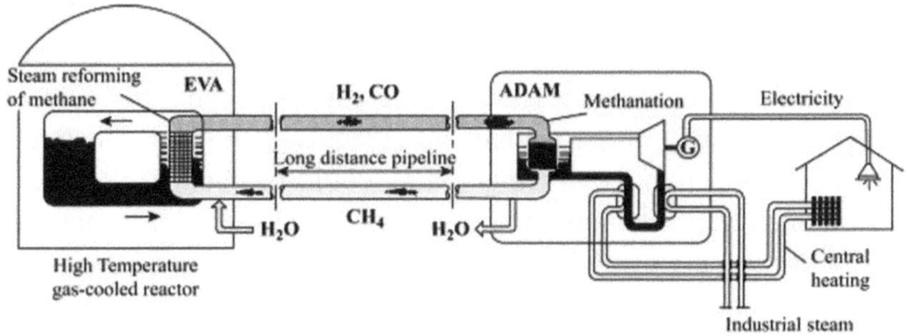

Figure 8.15
Schematic representation of the Adam and Eve concept. *From reference of footnote 6, reproduced with kind permission of Elsevier.*

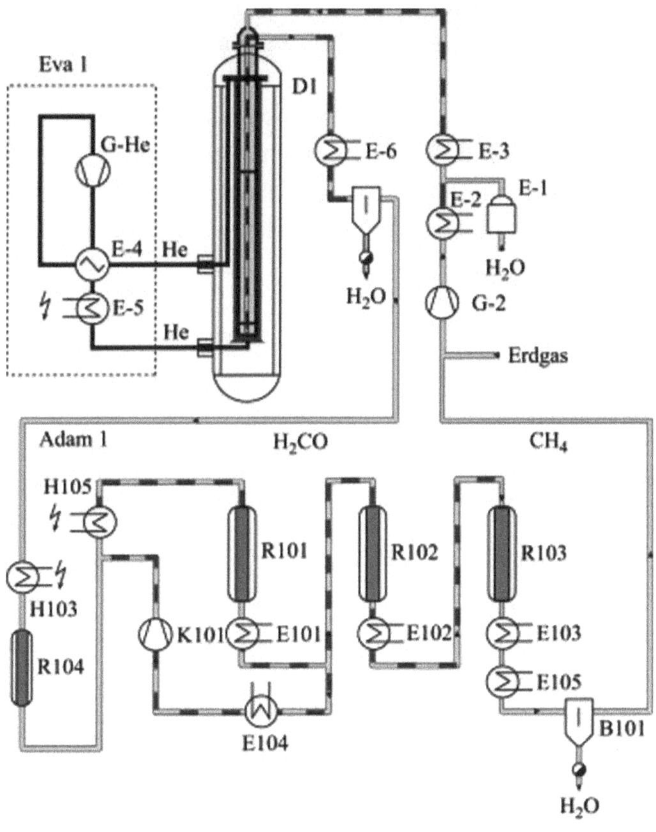

Figure 8.16
The Adam and Eve system, showing details of the reactors, feed systems, etc. *From reference of footnote 6, reproduced with kind permission of Elsevier.*

In the context of this chapter, the geometry of the Adam reactor is particularly informative. In order to achieve good conversions back to methane, a sequence of three reactors with interbed cooling of the type described earlier for the ammonia synthesis reaction was used. However, because the enthalpy of the methanation reaction is so high, the temperature rise in the first bed would be such that exit temperature would be about 800°C if an undiluted feed was used with an entrance temperature of about 300°C. In order to prevent such a large temperature increase, a recycle loop was introduced and this had the effect of decreasing the upper temperature to about 600°C. At the time when this work was initiated, the only methanation catalysts available commercially were those used for the methanation of traces of CO present in the hydrogen stream of an ammonia plant prior to entering the ammonia synthesis reactor. The temperature rise in such a methanation reactor is small and so the demands on the catalyst are relatively straightforward: sufficient activity at about 450°C with adequate thermal stability at that temperature to ensure that the catalyst does not deactivate during use over extended periods. The requirement for the Adam reactor was different: adequate activity at 300°C to start the reaction and sufficient thermal stability at 600°C to resist the severe sintering that would occur with standard methanation catalysts. The work led to the development, in collaboration with Haldor Topsøe, of new Ni-based methanation catalysts that exhibited the required activity and stability. Typical results for the temperature profiles for the three beds of Adam 1 using a Haldor Topsøe catalyst are shown in Fig. 8.17. The larger Adam and Eve II pilot plant operated for some time in the

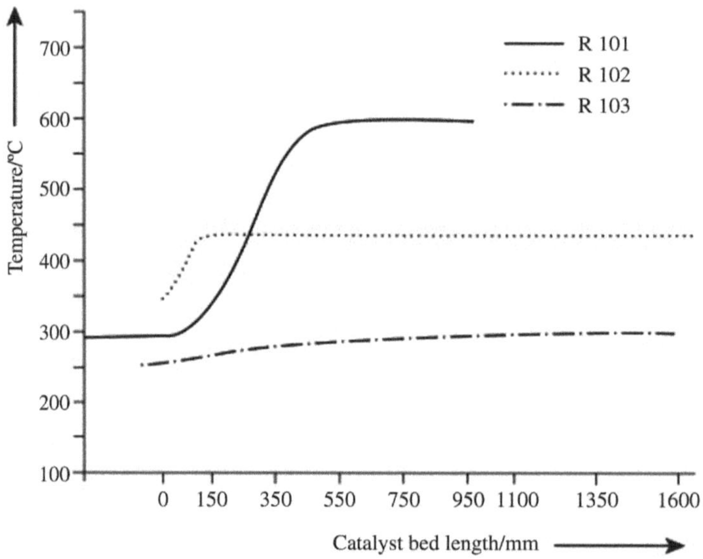

Figure 8.17
Temperature profiles in the Adam 1 reactor. *From reference of footnote 6, reproduced with kind permission of Elsevier.*

early 1980s, producing sufficient heat for the needs of a small town. However, political pressures led to the abandonment of the high-temperature nuclear reactor and hence of this process concept.[8] (See also Task 8.2.)

Task 8.2 The Chemical Heat Pipe Concept Applied to Solar Energy

The idea behind the Adam and Eve system has more recently been adopted in work aiming to harness sunlight using CO_2 reforming of methane as the endothermic reaction. The reforming reactor is placed at the focus of the heliostat and the syngas produced is piped to where the energy is needed. Examine the literature to learn something more about the use of such systems in solar applications using the paper by Levy et al., Sol. Energy 50 (1993) 179–189 as a starting point. See if you can find out about other conceptual designs for chemical heat pipes.

(This concept is handled in more detail in Chapter 11: Production of Hydrogen and Syngas From Methane and Some Other Reactions of Methane.)

[8] There is some evidence that pressure is currently mounting to recommence the development of this type of nuclear reactor. Work has also been carried out on the development of similar concepts for the utilization of solar energy (see Task 8.2), a topic considered further in Chapter 11, Production of Hydrogen and Syngas From Methane and Some Other Reactions of Methane.

PART II

Current Applications of Heterogeneous Catalysis

CHAPTER 9

CO_2 and Energy

Chapter Outline

9.1 Introductory Remarks

Having in the first part of this book introduced the basic concepts of heterogeneous catalysis, touching briefly on some of the applications of this important discipline, the object of the second part is to examine some of the work on a variety of topics that are of greatest current academic and industrial interest. The aim is to give the reader a starting point for further self-study on each of the topics covered rather than going into every aspect of all of them and each chapter includes a wide range of tasks to encourage this approach. The emphasis—working on the assumption that you will learn much more by reading the original articles written by authors with hands-on experience than from a dry textbook—is on encouraging you to read the relevant literature. Wherever possible, you are given links to review articles that summarize the knowledge of the author or authors of the reviews of the fields in question; however, occasionally you will be pointed in the direction of a particularly important original paper. Do not try to read and understand each and every detail of anything you read—as each time you read a paper you will find something different of interest; try instead to distill out the most relevant points relevant to your current search.

This chapter deals with the important topic of energy production from fossil fuels and the problem of the resultant CO_2 emissions. It begins with a general discussion of the various fossil raw materials (wood, coal, oil, and natural gas) used as sources of energy and also of chemicals. It then considers the role that is played by heterogeneous catalysis in some aspects of the production/upgrading and use of these fossil fuels for energy-related purposes. Particular attention is given to the role of combustion processes in improving energy efficiency and reducing CO_2 emissions, introducing the ideas behind catalytic combustion and examining some of the important work in that area.

Contemporary Catalysis.
DOI: https://doi.org/10.1016/B978-0-444-63474-0.00009-6

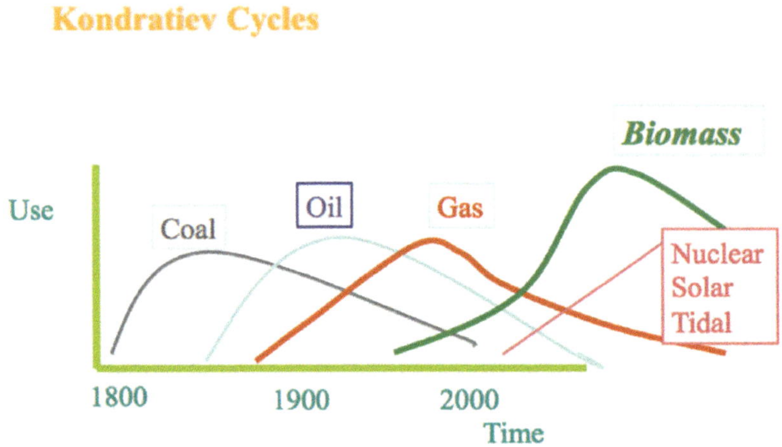

Figure 9.1
Kondratiev cycles for the use of various feedstocks.

Biomass and fossil fuels have been used as sources of energy for many centuries and the increase of the use of the combustion of fossil fuels since the period of the Industrial Revolution is blamed for the steadily increasing levels of greenhouse gases in the atmosphere and the associated global warming that has occurred. The Russian economist, N.D. Kondratiev, suggested in 1925 that, in addition to the more easily seen shorter term changes, economic cycles spanning approximately half a century also have an important effect. Since then, G.F. Ray has suggested that these cycles were tied to changes in energy usage.[1] Such cycles related to energy usage are illustrated schematically in Fig. 9.1.

This figure shows how the importance of the various fuels has gone through successive maxima since about the time of the Industrial Revolution. It is probably worth pointing out at this stage that much of the work on catalysis since 1900 has tended to follow similar cycles, the occurrence of these depending on the current importance of one or another fuel. Some aspects of the use of each of these fuels will now be considered in turn.

9.2 Wood and Biomass

Prior to the Industrial Revolution, biomass (largely wood) was the main fuel in use and, indeed, man was responsible for the disappearance of a large proportion of the tree-cover of much of the populated world by burning wood for heating and cooking. It was also used as a building material. (See Fig. 9.2 for an example of deforestation. In the Burren area shown, there is geological evidence that wood was also used for the construction of dwellings.)

[1] Kondratiev's paper was first published in 1925 (Archiv für Sozialwissenschaft und Sozialpolitik, Tübingen) but was later published in English (Review of Economics and Statistics, November 1935, reproduced in Lloyds Bank Review, No. 129, July 1978). See also G.F. Ray, Energy Economics, January 1983, 3–8.

Figure 9.2
A view of part of the Burren, Co. Clare, Ireland, an area that was deforested in c. 4000 BC by Neolithic man who used the wood for fuel.[2]

As time progressed, wood was also used as a source of energy for the smelting of metals (e.g., iron and bronze). Wood-burning stoves were first patented in Strasbourg in 1557, well before the Industrial Revolution, and are still in use.[3] Modern wood-burning stoves are constructed in such a way as to improve combustion efficiency and to reduce undesirable emissions. Some (Fig. 9.3) incorporate catalytic converters similar to those used for air purification applications or in car exhaust clean-up that will be discussed further in Chapter 13, Environmental Catalysis; they are also used for natural gas combustion (Section 9.4). These stoves are claimed to be significantly more efficient than noncatalytic ones but there can be problems associated with the operation of the catalyst such as rapid deterioration due to poisons in the flue gas, etc., and so the catalytic components may have to be replaced relatively frequently. Remember that it is to be expected that the stove will be operated by a relatively unskilled user! As we will see later, similar problems can exist with automobile exhaust catalysts. (See also Task 9.1.)

The wood used in modern wood-burning stoves is usually in pellet form and these pellets can be made from a variety of different types of wood. The pellets are very dense and generally have very low moisture contents that make them suitable for continuous feeding methods enabling automatic operation of the combustion system. They can also be made from other biomass material such as empty fruit bunches, palm kernel shells, coconut shells, and the branches of

[2] The Burren is a glacio-karst landscape that was in the Mesolithic period (8000–7000 BC) covered by a mixture of deciduous, pine and yew trees. For further information, read an article on the Burren in Wikipedia or find similar articles on the same subject that will be found by doing a Google search.

[3] We will not discuss the current use of biomass combustion for energy purposes any further in this section as little has changed in that regard over the centuries. The combustion of biomass in the form of wood and other crops (as well as of coal, see Section 9.3) is still used widely, particularly in developing countries, and this contributes quite significantly to worldwide CO_2 emission figures. However, as biomass is regenerable, such emissions are not considered to contribute to increasing global CO_2 emission figures.

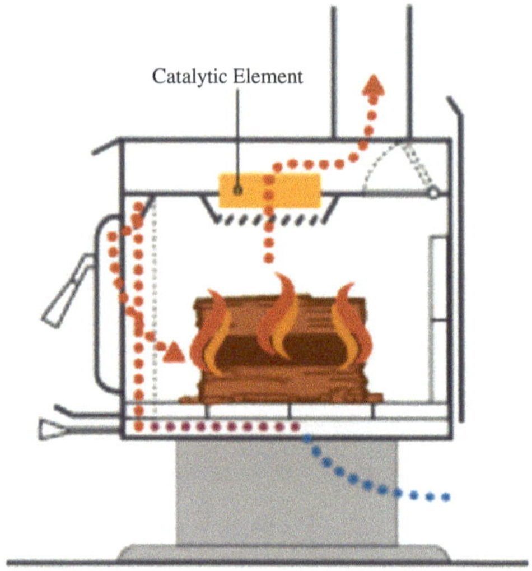

Figure 9.3
A catalytic wood-burning stove showing the various flows of gases: combustion air with preheating system and flue gas together with a damper to allow the catalyst to be by-passed if necessary during start-up. *Diagram from the US EPA Burn Wise website.*

Task 9.1 Catalytic Wood-Burning Stoves

Carry out an Internet search to find out more about wood-burning stoves fitted with catalytic converters. These catalysts will be exposed to very variable circumstances and will be required to start working from cold. See what you can find out about the types of catalyst used and their stability in use.

trees that result from logging operations and are not used for other purposes. As the CO_2 produced during combustion originates from a renewable source, it does not contribute to an increase in CO_2 emissions and so wood-burning is considered to be a "green technology." Other "green" uses of biomass will be discussed in Chapter 15, Catalysis in Biomass Conversion.

9.3 Coal

From about 1800 onward, coal became the main source of energy and many different chemicals were also produced from it. As in the case of wood, much was simply burnt in relatively inefficient grates and the associated emissions (high in SO_2 and NO_x) were major contributors to the creation of smog in major cities. Air-pollution regulations were introduced in the years around the middle of the 20th century and this lead to the more widespread use of the so-called smokeless fuels.

Coal was also used as a source of other products. One of the most important energy-related processes, to be found in every large center of population, was coal gasification to produce

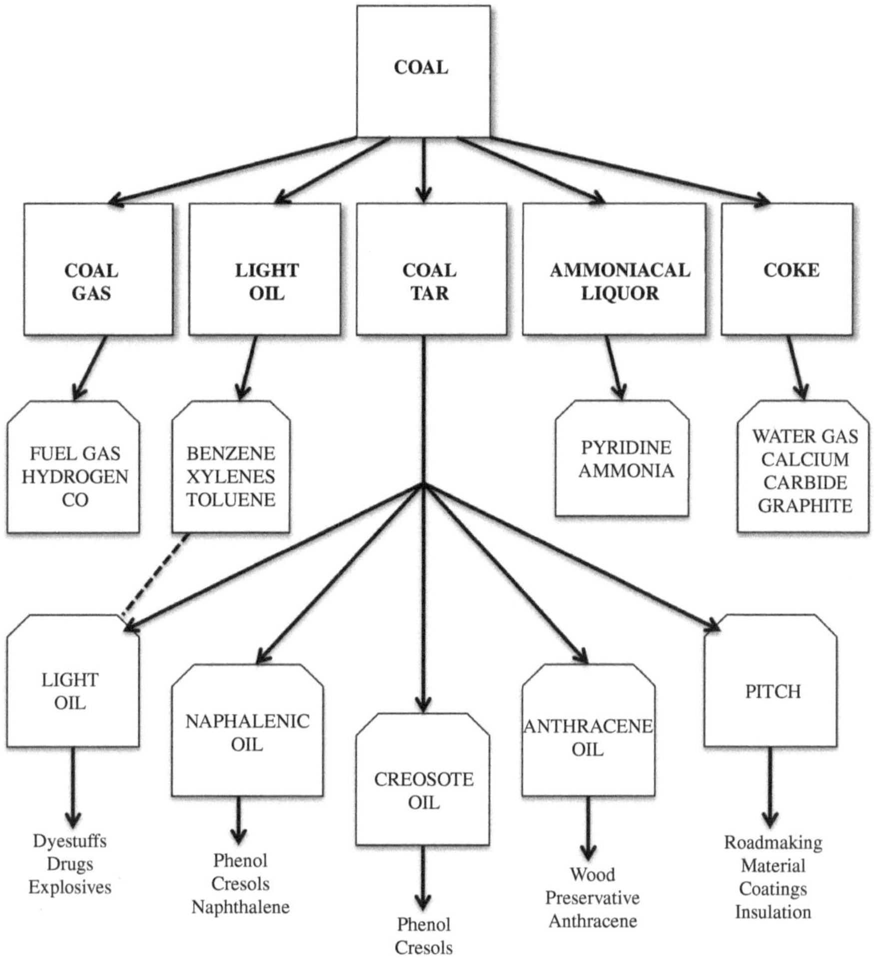

Figure 9.4

Some of the chemicals that were produced by the destructive distillation of coal at about 1100°C. This is a simplified version of one that appeared in a book by A. Lowry.[4] *See http://www.hatheway.net/images/wastepicts/big_tar_chem_furnas_1942.large.jpg.*

a towns gas, a mixture containing in a typical case 50% hydrogen, 35% methane, 10% carbon monoxide, and 5% ethylene. The process also produced tars, coke, sulfur, and ammonia; the tars contained compounds that could be converted to many different products including drugs and dyestuffs. This was the main source of the industrially important chemicals in use at that time. The many different products that could be derived from such plants are shown schematically in Fig. 9.4.

[4] Dr. Hatheway describes his diagram as follows: The wide variety of residuals resulting from destructive distillation of coal, including the processes resulting in manufacture of artificial gas. The chart was produced in 1923 by coal-combustion expert Alexander Lowy and appeared in several editions of Rogers' Industrial Chemistry, edited in later revisions by C.C. Furnas.

Towns gas plants were notoriously "dirty" and they created serious urban air contamination problems as mentioned earlier. It was only in the 1960s, when natural gas became more generally available, that these coal gasification plants were phased out and alternative routes to some of the chemicals hitherto produced from coal tar were developed. Coal gasification was also used for the production of syngas (a mixture of CO and hydrogen) using Lurgi gasifiers, the syngas then being used for methanol production or in Fischer–Tropsch plants (see Chapter 12: Catalytic Reactions Involving Syngas, Hydrogen, or Carbon Monoxide for the Production of Intermediates and Chemicals). Although the gasifiers used for towns gas and syngas preparation did not themselves require the use of any catalytic processes, catalysis was used in many downstream steps such as the conversion of SO_2 to sulfur by the Claus process or the production of hydrogen-rich gases by carrying out the water-gas shift reaction:

$$CO + H_2O \rightarrow CO_2 + H_2$$

We will discuss the importance of the water-gas shift reaction in relation to natural gas processing in Chapter 10, The Production of Energy Carriers From Oil; it also contributes to many other reactions as will be discussed in later chapters.

BOX 9.1 Environmental Aspects of Gas Manufacture

It should be noted here that urban gas manufacturing plants were probably among the worst examples of environmentally damaging installations. Not only were the gaseous emissions from such plants contributors to the formation of smog and the cause of many health problems but the tars formed at various stages in the process contained myriads of different molecules. Some of these were of value but many were regarded as wastes and just discarded. As a result, the surroundings of old gas works are frequently still seriously contaminated. Needless to say, environmental concern did not feature at the time when these plants were in their heyday and so there are many regions where remediation is still needed. For further information on the residual problems from such plants, the interested reader should consult the website: www.hatheway.net.

Task 9.2 Products From Towns Gas Plants

Use a suitable Internet search to find out more about the chemical industry based on coal gasification as depicted in Fig. 9.4, particularly from the light oil fraction containing benzene and other aromatic compounds. Pay particular attention to any catalytic processes used in converting these products to other useful chemicals. What is the current status of such processes?

As a result of the need in the period leading up to the World War I for liquid fuels, Freidrich Bergius developed in 1913 a hydrogenation process using coal as raw material. This operated under pressure, with a catalyst, producing liquid mixtures of hydrocarbons usable as fuels and lubricants; the original system used iron sulfides present in the coal as catalyst but more recent developments have used other materials such as tungsten or molybdenum sulfides. This work led later to the construction of a number of plants producing these fuels for use during World War II. More recently, there have been a number of research programs aimed at developing coal to liquid technologies and agencies in the United States and other countries are still funding a large number of such projects. An essential feature of each of these is that the process developed has to contain a method of CO_2 sequestration since very significant quantities of this greenhouse gas are generated at various stages, largely as a result of the need for process heat. (See also Task 9.3.)

Task 9.3 Coal Liquefaction

Search for details of recent coal liquefaction processes and find out what you can about the catalysts used in these processes. Pay particular attention in your search to recently funded projects. Consider also any developments that have occurred in the associated CO_2 capture technologies.

In the 1920s, Frans Fischer and Hans Tropsch developed what is now known as the Fischer–Tropsch process, an indirect route to produce hydrocarbon liquids from coal. This process uses Lurgi technology to gasify the coal to produce syngas (a mixture of CO and H_2) and then converts this mixture to hydrocarbons under pressure using catalysts containing Co or Fe. Modern developments associated with the process have involved the use of syngas made from natural gas. We will therefore return to this topic in Chapter 12 when discussing processes based on natural gas conversion, describing some of the history of the development of the process and its current status.

9.4 Oil

With the discovery of abundant reserves of crude oil and the development during the early part of the 20th century of the methods used for its extraction, fractionation, and upgrading (see Chapter 1), oil soon became the main source of both energy and chemicals and as a result the modern chemical and refinery industries based on crude oil still predominate. Some of the fractions of the crude oil separated by distillation are used directly for energy production without any purification (e.g., fuel oil for used in ships); such uses have associated with them serious undesirable emissions, just as with coal combustion, and so further measures have to be taken to clean up the flue gases formed. Although, as we will see in Chapter 15 there has recently been a renewed interest in the use of biomass as a

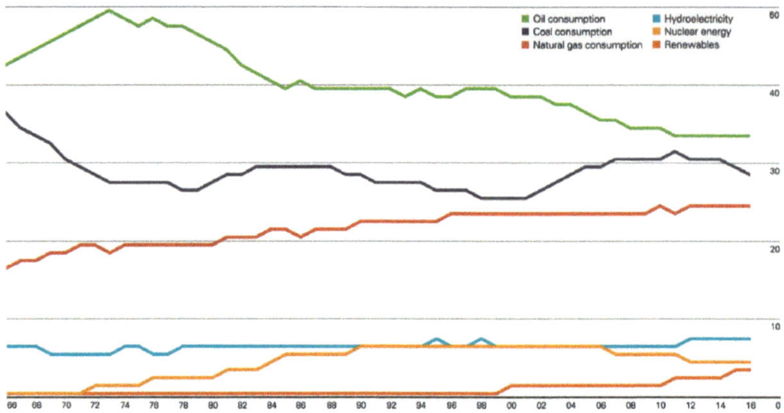

Figure 9.5

The relative shares of the global energy consumption per energy type for the years 1966–2016. *Reproduced with kind permission from the BP Statistical Review of World Energy, June 2017; www.bp.com/statistical review. Note that this review is updated every year. An app is available that makes all the data available in the review accessible for calculations, etc.*

source of renewable energy and also of chemicals, undoubtedly the use of nonrenewable fossil resources (oil and gas) for energy production will continue for several more decades. Such a situation is likely to prevail at least until new technologies are developed and introduced for the use of renewable energy and for the production of biomass-based chemicals. In this context, it should be recognized that there is only a finite reserve of hydrocarbons and that it is thus vital that these fossil fuels are used as efficiently as possible with minimal emission of greenhouse gases. As will be seen in Chapter 10 where we discuss processes used for oil conversion to important products, hydrocarbon conversion is an area in which catalysis plays an extremely important role. It is interesting to note that the *relative share* of oil in world energy usage (Fig. 9.5) is dropping steadily even though the proven reserves are still increasing (Fig. 9.6), albeit with changes in relative regional distributions. Fig. 9.5 also shows that the share of the energy market occupied by natural gas has increased steadily over the period covered by the report from which these figures were obtained and that there is a clearly discernable increase in the use of renewables for energy production. We will return to both of these topics in the following section.

9.5 Gas

Fig. 9.1 shows the use of natural gas (predominantly methane but also containing small proportions of volatile C_2–C_4 hydrocarbons) peaking around the year 2000 but the maximum as drawn is probably slightly premature (as shown in Fig. 9.5); this is because

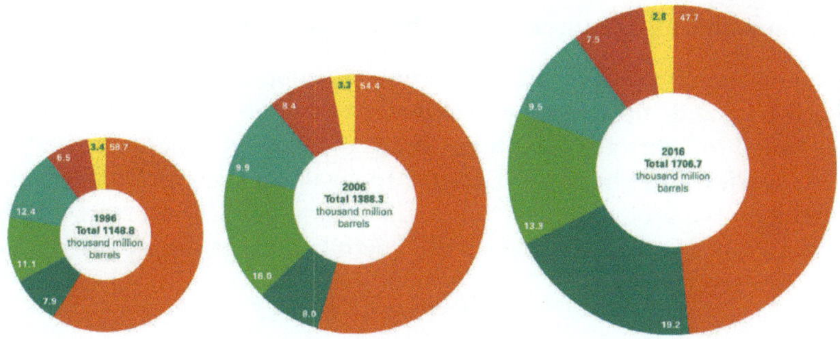

Figure 9.6

The distribution of proven oil reserves per world region at 10-year intervals from 1996 to 2016. *Reproduced with kind permission from the BP Statistical Review of World Energy, June 2017; www.bp.com/statistical review.*

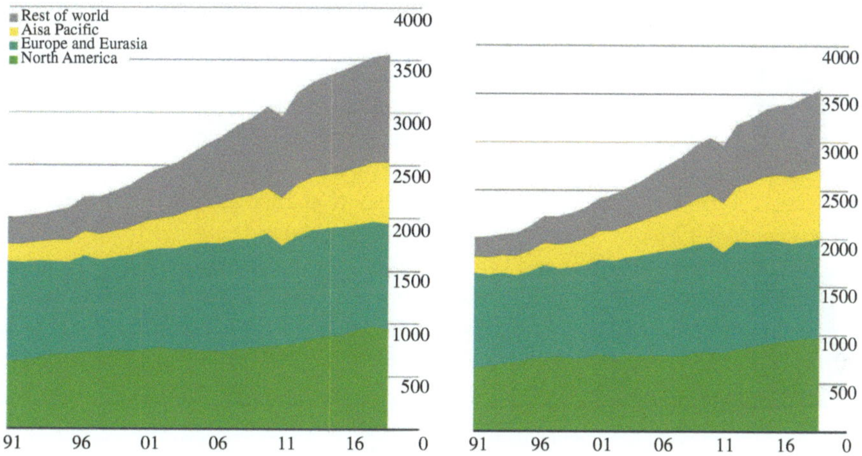

Figure 9.7

The relative shares of natural gas production and consumption for the years 1991–2016. *Reproduced with thanks from the BP Statistical Review of World Energy, June 2017; www.bp.com/statistical review. Note that this review is updated every year. An app is available that makes all the data available in the review accessible for calculations, etc.*

there is still a slight but continuous rise in the use of natural gas for energy purposes. This continuing increase is most probably due to the fact that fracking technologies have been introduced in a number of countries, particularly the United States, since about 2000. This newly applied technology has changed the situation very considerably and the United States has become a major supplier rather than an importer of natural gas. As shown in Fig. 9.7, this growth of both the production and consumption of natural gas has occurred in most of

BOX 9.2 Methanation as a Source of Synthetic Natural Gas

In the 1970s, many states in the United States required supplies of natural gas as they had no resources of their own and, although the Gulf states had good supplies, the cost of transportation added to interstate taxes made it attractive for the more northerly states to seek alternative supplies. This shortage of gas led to significant research as well as to development work on the production of "synthetic natural gas" (SNG) by the steam reforming of the naphtha fraction of crude oil, followed by methanation of the syngas produced:

$$C_nH_{2n+2} + nH_2O \rightarrow nCO + (m/2 + 1)H_2 \tag{1}$$

$$CO + 3H_2 \rightleftarrows CH_4 + H_2O \tag{2}$$

with, where necessary, some contribution from the water-gas shift reaction:

$$CO + H_2O \rightleftarrows CO_2 + H_2 \tag{3}$$

to supply the extra hydrogen needed to give the allover reaction:

$$C_nH_{2n+2} + \frac{(n-1)}{2}H_2O \rightarrow \frac{(3n+1)}{4}CH_4 + \frac{(n-1)}{4}CO_2 \tag{4}$$

We have already discussed in Chapter 1, some of the history of steam reforming and in Chapter 4, we considered some of the types of catalyst with high nickel content that have been developed for this reaction. (The methanation reaction was also discussed in relation to the Adam and Eve process in Chapter 8). For the production of SNG, the reaction temperature needs to be well controlled in such a way that methane is the favored product at the end of the reactor; a balance needs to be reached between the need for a sufficiently high temperature for reaction (1) to proceed but a low enough temperature for methanation (reaction 2) to be favored. As indicated in the main text, the shortage of methane of the 1970s has now been remedied, so much so that the United States has now become a net exporter of natural gas.

the regions of the world; only in Europe and Eurasia have both production and usage been relatively constant. (See also Boxes 9.2 and 9.3; Tasks 9.4 and 9.5.)

One of the most significant uses of natural gas is for heating purposes. This usage occurs because, when natural gas became readily available in the 1960s, it was able with the minimum of trouble to replace towns gas.[5] Natural gas was introduced at a time when environmental pressures were beginning to mount and so the combustion of natural

[5] One of the problems encountered when towns gas was substituted by natural gas was that the latter was much drier. This led to significant leaks in gas pipelines which had been constructed to transport the very much more humid towns gas. The leaks were so serious that the losses of gas to the atmosphere became quite high, up to 20% by volume.

BOX 9.3 Cyclic Reforming of Naphtha for Production of Natural Gas

The so-called cyclic reforming was carried out for many years as a means of producing towns gas from naphtha, a fraction of crude oil. A cyclic reforming plant is shown very schematically in the following figure.

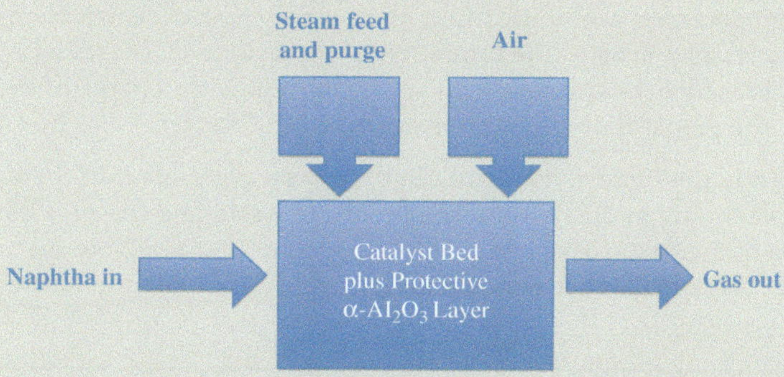

The reaction is based on steam reforming of naphtha, a hydrocarbon feed of relatively high molecular weight (obtained by distillation of crude oil).[a] The reaction is similar to that described by Eqs. (1)–(4) of Box 9.2 but now the temperature is much higher and so the product gas contains much more CO and H_2 and much less CH_4. (Reaction 2 occurs as an equilibrium and so little methane is formed at the higher temperatures of operation used in cyclic reforming, about 725°C.) The reaction cycle employed is as follows:

1. The catalyst bed is heated by a flame from the naphtha feed mixed with air. This flame is played on the surface of a protective α-Al_2O_3 layer above the catalyst bed.
2. Once the bed has reached the desired temperature, the air/naphtha feed is turned off and the system is purged with steam.
3. The naphtha is then fed to the steam feed and reforming occurs. As the reaction is endothermic, the temperature drops steadily and the cycle is terminated when the conversion rate is low.
4. The system is purged once more with steam prior to starting the cycle once more.

As the reaction is accompanied by very significant laydown of carbon on the catalyst (a point to be discussed in more detail in Chapter 11), this carbon is also burnt out during the first step of the next cycle, this adding to the heating effect of burning the naphtha.

The catalyst used for these plants is a relatively simple Ni-containing material supported on a porous α-Al_2O_3 support that is in the form of large Raschig rings; the Ni content is about 5 wt% and this is achieved by simple impregnation of the support material with $Ni(NO_3)_2$, followed by calcination, the resultant material being reduced in situ by exposure to the (reducing) reaction mixture. The most important aspect of the catalyst is that it must be thermally and mechanically very strong since it must withstand significant thermal shocks: increases and decreases in temperature as well as mechanical stress due to the deposition and removal of carbon. The catalyst lifetime is relatively short and so ease of replacement of the bed is important.

(*Continued*)

BOX 9.3 Cyclic Reforming of Naphtha for Production of Natural Gas (Continued)

It is interesting to note that it was also possible to operate cyclic reforming plants with a feed of natural gas in order to produce a towns gas. This was necessary for several years in the early days of the availability of natural gas before the gas-burning devices of domestic users had been converted to burn natural gas. Operation of the plants with methane as feed required higher temperatures and this caused more rapid catalyst deactivation due to sintering of the nickel particles as well as to the formation of nickel aluminate (spinel). This problem was solved very successfully in at least one plant by introducing a charge of the Ni-Al-La/ α-Al_2O_3 catalyst, the preparation of which is described in Chapter 4 and Box 4.4.

[a]This box is written in the present tense although it is very probable that there are few if any cyclic reforming plants still in operation throughout the world and the past tense might therefore be more appropriate. This material is included as it shows how such a plant worked and also because it is possible that the technology might be adapted in future for other purposes.

Task 9.4 Hydraulic Fracking and the Upturn in Natural Gas Production

Carry out a web search on the topic of fracking, learning what you can about the current status of this technique. Pay particular attention to the potential environmental problems associated with the technique as well as to the current views on the subject in the United States and Europe. How has fracking changed the status of the gas industry in the United States?

Task 9.5 SNG Production

Use the following article to find more recent articles on steam reforming and methanation: J.R.H. Ross, Metal Catalysed Methanation and Steam Reforming, Catalysis (Specialist Periodical Reports, Royal Society of Chemistry, London), 7 (1985) 1–45. See if you can identify any articles that describe how hydrogen produced by either solar or wind power can be used to methanate CO_2 as a means of sequestering this greenhouse gas.

gas for energy purposes took over many applications previously filled by coal and even oil.[6] New approaches were also developed, particularly relevant to the present discussion being work on catalytic combustion. (See also Box 9.4.) Catalytic (flameless) combustion takes place at much lower temperatures than those encountered in

[6] Another important reason for the current usage of natural gas is that when it is burnt, the amount of CO_2 produced per unit of energy is lower than that from the combustion of a heavier hydrocarbon or of coal, this being due to the higher hydrogen content per atom of carbon.

BOX 9.4 Catalytic Combustion of Methane

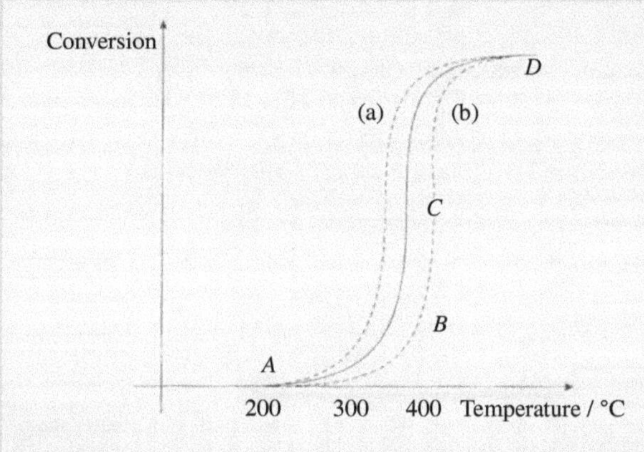

One of the most commonly used catalyst for methane combustion is Pd supported either on alumina or on some material containing alumina such as mullite or cordierite. The figure (reproduced with some modification from J.H. Lee and D.L. Trimm, Fuel Process. Technol. 42 (1995) 339–359) shows several features of the reaction occurring over a typical catalyst formulation. The *solid curve* shows schematically typical behavior observed over a Pd/alumina catalyst, the conversion being plotted as a function of the temperature. Region A occurs at a typical "light-off" temperature of such a catalyst, around 200°C. The reaction then proceeds under kinetic control (region B). However, as the reaction itself is highly exothermic, it is difficult to control the temperature and this then rises so that the behavior (region C) is typical of mass transfer control (internal diffusion limitations). Finally, in region D, at c. 100% conversion, the reaction becomes controlled by the rate of supply to the external surface ("reactant depletion"). The temperature scale shown is only approximate but it can be seen that depletion occurs at c. 400°C. The upper temperature is well below the range in which the formation of NO_x occurs and is also below the range of temperature in which gas-phase radical reactions are generally observed.

There is a body of evidence to show that the active surface of a Pd catalyst is (at least partially) oxidized during methane oxidation. The *dotted curve* (A) represents results obtained after the data for a freshly reduced catalyst (*solid curve*) that had been pretreated in the oxygen-containing reaction mixture before the measurements, there being a significant improvement in the catalytic behavior. The results shown by the *dotted curve* (B) illustrate another aspect of all work with heterogeneous catalysts: a significantly lower set of activities resulting from poisoning by one of the components of the gas feed used, in this case 100 ppm H_2S. Under the reaction conditions used in the work shown, adsorbed sulfate species were formed that could then be removed once more by thermal treatment in a nitrogen atmosphere or under vacuum.

(*Continued*)

BOX 9.4 Catalytic Combustion of Methane (Continued)

The review by Lee and Trimm referred to earlier gives an excellent description of some of the work reported in the literature on methane combustion; the effects of activation and of H_2S addition shown in the figure earlier are from results originally reported by L.J. Hoyos, H. Praliaud, M. Primet (Appl. Catal. A 98 (1995) 125–138). Another useful review article on methane combustion over Pd-based catalysts is by D. Ciuparu (Catal. Rev. Sci. Eng. 44 (2002) 593–644).

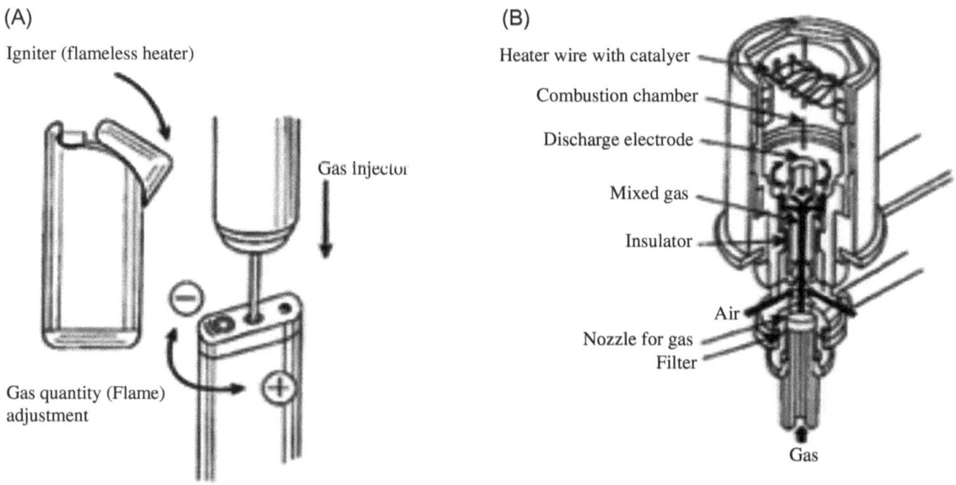

Figure 9.8

A catalytic lighter. *Reproduced from A. Nishino, Catal. Today 10 (1991) 107–118, with kind permission of Elsevier. No better figure available on* Science Direct.

homogeneous (flame) combustion, the advantage of this being that the formation of NO_x (a mixture of N_2O, NO, and NO_2) is much reduced, so much so that no flue-gas treatment is necessary. Catalytic combustion has many other advantages. For example, if a reactor wall is coated with a suitable combustion catalyst, the evolution of heat can be restricted to the region of that wall and so heat transfer is much more efficient. This is particularly relevant in microreactor applications (see Chapter 6: Catalytic Reactors and the Measurement of Catalytic Behavior). As the combustion is flameless, it can also be applied in situations in which flammability would cause the danger of fire or even explosion, for example, in paint drying applications in which the evaporating solvent is flammable. A novel application is shown in Fig. 9.8 that shows a catalytic combustor for

Task 9.6 Catalytic Combustion

Carry out a web-based search using catalytic combustion as search term. You will find that Willian C. Pfefferle of Engelhard Corporation is mentioned as one of the main scientists behind the development of catalytic combustion in the 1970s. Follow up leads to other companies that were also involved with developing catalysts for this use. Distinguish clearly between fuel-lean and fuel-rich applications.

A highly cited review on the preparation of wash-coated combustion catalysts is that by M. Valentini, G. Groppi, C. Cristiani, M. Levi, E. Tronconi, P. Forzatti, Catal. Today 69 (2001) 307–314. Use this paper and citations to it to learn more about such catalysts and their use for catalytic combustion.

use in a cigarette lighter. In the case, the fuel is more likely to be butane than natural gas but the application is effectively the same. (See also Task 9.6.)

We will return in Chapter 11 to a discussion of the use of natural gas in the production of hydrogen or syngas mixtures and will then discuss in Chapter 12 the various processes in which hydrogen and syngas can be used. Before doing that, however, we will discuss in Chapter 10 the conversion of crude oil to automotive fuels and other products.

CHAPTER 10

Catalysis in the Production of Energy Carriers From Oil

Chapter Outline

10.1 Introduction

We have in the previous chapter considered the combustion of biomass, coal, oil, and natural gas as sources of energy and we also considered the gasification of coal to produce towns gas and syngas for Fischer–Tropsch synthesis of hydrocarbons. We now turn to ways in which oil is converted to give products that can then also be used as sources of energy and also of hydrocarbon-based products. We start by considering the refining of crude oil and the production of automobile fuels. A discussion of the production of fuels by the conversion of natural gas to various different products is then the subject of Chapter 11, Production of Hydrogen and Syngas From Methane and Some Other Reactions of Methane.

10.2 Crude Oil

In the early part of the 20th century, the chemical industry relied largely on coal as the main source of chemical products and it was only with the increasing supplies of crude oil available in the United States and later in the Middle East and Europe that the modern petroleum and petrochemical industries came into their own. The oil industry is now the world's largest business, supplying a vast proportion of the world's energy demands as well as the majority of the petrochemical intermediates needed for the production of a wide variety of the goods that we all now take for granted: polymers, pharmaceuticals, dyestuffs, etc. Catalysis plays an important role in almost all processes carried out in the oil industry.

Contemporary Catalysis.
DOI: https://doi.org/10.1016/B978-0-444-63474-0.00010-2

Many of these processes are now well established and so the amount of research carried out on them relative to that carried out in other areas of interest has decreased significantly in the last 20–30 years. Nevertheless, there is still a great need to improve the efficiencies of the processes used since even a small percentage increase in the efficiency of a process can give enormous economic benefit and gives a decrease in the allover consumption of crude oil. This chapter does not attempt to cover the whole of this very vast subject. Instead, the aim is to give a flavor of some of the most important catalytic reactions and to give pointers for further reading.

A typical crude oil contains a very wide range of different components, from methane and other low-molecular-weight hydrocarbon molecules to heavy components ranging up in molecular weight to tars and waxes. These fractions are separated from one another by distillation before going for further purification and upgrading steps. Typical components are summarized in Table 10.1 and the combustion properties of the different types of molecule are given in Table 10.2. Olefins are not found in crude oil but are formed from it by thermal cracking processes. The distillation and cracking processes are discussed briefly in the following section (10.3) before proceeding to descriptions of some of the catalytic processes used to increase the values of the research octane number (RON) and motor octane number (MON) of some of the fractions of crude oil.

Table 10.1: Typical components of crude oil

Boiling Point Range	Fraction	Typical Components
<0°C	Fuel gas	C_1, C_2
	Propane	C_3
	Butanes	C_4
0–145°C	Light naphthas	C_5–C_9
145–185°C	Heavy naphthas	C_{10}–C_{11}
185–220°C	Jet fuel/kerosine	C_9–C_{14}
220–380°C	Light gasoil	C_{13} upward
	Heavy gasoil	Up to C_{25}
380–550°C	Vacuum gasoil, etc.	C_{20}–C_{50}
>600°C	Vacuum residue, asphalt	C_{40}–C_{50+}

Table 10.2: Research octane numbers (RONs) and motor octane numbers (MONs) for the various components of crude oil and of the olefins formed by cracking[1]

Hydrocarbon	n-Paraffins	Isoparaffins	Naphthenes	Aromatics	Olefins
RON	V. low (−19)	89–100	Low (80)	>100	90
MON	V. low (0)	Low (<3)	77	>95	Low (75)

[1] The octane numbers indicate the antiknock properties of each component relative to the reference fuels, iso-octane (100) and n-heptane (0); RON values are obtained under mild conditions (ordinary driving conditions) and MON values are obtained under severe operating conditions (high speed, high temperature).

10.3 Crude Oil Distillation and Feedstock Cleaning

Fig. 10.1 gives a very schematic representation of the arrangement of a typical oil refinery and Fig. 10.2 gives a very simplified version. The feed to be treated, a crude oil, enters at the left-hand side where it is separated by distillation and vacuum distillation into a variety

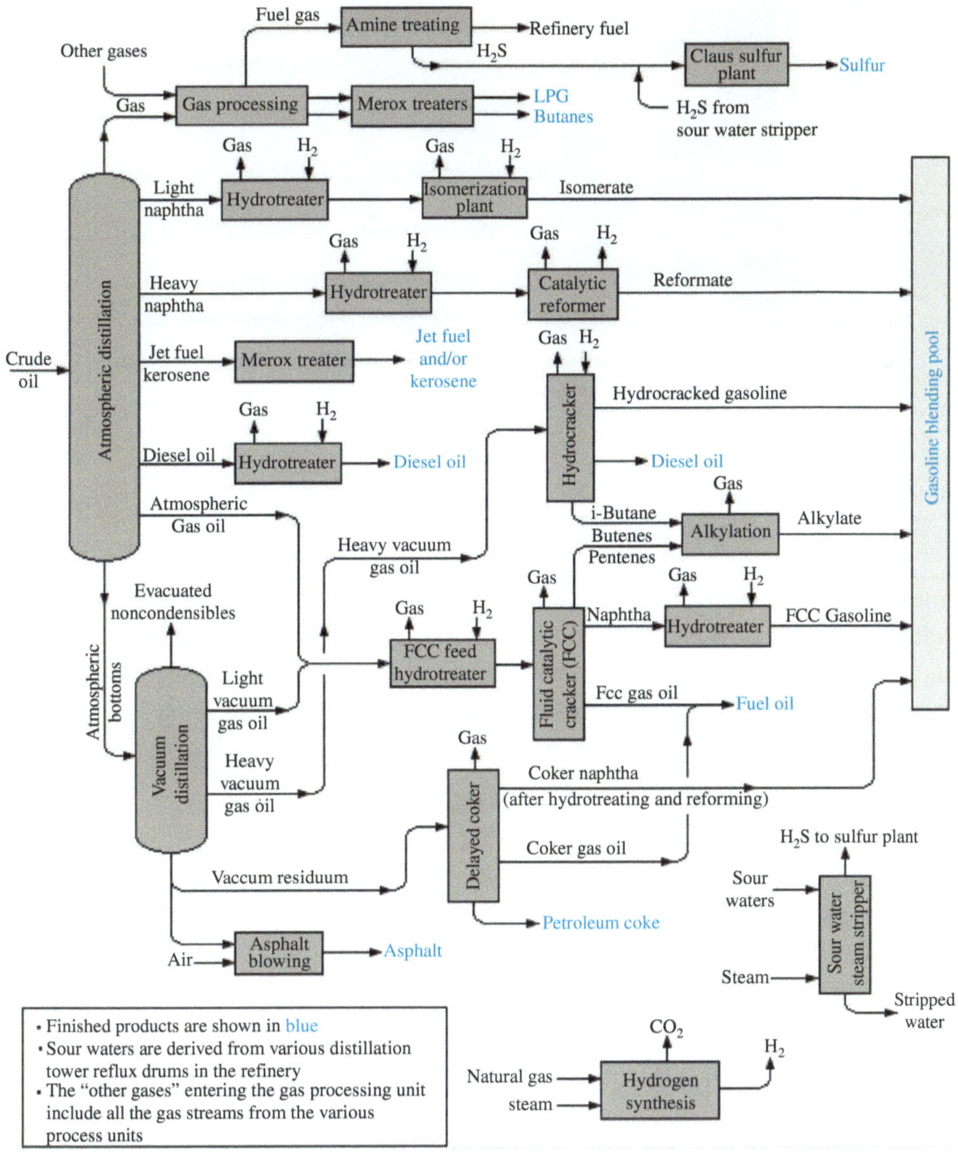

Figure 10.1

Schematic representation of the processes in a typical petroleum refinery. *http://en.wikipedia.org/wiki/File:RefineryFlow.png*

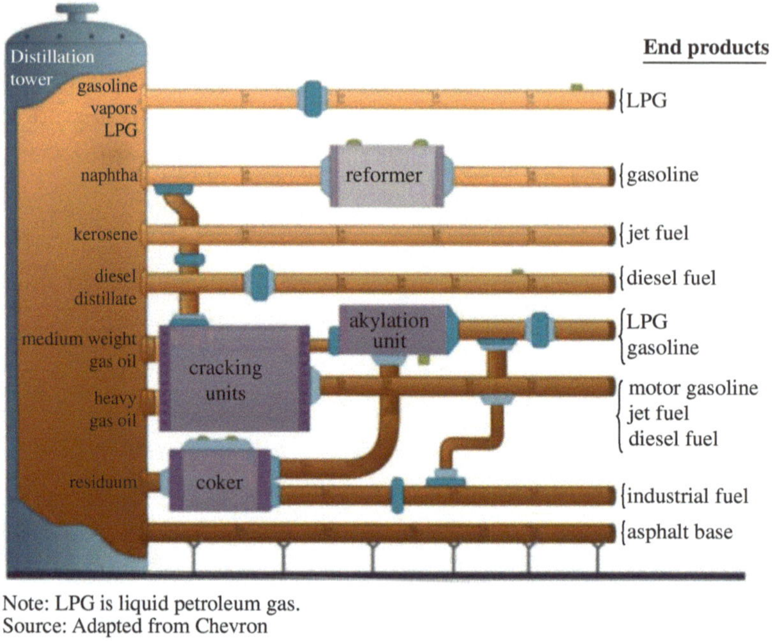

Figure 10.2
A simplified drawing showing most of the most important processes in a refinery. *U.S. Energy Information Agency, December 2017: www.eia.gov/*

of different fractions ranging from gases to asphalt. These are then treated further as appropriate by a range of other methods, some of which will be discussed further in the following text. It should be noted that the hydrogen required for a number of the individual steps is produced by steam reforming of natural gas, a topic to be discussed further in the next chapter (Chapter 11).

Fig. 10.3 shows the range of different products from the US oil industry in 2016, this being relatively indicative of the proportions of the different species produced worldwide. It can be seen that roughly 50% of the oil ends up as motor gasoline and that fuel oil is the next most important product. In other words, the vast majority of crude oil is used for energy processes, ending up with the emission of large quantities of CO_2. A very useful source of further information on the sources and uses of energy in the United States is the website of the US Energy Information Administration (www.eia.gov/).

Since the crude oil, depending on its origin, may contain a wide range of different components and impurities, each refinery will be different, dependent on that source. In the refinery depicted in Fig. 10.1, several of the fraction streams are passed through Merox units to remove mercaptans. The Merox process, introduced by UOP (http://en.wikipedia.org/wiki/Merox/), involves reaction of the mercaptans with alkali to give gives organic

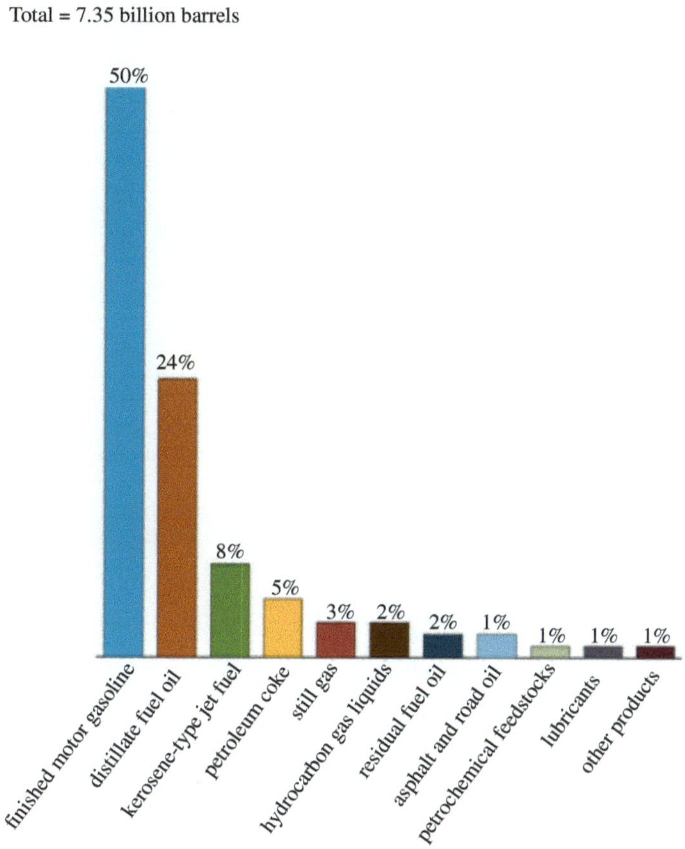

Figure 10.3

Net production of petroleum products in the United States in 2016. *U.S. Energy Information Administration, Petroleum Supply Annual, September 2017; www.eia.gov/*

disulfides that are then removed from the system. The sulfur in several other streams is removed by hydrotreating (see later). Hydrogen sulfide, which may be present in the gaseous fraction of the crude oil and also formed as a product of hydrotreating, is removed by amine scrubbing and subsequent treatment in a Claus plant (http://en.wikipedia.org/wiki/Claus_process/). As discussed in Box 10.1, more stringent requirements have led to the introduction of the SUPERCLAUS Process in which residual H_2S is selectively oxidized to sulfur (see the reference given in Box 10.1). (See also Task 10.1.)

10.4 Hydrodesulfurization

The term "hydrotreating" applies to a number of different processes that involve treatment of crude oil fractions with hydrogen. (The corresponding units in Fig. 10.1 are shown as "hydrotreaters"; see, for example, http://www.shell.com/business-customers/global-solutions/refinery-technology-licensing/hydrotreating.html.) These can be grouped

BOX 10.1 The Claus Process

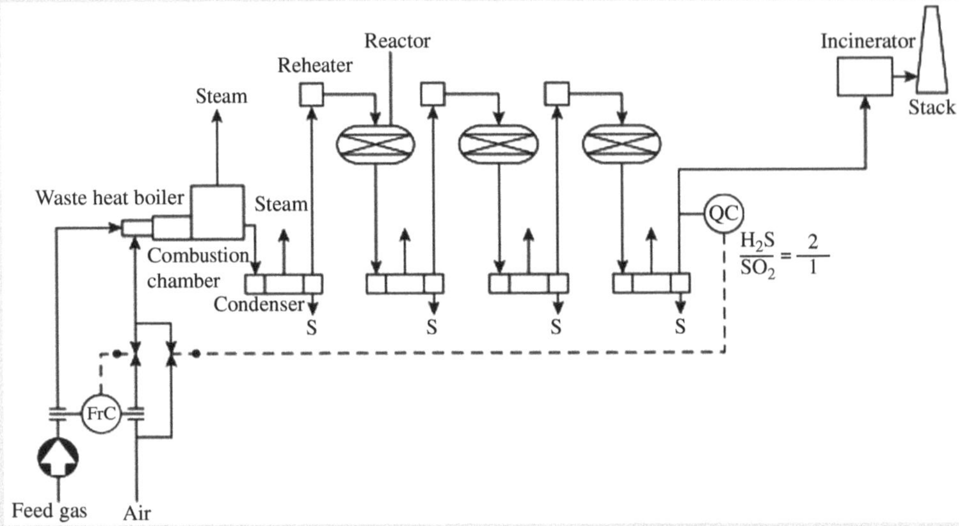

The Claus Process, depicted schematically earlier, is essentially the reaction:

$$2H_2S + SO_2 \rightleftarrows 3S + 2H_2O$$

this being catalyzed by alumina or titania.

The SO_2 required for the reaction is prepared in a preburner by the reaction:

$$2H_2S + 3O_2 \rightarrow 2SO_2 + 2H_2O$$

The allover reaction is therefore:

$$2H_2S + O_2 \rightarrow S_2 + 2H_2O$$

The first reaction is equilibrium limited and gives a conversion efficiency of only 95%–98%, leaving significant concentrations of H_2S in the gas stream. Many Claus plants have therefore been retrofitted by a Super-Claus reactor containing a catalyst comprising an α-alumina support impregnated with iron and chromium oxides. This catalyst selectively oxidizes the majority of the remaining H_2S to sulfur at 200–300°C without the formation of any further SO_2, giving an allover efficiency of 99%.

Source: *Figure courtesy of Elsevier Science Publishers, from P.F.M.T. van Nisselrooy and J.A. Lagas, SUPERCLAUS reduces SO_2 emission by the use of a new selective oxidation catalyst, Catal. Today 16 (1993) 263–271.*

Task 10.1 Recent Work on Catalysts for the Claus Reaction

Using the paper by Van Nisselrooy and Lagas shown in Box 10.1 as a source, carry out a citation search to locate the subsequent literature and find details of other catalysts which have been reported as active for the Claus reaction.

If you are using Scopus, you will find that these authors are listed as Van Nisselrooya and Lagasb!

into two subclassifications.[2] The first of these, often termed "hydropurification," covers the removal from the feedstock molecules of sulfur, nitrogen, oxygen, and/or heavy metals (e.g., Ni, V, As, etc.), the corresponding treatment processes being known as hydrodesulfurization (HDS), hydrodenitrogenation (HDN), hydrodeoxygenation (HDO), and hydrodemetallization (HDM). The removal of these components means that the product molecules have improved properties, do not emit contaminants on combustion, etc.; the products can also be further treated catalytically without poisoning the catalysts used. The second group of treatments, often referred to collectively as "hydroconversion," involves changes in the molecular weights and structures of the treated molecules in processes such as hydrogenation (HYD), hydrodearomatization (HDA), isomerization (ISM), and hydrocracking (HCG). We will first deal briefly with the process of HDS and will then in the following sections (10.5–10.7) discuss a number of different hydroconversion processes.

HDS is a process in which sulfur atoms are removed selectively from S-containing molecules such as thiols and thiophenes. The catalysts for HDS, generally based on the Co/Mo/alumina system, are often made using preformed γ-Al_2O_3 extrudates by a simple impregnation method involving, for example, solutions of cobalt nitrate or formate and ammonium molybdate. (See also Task 10.2.)

Task 10.2 Reviews on the Hydrodesulfurization Process

Grange and Vanhaeren [Hydrotreating catalysts, an old story with new challenges, Catal. Today 36 (1997) 375–391] list some 37 reviews on the subject of hydrotreatment published from 1957 onward. Using some or all of these as search terms in Scopus, identify and list in tabular form some more recent reviews, giving a brief description of the content of each. Using the Grange and Vanhaeren review and some of the more recent reviews, discuss the various models postulated for the active sites involved in catalysts for the HDS process.

One of the simplest S-containing molecules used for model studies is thiophene. Hargreaves and Ross examined the relative rates of the HDS of thiophene and tetrahydrothiophene as well as of the HYD of butadiene and but-1-ene over a catalyst with typical Co and Mo contents that had been presulfided.[3] They showed that the results could be explained by the scheme shown in Fig. 10.4.

[2] See, for example, P. Grange, X. Vanhaeren, Hydrotreating catalysts, an old story with new challenges, Catal. Today 36 (1997) 375–391.

[3] A.E. Hargreaves, J.R.H. Ross, An investigation of the mechanism of the hydrodesulfurization of thiophene over sulfided Co-Mo/Al_2O_3 catalysts. II. The effect of promotion by cobalt on the C-S bond cleavage and double bond hydrogenation/dehydrogenation activities of tetrahydothiophene and related compounds, J. Catal. 56 (1979) 363–376.

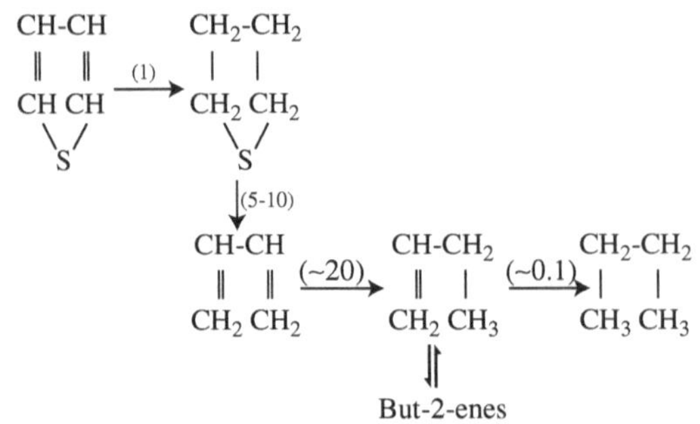

Figure 10.4

Scheme for thiophene hydrogenation to butenes and butane; the relative rates are given above the arrows for a catalyst containing 4.9 wt% CoO and 11.6 wt% MoO_3 with a balance of γ-Al_2O_3. The catalyst was presulfided in H_2S at 573 K for 17 h and then in H_2 at 521 K for 1 h.[3] *Reproduced with kind permission from Elsevier.*

The authors argued from their results that the parallel scheme in which the first product is but-1-thiol (i.e., ring opening before desulfurization) did not occur to any appreciable extent and that the predominant route is the HYD of the double bonds in adsorbed thiophene species prior to the cleavage of C-S bonds in the adsorbed tetrahydrothiophene to give but-1-ene as the predominant preliminary product. These conclusions contrasted with those reached earlier by Amberg and his colleagues, who thought that the formation of but-1-thiol was the first step in the process.[4] However, the difference between the two sets of conclusions may largely be due to the different experimental conditions used in the two studies. Although this work was carried out more than 30 years ago, there is still discussion on the sequence of surface reactions. For example, Henrik Topsøe and his colleagues at Haldor Topsøe, who have added a great deal of new information to the literature on the HDS reaction, have carried out a series of computations on the energetics of various possible surface steps on an idealized MoS_2 surface. (For further information on these computations, carry out Task 10.3.) The HDN and HDM processes are very similar to the HDS process, although the catalysts used may be slightly different. For example, HDN often uses an Ni—Mo—alumina catalyst and HDM requires catalysts with wide pores that are not poisoned too seriously by the deposition in them of the metallic impurities. (See also Tasks 10.4, 10.5 and 10.6.)

10.5 Isomerization

The aim of the majority of the steps following the hydrotreating step shown in Fig. 10.1 is to improve by the use of various catalytic reactions the octane numbers of the hydrocarbon fraction in question; this can be done, for example, by using isomerization (ISM) to

[4] See, for example, P. Dekisan, C.H. Amberg, Catalytic hydrodesulphurisation of thiophene. 5. Hydrothiophenes. Selective poisoning and acidity of catalyst surface, Canad. J. Chem. 42 (1964) 843.

Task 10.3 Computational Approach to HDS Mechanisms

Moses et al. (references below) have carried out a series of computations on the steps which occur during the HDS of thiophene over the active sites on MoS_2 and have shown the relevant importance of the various possible steps on S and Mo edge sites as shown in the following scheme.

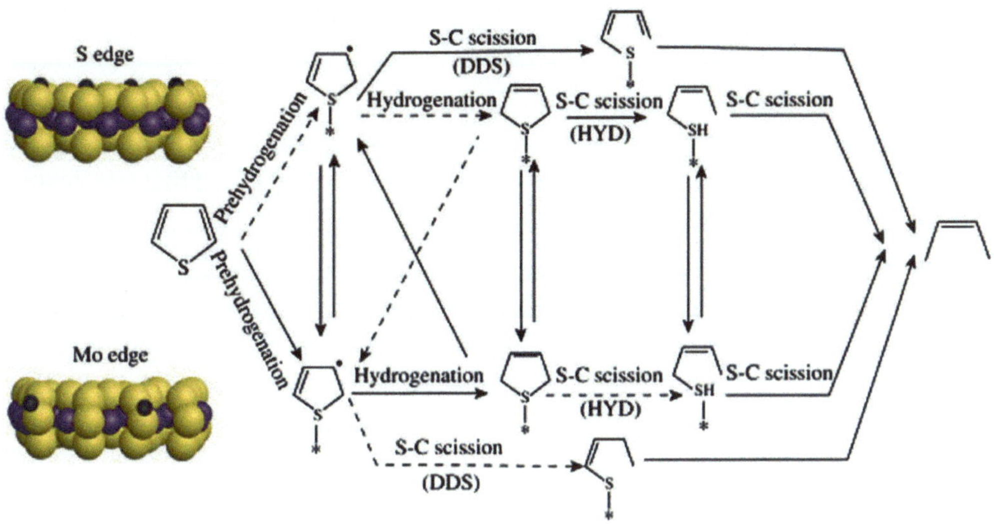

These authors have also shown in the corrigendum to the original paper that the S-C scission (DDS—direct desulfurization) route on S-edge sites at the top of the diagram is more important than the HYD route on Mo sites at the bottom. The predominant route is therefore likely to be prehydrogenation on the Mo edge sites followed by transfer to the S-edge sites and S-C scission on those sites. Use the references in the main paper to gain an understanding of the background to this work and the evidence for the site geometry discussed.

- P.G. Moses, B. Hinnemann, H. Topsøe, J.K. Nørskov, The hydrogenation and direct desulfurization reaction pathway in thiophene hydrodesufurization over MoS_2 catalysts at realistic conditions: a density functional study, J. Catal. 248 (2007) 188–203.
- P.G. Moses, B. Hinnemann, H. Topsøe, J.K. Nørskov, Corrigendum to the previously cited paper, J. Catal. 260 (2008) 202–203.

(*NOTE: It is not suggested that you attempt to understand all the details of the computational methods used, although some readers may find this very instructive.*)

Task 10.4 HDS, HDN, and HDM Catalysts and Processes

Using as a source reference the following very comprehensive review by E. Furimsky and F.E. Massoth examine the literature on HDS, HDN, and HDM catalysts and processes: "Deactivation of hydroprocessing catalysts", Catal. Today 52 (1999) 381–495. You should also pay attention to a more recent paper by the same authors on the HDN process: "Hydrodenitrogenation of petroleum", Catal. Rev. 47 (2005) 297–489.

Task 10.5 Commercial Suppliers of Hydroprocessing Plants and Catalysts

Carry out a web search for suppliers of plants and catalysts used for hydroprocessing. A useful starting point for part of your study will be the Shell web page http://www.shell.com/business-customers/global-solutions/refinery-technology-licensing/hydrotreating.html/ from which you will be led to the web pages of the Shell associate company, Criterion Catalysts. Attempt to identify other suppliers of both plants and catalysts and learn what you can about both the plants and the catalyst formulations on offer from these firms.

increase the degree of branching of the C–C chains in the hydrocarbon or by using HCG to decrease the chain length of the heavier fractions. The majority of the processes involve catalysis, the exception being thermal cracking.[5]

Light naphtha (see Fig. 10.1) contains a small proportion of butane and larger fractions of, in particular, n-pentane and n-hexane. As they stand, the latter two molecules have relatively little value as fuel additives because of low RON numbers.[6] However, if they are isomerized to give 2-methyl butane and 2,2-dimethyl butane respectively, RON values of 99 and 89 are achieved. The aim of the isomerization (ISM) step shown is therefore to increase the proportion of branched hydrocarbon species. While the ISM of alkenes is relatively easily achieved, that of the alkanes is not. Hence, it is necessary for the catalysts used to have two functions: dehydrogenation/hydrogenation and ISM. The former function is normally provided using a noble metal component, notably Pt, while the latter function is normally given by using a solid acid material. Hence the term: bifunctional catalyst. Early bifunctional catalysts (see Section 1.3.2) consisted of a strong acid such as $AlCl_3$ promoted by Pt but these have been phased out for environmental reasons; more recently, other acid components such as alumina, silica–alumina, zeolites, and sulfated zirconia have all been used, again promoted by Pt. Fig. 10.5 shows schematically some of the developments associated with the catalysts for the isomerization (ISM) reaction of n-pentane to isopentane, this showing the temperature range applied and the selectivity to isopentane for each type of catalyst.

[5] Thermal cracking is a noncatalytic process that is applied to convert various hydrocarbon fractions into lower molecular weight molecules. It was originally developed as a process for the production of liquid fuels but has been surpassed by catalytic processes such as FCC, a subject to be considered later in this chapter. The current usage of thermal cracking is for the production of olefins such as ethylene and propylene (for the production of polymers) from low-molecular-weight hydrocarbons such as naphthas, butane, propane, and ethane.

[6] As discussed in Section 10.2 earlier, the RON gives the relative ability of a molecule to resist reaction in an internal combustion engine in an oxygen-containing atmosphere until the conditions are at those required for ignition. Molecules with low values will give "knocking" or "pinking" while those with high values will not. The molecule iso-octane (2,2,4-trimethylpentane) is taken to have a value of 100. In the past, a molecule such as tetra-ethyl lead was added to the mixture to prevent knocking but this addition was phased out several decades ago.

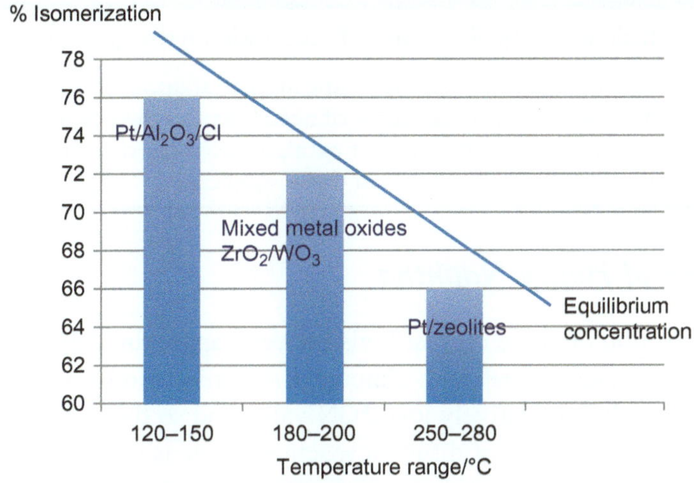

Figure 10.5

Selectivity to isopentane for the conversion of n-pentane over various different catalysts with the temperature range necessary for each; the selectivities approach the equilibrium concentrations at these temperatures. Pressure: 25–35 Bar; $H_2/C_5H_{12} = 1-2$. *Data from S.P. Srivastava, J. Hancsók, Fuel and Fuel-Additives, Wiley (2014).*

A typical feedstock has an RON value of about 70 and the most active of the catalysts shown in Fig. 10.5, the chlorinated Pt/Al_2O_3, will give a mixture of isomers with an RON value of about 84 in a single cycle mode of operation. If separation and recycling of the lower RON number (C_5 and C_6) components of the mixture obtained are then carried out with the same catalyst, an octane number of 91–93 can be obtained.

The operation of a bifunctional catalyst for the isomerization (ISM) of n-pentane (dehydrogenation and later rehydrogenation over the metal sites, ISM on the acidic oxide sites of the support) can be depicted as follows[7]:

$$nC_5H_{12} \rightarrow nC_5H_{10} + H_2(\text{on Pt}) \tag{10.1}$$

$$nC_5H_{10} + H^+ \rightarrow nC_5H_{11}^+(\text{on solid acid}) \tag{10.2}$$

$$nC_5H_{11}^+ \rightarrow isoC_5H_{11}^+(\text{on solid acid}) \tag{10.3}$$

$$isoC_5H_{11}^+ \rightarrow isoC_5H_{10} + H^+(\text{on solid acid}) \tag{10.4}$$

$$isoC_5H_{10} + H_2 \rightarrow isoC_5H_{12}(\text{on Pt}) \tag{10.5}$$

The isomerization (ISM) reaction is generally carried out in the presence of hydrogen as this helps retain catalyst activity over longer periods without there being a need for frequent regeneration as a result of coke formation. When required, coke burn-off can be achieved using a mixture of steam and air. (See also Task 10.6.)

[7] This scheme is taken from an excellent review on the subject by Y. Ono: A survey of the mechanism in catalytic isomerization of alkanes, Catal. Today 81(2003) 3–16.

Task 10.6 Other Industrial Applications of Catalytic Hydrogenation

Catalytic hydrogenation has many other important industrial applications. These include fat hydrogenation and the selective hydrogenation of acetylene to ethylene. Carry out a web search on these topics and identify the types of catalysts used for the processes. Examine some of the papers involving basic research on these subjects.

10.6 Reforming of Heavy Naphtha

Heavy naphtha, a fraction of crude oil that boils between approximately 70 and 200°C, contains mostly alkanes and cycloalkanes although there may also be a small fraction of aromatics. This mixture has a relatively low RON value and so the process is aimed at increasing that value by a mixture of different reactions such as dehydrogenation, isomerization (ISM), dehydrocyclization (ring closure followed by increasing the aromatic content by dehydrogenation) and some hydrocracking (HCG) to reduce the chain length of the heavier molecules present. As with the ISM process discussed earlier, the catalysts used must have a dehydrogenation function to produce olefins and an acid function to bring about thee ISM; the metallic function also gives some limited cracking of the heavier molecules which are also then isomerized. Much has been written about the catalysts, reaction schemes, and processes for naphtha reforming and the research literature on the subject is vast. According to Bond in his seminal book on "Metal-Catalysed Reactions of Hydrocarbons" (Springer, 2005), which the reader might peruse to gain a very full understanding of the subject, bimetallic catalysts such as Pt-Sn, Pt-Ir, and Pt-Re have had widespread use in hydrocarbon reforming. It is generally accepted that one of the main functions of the added metal promoter is to stabilize small Pt crystallites and thereby to decrease the chances of carbon deposition. As with light naphtha isomerization (ISM), the catalysts still need to be regenerated regularly. Performing Task 10.7 will give the reader a feel for the subject.

Task 10.7 Bimetallic Reforming Catalysts

Using the name of J.H. Sinfelt carry out a citation search on bimetallic catalysts and their uses. Select one of his most cited papers on the subject and then use this article to identify some of the other researchers active at the same time as Sinfelt. Ascertain whether or not research on this and related topics is still occurring.

Two important articles on the topic of bimetallic reforming catalysts from the period around 1980 are the following: F.M. Dautzenberg, J.N. Helle, P. Biloen, W.M.H. Sachtler, Conversion of n-hexane over monofunctional supported and unsupported Pt-Sn catalysts, J. Catal. 63 (1980) 119–128; and V. Ponec, Catalysis by alloys in hydrocarbon reactions, Adv. Catal. 32 (1983) 149–214. Use these references as source material to examine the work by these authors (and by others referred to in these articles or citing them) on the mechanisms of hydrocarbon reactions on bimetallic catalysts. Your work should aim to list the catalysts that have been studied and to gain an idea of the mechanisms of the reactions and of the effects of alloying on carbon deposition kinetics.

10.7 Fluid Catalytic Cracking

Catalytic cracking is now one of the most important processes practiced in petroleum refining as it allows the use of a much larger fraction of crude oil, converting relatively high molecular weight materials into high octane fuels. In its earliest development (http://en.wikipedia.org/wiki/Fluid_catalytic_cracking/), Friedel Crafts catalysts ($AlCl_3$) were used but the process using them was too expensive. In 1922, two Frenchmen named E.J. Houdry and E.A. Prudhomme set up a plant to produce 60 tons per day of petroleum from lignite using a catalyst discovered by Houdry, a Fuller's Earth containing silica–alumina. Following further work on oil conversion, Houdry was invited by the Vacuum Oil Company in 1930 to come to the United States and there he worked, first with a new company set up between the Vacuum Oil Company and Standard Oil of New York (Socony), the Socony-Vacuum Oil Company, which operated the first plant, producing 32,000 L per day. Then, working in collaboration with the Sun Oil Company, the first demonstration unit was operated in 1936 in Paulsboro, New Jersey, producing 10 times the volume. By 1940, there were 14 full-scale units in operation, these being fixed-bed semibatch systems. The FCC process as we know it arose from work in a consortium, Catalytic Research Associates (CRA), made up of Standard Oil of New Jersey, Standard Oil of Indiana, Anglo-Iranian Oil, Texas Oil, Dutch Shell, M.W. Kellogg, Universal Oil Products and IG Farben, this being based on work by Standard Oil of New Jersey prior to the depression years. The first plant based on the new technology, built by M.W. Kellogg Company, was commissioned in 1940 at the Standard Oil of New Jersey (later Exxon Mobil) Baton Rouge refinery in Louisiana.[8] Further details of some of the history of these important developments are to be found on the North American Catalysis Societies' website at http://www.nacatsoc.org/edu_info.asp.

Fig. 10.6 is a schematic representation of a typical FCC plant, showing the surrounding peripheral units as well as the reactor, while Fig. 10.7 gives more details of the central "riser" reactor itself. Fresh catalyst is fed to the riser section in which the acid-catalyzed cracking reaction takes place. The cracked product is then removed for further processing in the cyclone vessel. The catalyst, which now contains substantial quantities of coke, is then fed to the regenerator that is also fed with combustion air to burn off the coke. (Other additives can be added at this stage, these including deNO_x components designed to reduce any NO_x formed in the regenerator; see Harding et al., reference in Task 10.8.) The regenerated catalyst is then fed back to the riser where additional fresh catalyst is also added. In addition to having high activity for the cracking reaction, the catalyst must have high strength to resist degradation in the fluidized beds and to that end, many other constituents are added. (See also Task 10.8.)

[8] High octane fuel made in the early plants was provided to the RAF for use during the Battle of Britain and it has been claimed that the superior performance achieved assisted victory in that battle.

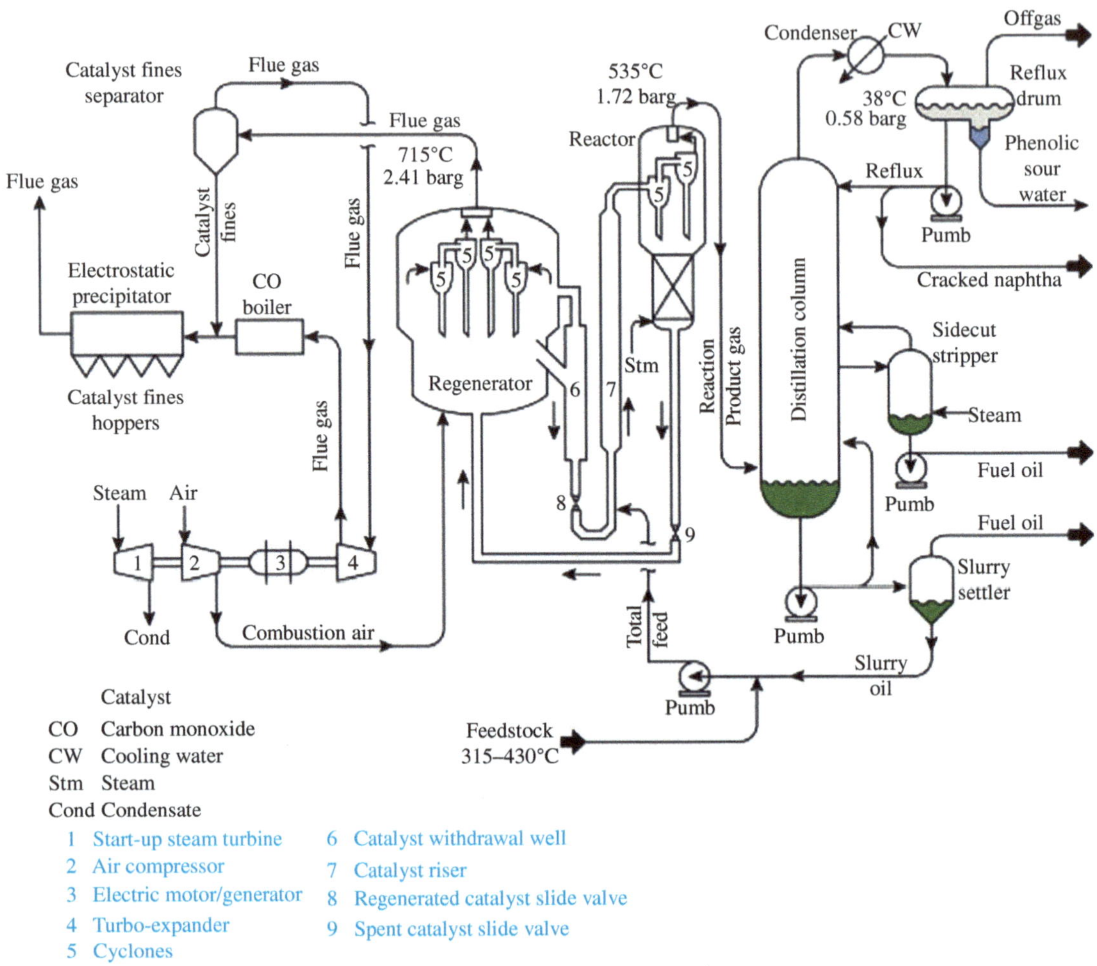

Figure 10.6

Schematic representation of a typical FCC plant. *Reproduced from Wikipedia: http://upload.wikimedia.org/wikipedia/commons/9/95/FCC.png*

Essential for the operation of a good cracking catalyst is the presence on the surface of the catalyst of strong acidic sites that can initiate the reaction by forming carbonium ions that then react further to form carbenium ions:

$$R\text{-}CH_2\text{-}CH_3 + H+ \rightarrow R\text{-}\overset{\displaystyle H\atop |}{\underset{/\ \backslash \atop H\ \ H}{C^+}}\text{-}CH_3 \rightarrow R\text{-}\underset{| \atop H}{C^+}\text{-}\ CH_3 + H_2 \tag{10.6}$$

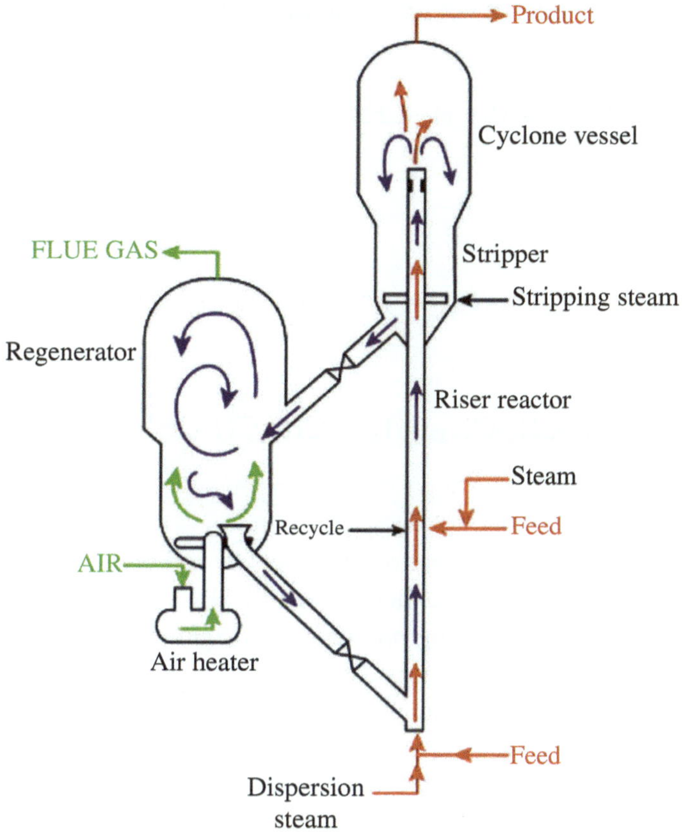

Figure 10.7
Schematic representation of a riser reactor. *W.-C. Cheng, C.T. Kim, A.W. Peters, X. Zhao, K. Rajagopalan, Cat. Rev. Sci. Eng. 40 (1998) 33–79.*

Task 10.8 FCC Catalysts

FCC catalysts are nowadays based on zeolitic materials together with binders and promoters of various sorts. Examples of the vast body of information available on the development of the catalysts used in the process are papers by P. O'Connor, P. Imhof, and S.J Yanik ["Catalyst assembly technology in FCC, Part 1. A review of the concept, history and developments", in Studies in Surface Science and Catalysis, M.L. Occelli, P. O'Connor (Eds.), 131 (2001) 299–310] and R.H. Harding, A.W. Peters, and J.R.D. Nee [New Developments in FCC catalyst technology, Appl. Catal., A 221 (2001) 389–396]. Use these as starting points to learn more about the types of catalyst used in the process, the nature of the active sites and their function, and the use of various additives to improve environmental and other aspects of the process.

The carbenium ions can then react yet further in many different ways including by: hydrogen transfer:

$$\underset{\substack{|\\ H}}{R\text{-}C^{+}\text{-}CH_3} + R'CH_2\text{-}CH_3 \rightarrow RCH_2\text{-}CH_3 + R'CH^{+}\text{-}CH_3 \qquad (10.7)$$

C–C bond scission:

$$CH_3 - CH^+ - CH_2 - R'' \rightarrow CH_3 - CH = CH_2 + R''^+ \qquad (10.8)$$

and isomerization:[9]

$$CH_3\text{-}CH^{+}\text{-}CH_2\,CH_2\text{-}CH_2\text{-}CH_2\text{-}CH_3 \rightarrow CH_3\text{-}\underset{\substack{|\\ CH_3}}{CH}\text{-}CH^{+} - CH_2\text{-}CH_2\text{-}CH_3 \qquad (10.9)$$

Once a carbenium ion is formed, a chain reaction sets in and both bond cleavage and isomerization (ISM) (forming the more thermodynamically stable products with secondary and tertiary carbon atoms) occur, giving products with significant improvements in octane ratings. For further details on the FCC process, the reader is referred to some of the excellent textbooks dealing with petroleum refinery processes and also to a wide variety of websites dealing with the subject. (See also Box 10.2, Task 10.9.) We return in Chapter 14 to a discussion of the conversion of hydrocarbons to valuable intermediates and chemicals by catalytic routes involving hydrogenation and oxidation.

BOX 10.2 Basic Research on Hydrocarbon Reactions

A wide range of basic research on the adsorption and reactions of hydrocarbons was carried out over many decades in the second half of the 20th century, this being inspired by the rapid industrial developments related to petroleum refining occurring during that period. A monumental work summarizing some of that work and drawing it together in a very readable form is the book by Geoff Bond entitled "Metal-Catalysed Reactions of Hydrocarbons" (Fundamental and Applied Catalysis, in: M.V. Twigg, M.S. Spencer (Eds.), Springer, 2005). A very extensive and informative chapter (Chapter 7, "Hydrogenation and dehydrogenation of organic compounds") in "Fundamentals of Industrial Catalytic Processes" by C.H. Bartholomew and R.J. Farrauto (2nd ed., Wiley, NJ, 2006) gives some valuable information on research on industrially important of hydrocarbon reactions as well as giving useful details of their use industrially.

[9] The isomerization (ISM) process on the acidic catalyst is similar to that occurring on the acidic support in bifunctional catalysis as described earlier in the section on isomerization (ISM).

Task 10.9 Exchange Reactions of Hydrocarbons Over Metal and Oxide Catalysts

When a hydrocarbon encounters a metal or oxide surface on which a reaction will occur, it first becomes adsorbed on that surface. Then, if the temperature is sufficiently high, it will undergo further interaction, producing new surface species that can then desorb to give products. The adsorption and the desorption of the hydrocarbon prior to any reaction occurring can be studied using deuterium exchange techniques to give important information on the way in which adsorption occurs. Carry out a citation study on hydrocarbon exchange experiments over both metals and oxides using as starting point a review by C. Kemball: Adv. Catal. 11 (1959) 223–262. Among other reviews given in this search, you will find one on adsorption and exchange of methane by metals by A. Frennet: Catal. Rev. 10 (1974) 37–38. Use the results of your search to learn more about the exchange technique and how the results are interpreted.

CHAPTER 11

Production of Hydrogen and Syngas From Methane and Some Other Reactions of Methane

Chapter Outline

11.1 Introduction

Fig. 11.1 shows schematically the importance of catalysis in the conversion of the natural fossil fuel resources, coal, oil, and gas, to some useful products. The *solid arrows* 1, 2, and 3 here represent the use of these resources for energy purposes, as discussed in Chapter 9: the supply of heat and electricity by combustion processes, these processes being noncatalytic. The catalytic combustion of natural gas is also shown for completeness as a *dashed arrow* (3). As discussed in Chapter 9, catalytic combustion has the advantage over homogeneous combustion that the emission of NO_x is minimized; this reaction can also be used effectively to remove traces of methane or other hydrocarbons from other gases as will be discussed in Chapter 13, Environmental Catalysis. The only other noncatalytic processes shown are coal gasification (*solid arrow* 4) and the cracking of oil fractions to give olefins (*solid arrow* 10). For completeness, the conversion of oil fractions to petrochemicals and fine chemicals by catalytic means is shown by the *dashed arrow* (14); steam reforming of oil fractions, largely naphthas, can also be used for syngas production but for reasons of clarity this is not shown in the figure. We will now discuss the various routes for the catalytic formation of syngas from methane (*arrow* 4). We will then discuss some of the potential routes for the conversion of methane to other products shown in the figure (*dotted lines, arrows* 11 and 13). Some of the uses of the syngas formed (*arrows* 6 and 7) will be discussed in Chapter 12, Catalytic Reactions involving Syngas, Hydrogen or Carbon Monoxide for the Production of Intermediates and Chemicals.

Contemporary Catalysis.
DOI: https://doi.org/10.1016/B978-0-444-63474-0.00011-4

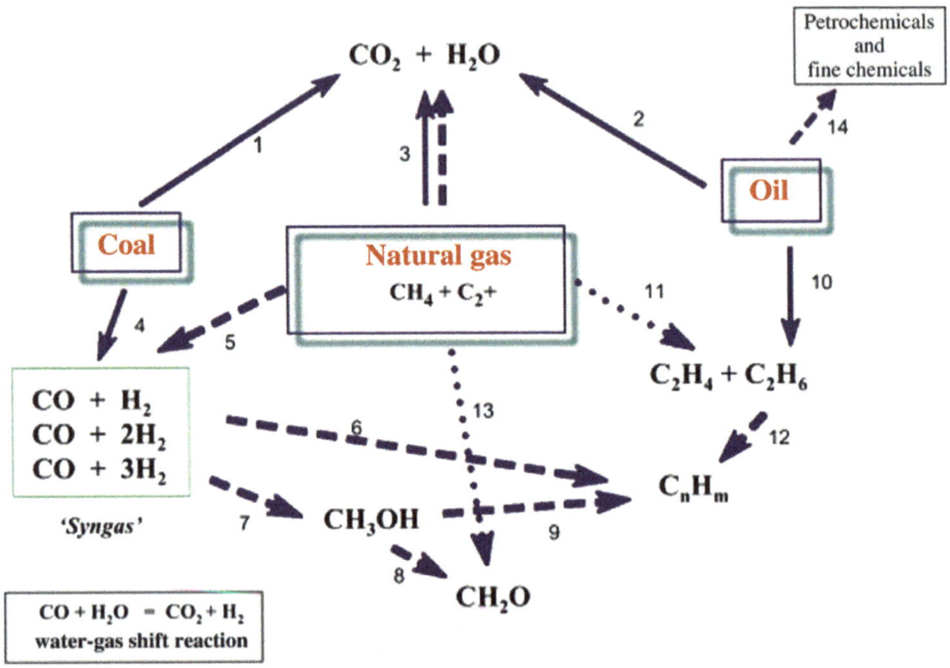

Figure 11.1
Schematic formulation showing the production of energy and chemicals from coal, oil and natural gas. The *solid arrows* represent noncatalytic routes, the *dashed arrows* represent established catalytic processes, and the *dotted arrows* represent possible future developments. (Route 14 represents the conversion of oil as discussed in Chapter 10: The Production of Energy-Carriers From Oil.).

11.2 Conversion of Natural Gas to Syngas

We have already discussed some aspects of the conversion of natural gas to syngas (Reaction 4) in relation to catalyst preparation (Chapter 4) and to its use in chemical heat pipes as used in the Adam and Eve Process (Chapter 8). The steam reforming of methane to give syngas is a mature technology and it is a subject that has been very well reviewed by Rostrup-Nielsen.[1] Some aspects of the reaction have also been reviewed by the present author.[2]

[1] J. Rostrup-Nielsen has published a number of papers on steam reforming. One of the most comprehensive reviews on the subject, entitled "Catalytic Steam Reforming", is published in Catalysis - Science and Technology, J.R. Anderson and M. Boudart (Eds.), Springer-Verlag, Berlin, Heidelberg, New York, and Tokyo, 5 (1984) 1–117. A later review describing recent developments is: "Steam Reforming and Chemical Recuperation", Catal. Today 145 (2009) 72–75. The interested reader is advised to carry out a comprehensive citation-based search using the name of this author as a starting point.

[2] J.R.H. Ross, "The Steam Reforming of Hydrocarbons", Surface and Defect Properties of Solids, M.W. Roberts and J.M (Eds.). Thomas, Specialist Periodical Reports, Royal Society of Chemistry, London, 4 (1975) 34–67; J.R.H. Ross, "Metal Catalysed Methanation and Steam Reforming", Catalysis, Edited by G.C. Bond and G. Webb, Specialist Periodical Reports, Royal Society of Chemistry, London, 7 (1985) 1–45; see also Natural gas reforming and CO_2 mitigation, Catal. Today 100 (2005) 151–158.

As shown in Fig. 11.1, route 5 encompasses three different methods of obtaining syngas from methane:

$$\text{Steam reforming: } CH_4 + H_2O \rightarrow CO + 3\,H_2, \quad \Delta H^\circ = -206.2\ \text{kJ} \tag{11.1}$$

$$\text{Partial oxidation: } CH_4 + 1/2\,O_2 \rightarrow CO + 2\,H_2, \quad \Delta H^\circ = -36\ \text{kJ} \tag{11.2}$$

$$\text{``Dry'' reforming: } CH_4 + CO_2 \rightarrow 2CO + 2\,H_2, \quad \Delta H^\circ = -247.4\ \text{kJ} \tag{11.3}$$

All three reactions are shown as being irreversible but they are all in fact, to a greater or lesser extent, equilibrium reactions and can only be considered as being irreversible if carried out at high temperatures[3]. Further, they are complicated by the fact that the water–gas shift equilibrium reaction (11.4) can also take place, altering in most cases the final gas composition [particularly in reactions (11.1) and (11.3) and, to a lesser extent, in reaction (11.2)] if some conversion to give CO_2 also occurs:

$$CO + H_2O \rightleftarrows CO_2 + H_2, \quad \Delta H^\circ = -41.2\ \text{kJ} \tag{11.4}$$

One final and very important complication is the lay-down of carbon on the catalysts by two different reactions:

$$\text{Boudouard reaction: } 2CO \rightleftarrows C + CO_2, \quad \Delta H^\circ = -172.4\ \text{kJ} \tag{11.5}$$

and

$$\text{Methane decomposition: } CH_4 \rightleftarrows C + 2\,H_2, \quad \Delta H^\circ = 74.9\ \text{kJ} \tag{11.6}$$

Carbon deposition can be largely avoided (thermodynamically) in the steam reforming [reaction (11.1)] by operating in the presence of an excess of steam. However, it cannot be avoided thermodynamically in the process of dry reforming and so only kinetic control (developing a catalyst that does not favor carbon deposition) has any chance of success in dry reforming. In partial oxidation, the problem of carbon deposition is less significant as long as the conditions are chosen such that any carbon formed can immediately be burnt off by even low partial pressures of oxygen in the feed. Although "dry reforming" gives an attractively low H_2/CO ratio in the products of approximately unity, the reaction is not currently practiced commercially, at least partially because of this carbon deposition problem.[4] (See also Box 11.1.)

[3] It is generally accepted that the partial oxidation reaction (11.2) occurs in two steps, the oxidation of part of the methane to give CO_2 and H_2O in the early part of the catalyst bed, this being followed by reactions (11.1) and (11.2) that convert the rest of the methane; oxidation occurs exothermically (irreversibly) at the entrance of the catalyst bed and then the slower endothermic (and reversible) reactions occur further down the bed.

[4] Several processes such as the Midrex, Armco, and Nippon Steel Purofer processes use the reaction to produce syngas for the reduction of iron ore; these processes recycle the gases exiting the ore reduction step and so the feed contains both CO_2 and water. (http://www.kobelco.co.jp/p108/dri/e/dri04.htm/). Haldor Topsøe have operated a dry reforming pilot plant using undiluted CO_2 feed but have not commercialized the technology.

BOX 11.1 Carbon Filament Growth

It is well established that the carbon that is formed on catalysts by the carbon-forming reactions discussed here is in the form of filaments. Much research work has been carried out on the mechanism of filament formation and on the use of the filaments formed, for example, as catalysts supports and adsorbents. Reaction (11.6) has also some potential as a source of CO-free hydrogen, even though the operation of such an approach would depend on there being a parallel economical use for the carbon formed. A recent review on this subject deserves attention: "Methane decomposition to CO_x-free hydrogen and nano-carbon material on group 8–10 base metal catalysts: A review", Yongdan Li*, Douxing Li, Gaowei Wang, Catal. Today 162 (2011) 1–48.

The current author has suggested that the formation of carbon during the CO_2 reforming reaction (see later) would be an energetically favorable way sequestering CO_2. ["Natural gas reforming and CO_2 mitigation", Catal. Today 100 (2005) 151–158.]

The arrangement of a typical steam reforming plant for the production of hydrogen or syngas from methane is shown schematically in Fig. 11.2 and a view of such a plant, showing clearly the scale involved, is shown in Fig. 11.3.

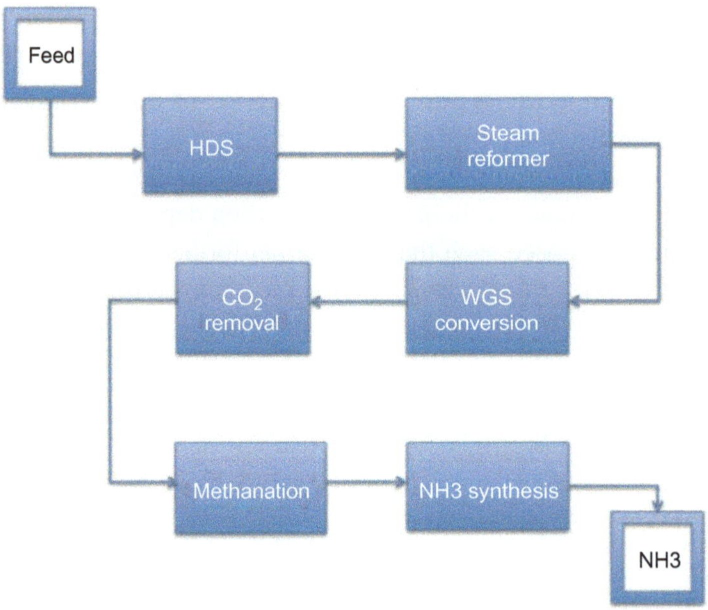

Figure 11.2
Schematic block diagram of a typical ammonia plant.

Figure 11.3
A view of a typical steam reforming unit. *Reproduced from C. Agrafiotis, H. von Storch, M. Roeb, C. Sattler, Renewable Sustainable Energy Rev. 29 (2014) 656–68, with kind permission of Elsevier.*

The following are some of the most important aspects of the steam reforming process:

- All systems must be preceded by sulfur removal using a hydrodesulfurization (HDS) unit.
- The most commonly used catalysts in the steam reforming reactor contain nickel on a refractory support, some of which have been discussed in previous chapters (Chapters 4 and 10); more expensive materials containing noble metals (Ru or Rh) can also be used.
- In order to avoid pressure drops through the catalyst bed, the active catalyst is normally supported on porous ceramic Raschig rings or even more complex shapes and the heat and mass transfer effects are such that the effectiveness factor of the catalyst is 10% or less.

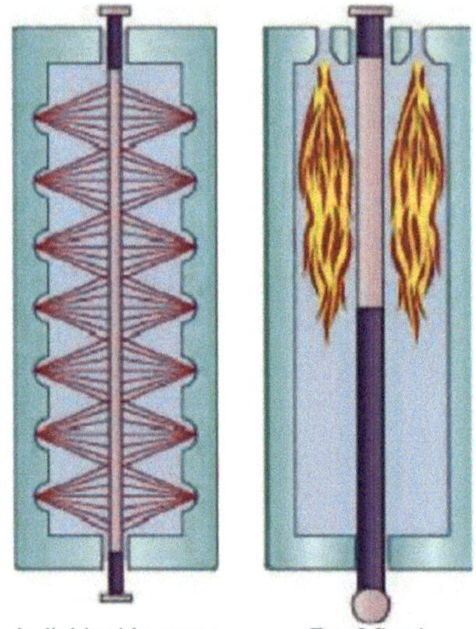

Figure 11.4
The arrangement of the heaters (fed by natural gas) in two types of steam reforming reactors. These use either individual horizontal burners that heat the tubes evenly or parallel roof-fired burners that heat the top of each tube. *Reproduced from C. Agrafiotis, H. von Storch, M. Roeb, C. Sattler, Renewable Sustainable Energy Rev. 29 (2014) 656–682, with kind permission of Elsevier.*

- The most commonly encountered reactors are composed of a series of parallel stainless steel tubes (typically 10 m or more in length and external diameter 10–15 cm) containing the catalyst; these tubes are capable of being heated to very high temperatures (typically 950°C at the bed entrance) by natural gas flames directed either horizontally or vertically past the reactor tubes (see Fig. 11.4).
- Prereformers are often included, these operating at a temperature somewhat lower than that of the main reformers and containing a catalyst that is more active than the methane reforming catalyst, this converting any other components of the natural gas into methane prior to entering the main reactor.
- The product composition is close to equilibrium as determined by thermodynamic calculations (see Fig. 11.5). The conversion does not reach completion until a temperature of 900–1000°C for reaction at atmospheric pressure and the conversion becomes lower with increasing pressure; as the plant is normally associated with either an ammonia plant or a methanol plant, the pressure will be even higher (above 25 atm.).[5]

[5] It should be noted that the thermodynamics for the dry reforming reaction are even less favorable than those for steam reforming, the temperatures required for complete reaction being some 200°C higher.

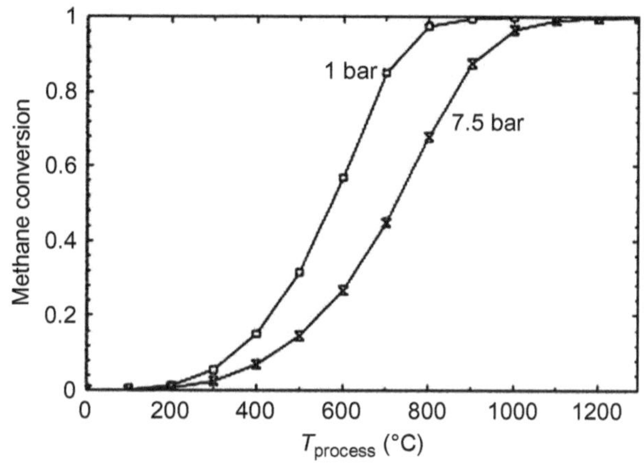

Figure 11.5
Methane conversion as a function of temperature at two pressures (1 and 7.5 bar). *Reproduced from C. Agrafiotis, H. von Storch, M. Roeb, C. Sattler, Renewable Sustainable Energy Rev. 29 (2014) 656–682, with kind permission of Elsevier.*

- Some plants operate with the addition of pure oxygen, this addition supplying some of the overall heat of the reaction by the partial oxidation reaction:

$$CH_4 + 1/2O_2 \rightarrow CO + 2H_2 \tag{11.2}$$

The H_2/CO ratio of the product gas is thus reduced. Although the efficiency of the reaction is improved by the addition of O_2, there is a significant extra cost imposed by the need for an air separation unit.

Depending on the end uses, different allover plant designs are used. For example, for hydrogen production for use in ***ammonia synthesis***, water–gas shift reactors and a tail-gas methanation reactor are included to treat the exit gas (see Fig. 11.2). The exothermic water–gas shift equilibrium is generally attained using two reactors: the first contains a so-called high-temperature shift catalyst (most frequently iron oxide promoted with about 10% chromium oxide) operated at between 310°C and 450°C; the second contains a Cu/ZnO/Al_2O_3 material (similar to that used for methanol synthesis) operated in the temperature range 200–250°C. Both reactors generally operate at high pressures (up to 80 atm.). Any residual CO (a poison for the ammonia synthesis catalyst) is removed by methanation using an Ni catalyst with high Ni content. The temperature rise in the methanation reactor will be small because only small quantities of CO are present. Hence, catalysts that are much more active than those for the steam reforming reaction are used since the possibility of sintering due to temperature increases is not a serious problem. Finally, the ammonia synthesis reaction takes place in a reactor assembly such as one of the various conformations described in Chapter 8. The catalyst used for the synthesis is generally a promoted iron-based material but Ru-containing catalysts may also have been applied (see Chapter 1).

For methanol synthesis, the allover design of the reaction system is simpler as all that is required is that the CO/H_2 ratio is adjusted to a value of 2 by the addition of the required amount of CO_2 with the application of the reverse water–gas shift reaction. In principle, the required CO/H_2 ratio could also be achieved by adding CO_2 to the CH_4/H_2O feed to the primary reactor. For the production of syngas for use in Fischer–Tropsch synthesis (see later), autothermal reformers are added, these being fed with a mixture of steam and pure oxygen.[6] (See also Task 11.1.)

Task 11.1 Tri-Reforming of Methane

Chunshan Song and Wei Pan (Catal. Today 98 (2004) 463–484) have described a "tri-reforming process" based on the combination of steam reforming, dry reforming and partial oxidation in a single reactor to give H_2/CO ratios in the range of 1.5–2.0. Find out more about this concept and also about the catalysts tested by Song and Pan. Using Scopus or some other citation tool, examine the citations to this paper to find out what other work of this sort has been carried out.

The methane reformer is the most costly part of a full-scale commercial ammonia or methanol plant due to the very high specifications required for the reactor tubes and the furnace arrangement. As indicated earlier, the heat required is provided by flame combustion of extra methane, this being equivalent to approximately 25%–30% of the total feed to the reactor. Each plant is a major producer of CO_2 and hence alternative methods of providing the reaction heat have the potential of contributing significantly to the reduction of greenhouse gas emissions. Examples of such approaches are the use of nuclear heat or solar energy (see Chapter 8) and research into these approaches continues. There are also many developments in the use of reforming reactors on different lower scales, for example, for producing hydrogen at fueling stations rather than shipping it by tanker to these stations. Fig. 11.6 shows the arrangement of the fuel processing unit of such a system. The section SREF and CC contains both a catalyst for steam reforming and one used for energy production by catalytic combustion. The product gas then passes to the water gas shift reactor section.

[6] P. Bakkerud, J. Gøl, K. Aasberg-Petersen, I. Dybkær, Preferred synthesis gas production routes for GTL, Stud. Surf. Sci. Catal. 147 (2004) 13–18. Other papers in the same volume discuss in detail other aspects of Gas to Liquids (GTL) technology.

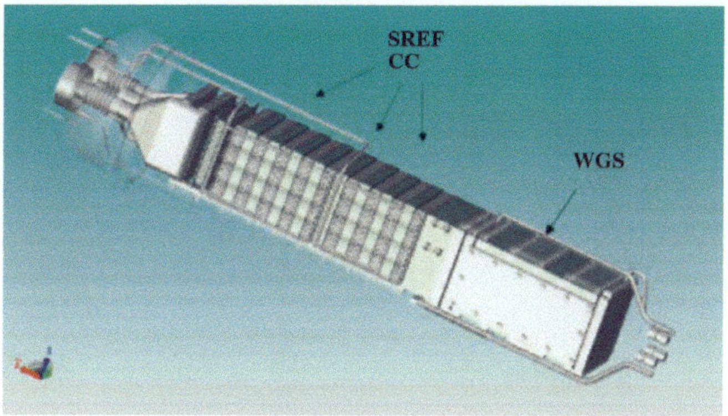

Figure 11.6
A small-scale steam reforming unit with integrated catalytic combustion for energy supply. Reproduced from P.C. Hultenberg et al., Int. J. Hydrogen Energy, 33 (2008) 1266, with kind permission of Elsevier.

Even smaller units based on microreactor technology (see Chapter 6) are also available and various types of catalyst for use in such small units have been described, particularly in the literature on solar reformers. A review by Agrafiotis et al.[7] describes some of the various confirmations that have been tested for solar reforming. Fig. 11.7, taken from this review, illustrates some of the aspects of the technology used and includes photographs of various catalyst conformations that have been tested, while Fig. 11.8 shows two different solar tower arrangements.

The reaction of choice in many studies of solar reformers has been dry reforming [Eq. (11.3)] rather than steam reforming [Eq. (11.1)]. As indicated earlier (and discussed further later), the dry reforming reaction generally requires significantly higher temperatures than does steam reforming. Further, the problem of carbon deposition becomes much more significant under dry reforming conditions. For this reason, much work has concentrated on the use of noble metals in place of Ni as catalytically active metal of choice. Even though the reaction operates under conditions in which C-formation is thermodynamically favored, the deposition reactions over the noble metals is kinetically less likely (although not totally excluded). The metal that has received most attention in work on solar applications is Rh for which carbon deposition appears to cause minimal problems. However, there are still problems with catalyst sintering under operation conditions. One way of avoiding this problem is to operate at lower temperatures and finding a way in which the thermodynamic limitation is less important. Fig. 11.9 shows one such concept in which a parabolic trough solar reflector operating at about 600°C, using circulation of a molten salt as heat transfer

[7] C. Agrafiotis, H. van Storch, M. Roeb, C. Sattler, Solar thermal reforming of methane feedstocks for hydrogen and syngas production—A review, Renewable and sustainable energy reviews, 29 (2014) 656–682.

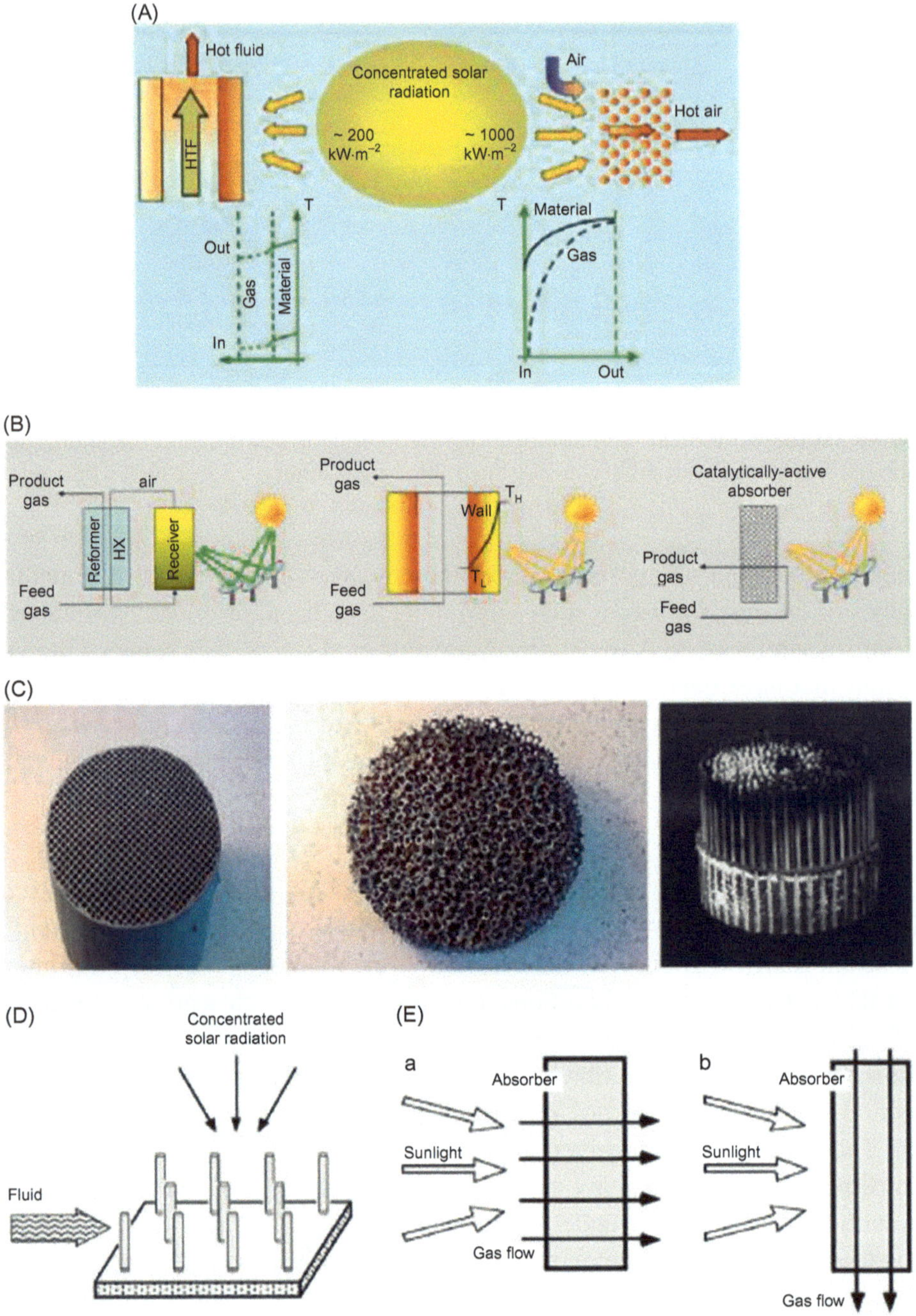

Figure 11.7

(A) Operational arrangements for indirectly heated tubular (left) and directly heated (right) solar receivers; (B) operation of various types of solar receiver; (C) different types of catalytic elements: monolithic ceramic, honeycomb, and pin-finned structures; (D) different conformations of the catalytic reactor tubes. *Reproduced from Agrafiotis et al., see footnote 7; with kind permission of Elsevier.*

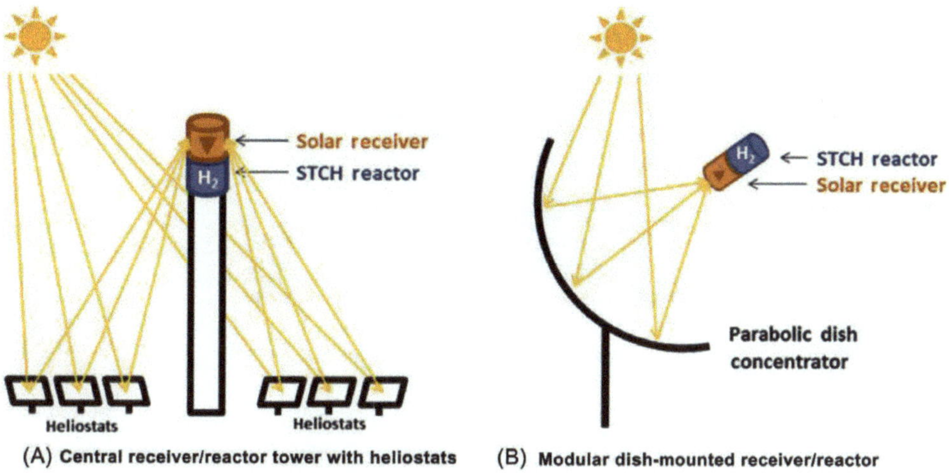

Figure 11.8
Two concepts for the use of solar power to drive methane steam reforming: (A) a series of heliostats focus sunlight onto a central reactor tower; (B) a parabolic dish concentrator is used to focus light onto a reactor module. *From www.energy.gov.*

medium, is used in combination with a reactor in which hydrogen is diffused through a membrane, thus shifting the dry reforming equilibrium toward higher conversions.[8]

Yet another application of steam reforming of methane is to be found in the area of fuel-cell technology with the molten carbonate fuel cell (MCFC). The MCFC operates at a temperature of about 650°C, the electrolyte providing electrical conductivity between the cathode (typically Ni/NiO) and the anode (typically Ni/Cr) being a molten carbonate mixture impregnated on a porous $LiAlO_2$ matrix (see Fig. 11.10). Each anode/electrolyte/cathode layer is separated from the next in a stacking arrangement by a corrugated separator plate. Instead of feeding the fuel cell with hydrogen, it is possible to use methane that is then steam reformed on suitable catalyst particles placed within the channels of the separator plate on the anode side of the assembly. The steam/methane mixture is then admitted to the catalyst compartment and $O_2 + CO_2$ is admitted to the cathode side. One of the problems associated with the operation of this type of fuel cell is that there is diffusion of alkali metal ions from the electrolyte layer to the catalyst where they then poison the steam reforming activity. Berger et al.[9] have made a detailed study of this phenomenon, showing that surface diffusion of alkali metal ions is involved in the migration of the

8 S.A.M. Said, M. Waseeuddin, D.S.A. Simakov, A review on solar reforming systems, Renewable Sustainable Energy Rev 59 (2016) 149–159.

9 R.J. Berger, E.B.M. Doesburg, J.G. van Ommen, J.R.H. Ross, Nickel catalysts for internal reforming in molten carbonate fuel cells, Appl. Catal., A 143 (1996) 343–365.

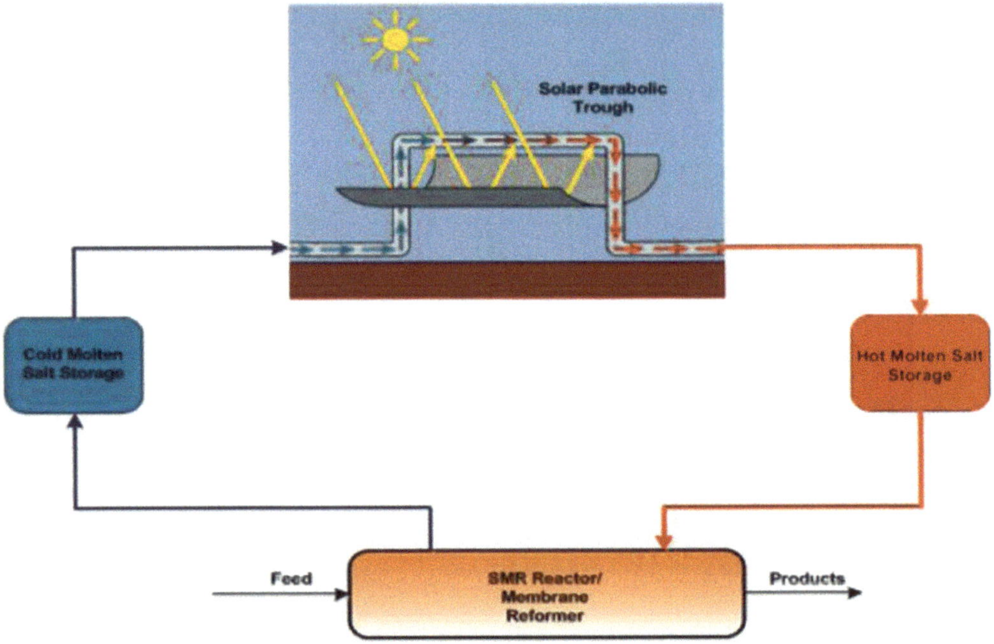

Figure 11.9
Solar parabolic troughs combined with membrane reactors that operate at about 600°C. *Reference in footnote 8; reproduced with kind permission of Elsevier.*

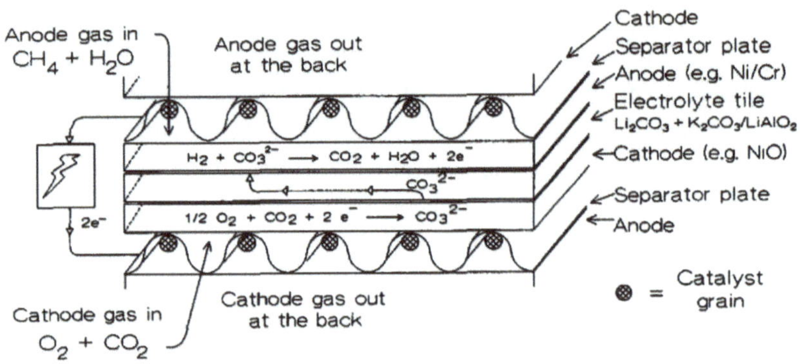

Figure 11.10
A cross-sectional view of one of the repeating units of an internal reforming molten carbonate fuel cell that operated at about 650°C. *Reference in footnote 9; reproduced with kind permission of Elsevier.*

poisoning metal species toward the catalyst. Deactivation of the catalyst in minimized if the outer surface of the catalyst particle is capable of adsorbing most of the alkali ions, leaving adequate active material to carry out the reforming reaction. It was found that either a coprecipitated Ni/alumina catalyst containing relatively low amounts of Ni (or an equivalent material an equivalent material prepared by deposition precipitation) and therefore containing sufficient alumina on which the adsorption of alkali ions could occur was capable of resisting poisoning. (See also Boxes 11.2 and 11.3; also Task 11.2.)

BOX 11.2 A Hydrogen Economy?

There are many uses for hydrogen other than in ammonia production, these including its participation in a variety of processes in petroleum refining. (See Chapter 10: The Production of Energy-Carriers From Oil.) One area in which there are currently many new developments, largely with a view to reducing greenhouse emissions, is that of the use of hydrogen as fuel for traction and other purposes, both in more conventional combustion systems and in fuel-cell powered vehicles. There have been many publications as well as related conferences and workshops on these topics. The widespread use of hydrogen as a fuel requires a well-established infrastructure and suitable storage facilities both for fueling stations and for the vehicles. This has led to very significant research and development efforts devoted to these subjects. The hydrogen can either be produced by steam reforming combined with on-site sequestering of all the CO_2 formed or by electrolysis with electricity produced by a sustainable renewable route (wind, solar, biomass, etc.). A recent (2017) report by an industrial grouping composed of producers and consumers of hydrogen energy, the Hydrogen Council, entitled "Hydrogen - Scaling Up" and sub-titled "A sustainable pathway for the global energy transition" provides a much useful background material to activity on this subject; see www.hydrogencouncil.com. This report provides particularly valuable information on the uses of hydrogen in different areas of the economy. The following diagram is an example of a part of the content and shows the vision of the Energy Council in relation to the recycling of captured carbon to produce various chemicals, in this specific case methanol.

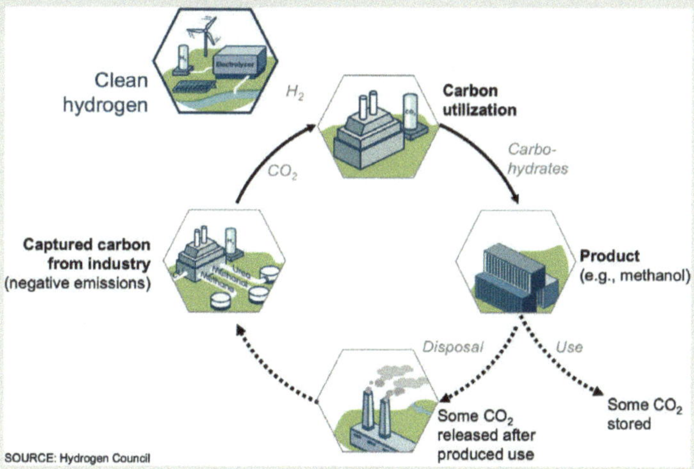

Task 11.2 Hydrogen as a Fuel for Vehicles

Read the report from the Hydrogen Council referred to in Box 11.2 and also perform a search on hydrogen-powered vehicles and the current status of the introduction of hydrogen fuel stations. What are the potential problems associated with the widespread usage of hydrogen as a fuel? Are battery-powered vehicles more likely to be used in the near future?

BOX 11.3 Some Additional Material on CO_2 Reforming Catalysts and Processes

As indicated in the main text, the dry reforming of methane has been a subject of very considerable interest both scientifically and technologically for many years. Some of the earliest work was reported in meetings of the Natural Gas Conversion Symposia, events that have been held approximately every two years since a meeting held in Oslo in 1990; this meeting was the successor to a series of international meetings on C1 reactions. The first four papers on Dry Reforming in these symposia were given at the third symposium 1993, these being followed by 8 and 14 papers in the 1995 and 1998 symposia, respectively. [The proceedings of these—and most of the subsequent meetings—have been published in the Elsevier book series, Studies in Surface Science and Catalysis (Volumes 61, 107, 119, respectively)].

There have also been a number of well-cited reviews on the subject, some of which are listed below with an indication of their contents.

Authors	Contents of Review	Source
S.B. Wang G.Q.M. Lu G.J. Millar	Identifies areas for future research: effect of metal-support interaction on activity; effects of promoters; reaction kinetics and mechanism; pilot reactor performance and scale-up. Ni on some supports gives carbon-free operation.	Energy Fuels 10 (1996) 896–904
Y.H. Hu E. Ruckenstein	Reviews catalysts studied, with emphasis on inhibition of C deposition. (Also reviews partial oxidation.)	Adv. Catal. 48 (2004) 297–345
J.H. Edwards A.M. Maitra	Rh and Ni catalysts most promising; MgO, Al_2O_3 or combinations good supports.	Stud. Surf. Sci. Catal. 81 (1994) 291–296
J.R.H. Ross A.N. van Keulen M.E.S. Hegarty K. Seshan	Review on the development of Pt/ZrO_2 catalysts	Catal. Today 30 (1996) 193–199
J.R.H. Ross	Review on dry reforming and CO_2 mitigation	Catal. Today 100 (2005) 151–158

(Continued)

BOX 11.3 Some Additional Material on CO_2 Reforming Catalysts and Processes (Continued)

Authors	Contents of Review	Source
X.E. Verykios	Ni/La_2O_3 catalysts very stable; $La_2O_2CO_3$ species critical to good operation	Int. J. Hydrogen Energy 28 (2003) 1045–1063
J.N. Armor	Review largely on steam reforming but also including CO_2 reforming, particularly limitations due to C deposition and also costs of supplying pure CO_2.	Appl. Catal., A 176 (1999) 159–176
J. Ma, N. Sun, X. Zhang	Review of CO2 conversion including CH4 + CO2 over Ni supported on a wide variety of supports.	Catal. Today, 148 (2009) 221–231

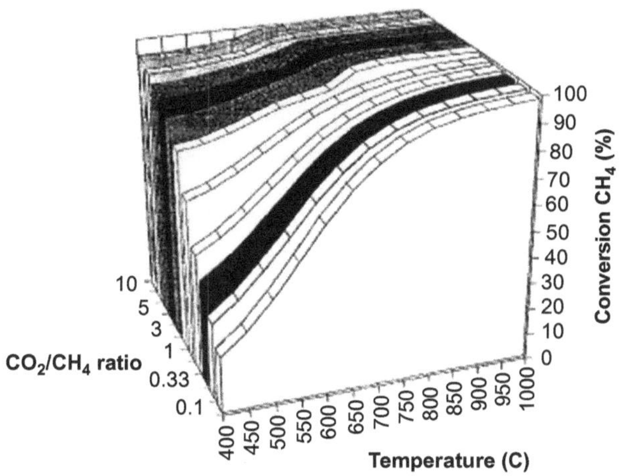

Figure 11.11

Thermodynamically calculated conversions of methane as a function of temperature for a series of different feed ratios. Calculations carried out using HSC chemistry (version 1.10, Outokumpu Research, Finland). *Reproduced from Catal. Today 100 (2005) 151–158, with kind permission of Elsevier.*

Dry Reforming. As already discussed, CO_2 reforming of methane ("*Dry Reforming*", reaction 11.3) offers an alternative route to the formation of syngas to give mixtures with CO/H_2 ratios of approximately 1.0. The reaction requires higher temperatures than does steam reforming to give full conversion; Fig. 11.11 shows the effect of temperature and CH_4/CO_2 ratio on the conversions to be expected thermodynamically. Further, the reaction occurs well within the region of operation in which carbon deposition is thermodynamically favored (see Fig. 11.12), especially if the water–gas shift reaction is fully in equilibrium,

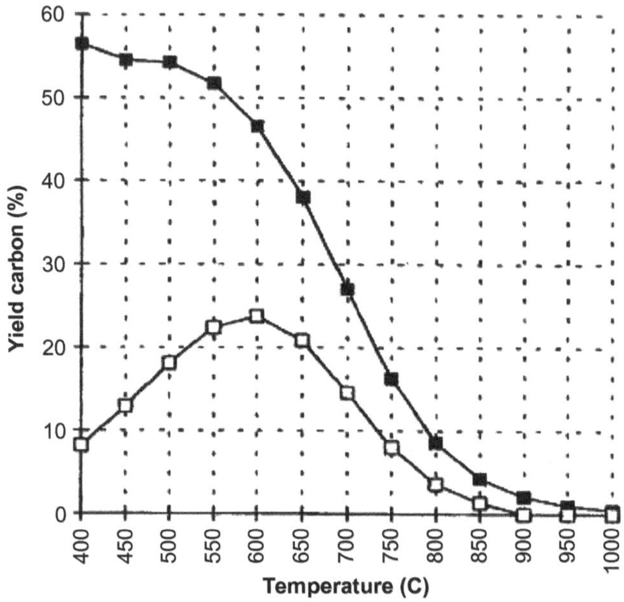

Figure 11.12
Thermodynamically calculated proportions of carbon formed under CO_2 reforming conditions with $CH_4/CO_2 = 1.0$; (■), with the reverse water–gas shift reaction in equilibrium; (□), without the reverse water–gas shift reaction. *Reproduced from Catal. Today 100 (2005) 151–158, with kind permission of Elsevier.*

and so any catalyst to be used must resist C deposition by kinetic means; carbon deposition becomes less pronounced at high temperatures but is still very significant. Despite these problems, CO_2 reforming of methane has been a "hot subject" for some years and continues to be so: the reaction and its catalysts have attracted a great deal of academic interest, despite the lack of industrial application. This is partly because of the misconception that the reaction can be used to remove unwanted CO_2. Even if catalysts could be found that would resist carbon deposition over the long periods necessary for industrial operation of the process, the potential worldwide use of any syngas formed would be extremely small in comparison with the vast amounts of CO_2 formed annually from even a small number of power plants. Nevertheless, the search for effective catalysts has led to some very useful academic research and a great deal of work has been carried out in an attempt to find how to make stable nickel-containing materials with very high dispersions (small Ni particles). However, many of these materials-oriented studies have been carried out in combination with catalytic tests carried out under relatively mild conditions: while nickel catalysts can generally be used for steam reforming applications, they generally do not resist carbon deposition for more than a relatively short period of operation under realistic CO_2 reforming conditions ($CH_4/CO_2 \approx 1.0$ and very high temperatures).[10] (See also Task 11.3.)

Task 11.3 Catalysts for Dry Reforming of Methane

Some aspects of the CO_2 reforming of methane are discussed in reviews entitled "Natural gas reforming and CO_2 mitigation" by J.R.H. Ross [Catal. Today 100 (2005) 151–158] and "A review of dry (CO_2) reforming of methane over noble metal catalysts" by J.J. Spivey [Chem. Soc. Rev. 43 (2014) 7813–7837]. Using the references in these articles together with citing articles found using Scopus or another citation indexing tool, do an up-to-date survey of the catalysts which have been examined for use for dry reforming. You can also, if you so wish, start from some of the references given in Box 11.3.

In contrast to the work on nickel catalysts, it has been reported that a variety of noble metal catalysts work reasonably well for extended periods. This is probably due to the fact that carbon can only form on the exterior of the noble metal crystallites, this being in contrast to nickel that dissolves carbon in the bulk. There are some discrepancies in the reported orders of activities. For example, using zirconia as a support, Van Keulen et al.[11] reported that the relative activities were in the order:

$$Rh > Pt > Ir, Pd > Ru$$

but the Pt sample performed better the other materials when the sample was heated in nitrogen to 800°C before admitting the reaction mixture at that temperature. Other supports have been reported to give slightly different orders of activities. For example, for MgO-supported materials, Rostrup-Nielsen[12] and Bak Hansen and Qin and Lapszewicz[13] reported the order:

$$Ru, Rh > Ir > Pt, Pd$$

Van Keulen and his colleagues tested the Pt/ZrO_2 sample at various temperatures up to 800°C and found that there was little evidence of catalyst ageing at that temperature.

[10] The present author has reviewed some of the many recent papers on the dry reforming of methane in "Transformation and Utilization of Carbon Dioxide", B.M. Bhanage and M. Arai (Eds.), Springer (2014): J. R.H. Ross, "Syngas Production Using Carbon Dioxide Reforming: Fundamentals and Perspectives", pp 131–161. In this article, it is argued that much research on this type of catalyst seems to ignore the sort of conditions under which a dry reforming catalyst would have to work, particularly the high temperatures required for an adequate conversion of methane.

[11] A.N.J. van Keulen, M.E.S. Hegarty, J.R.H. Ross, P.F. van der Oosterkamp, Stud. Surf. Sci. Catal. 107 (1997) 537–546

[12] J.R. Rostrup-Nielsen, J.-H. Bak Hansen, J. Catal. 144 (1993) 38.

[13] D. Qin, J. Lapszewicz, Catal. Today 21 (1994) 551.

Fig. 11.13 shows the results of a prolonged test in the temperature range of 650–700°C that indicates that there was some ageing under these conditions. The breaks in the curves coincided with in most cases a bypass of the reactor while the methane supply was replenished but the increase at point 5 coincided with a period during which the catalyst was exposed to pure CO_2 for a time, this treatment presumably removing some deposited carbon from the surface. The test was then continued for a further 1000 h at different temperatures up to 800°C (results not shown) after which there had been relatively little additional change in the behavior of the catalyst when the temperature was changed once more to 650°C.

Fig. 11.14 shows a model of the situation existing on the surface of a Pt/ZrO_2 catalyst under dry reforming conditions. In this scheme, methane adsorbs on the surface of the Pt crystallites of the catalyst and dissociates there, liberating hydrogen; the carbon formed

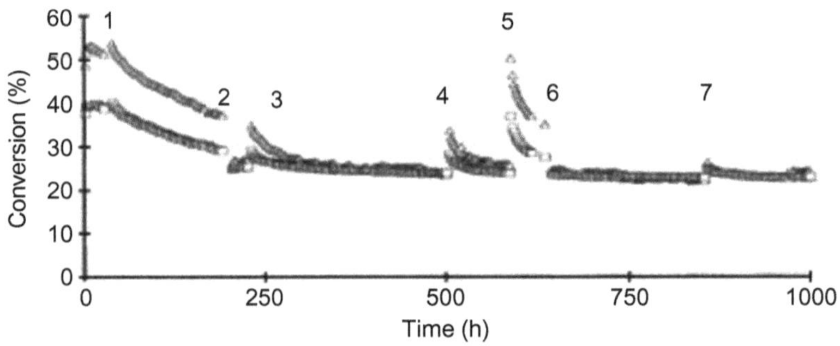

Figure 11.13
CH_4 and CO_2 conversions as a function of time on stream for a 0.5 wt% Pt/ZrO_2 catalyst. The temperature over the length of the catalyst bed was 650–700°C; GHSV = 16,000 h^{-1}, $CO_2/CH_4 = 2.0$. *Reproduced with permission of Elsevier from Studies in Surface Science and Catalysis; see footnote 11.*

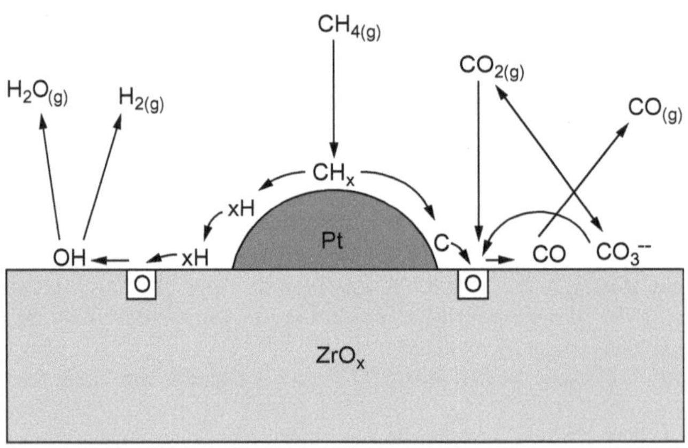

Figure 11.14
A model for the CO_2 reforming of methane over a Pt/ZrO_2 catalyst. *Reproduced from A.M. O'Connor, F.C. Meunier, J.R.H. Ross, Stud. Surf. Sci. Catal. 119 (1998) 819, with kind permission of Elsevier.*

diffuses to the metal-support interface where it picks up an oxygen entity from the support. Finally, the support is re-oxidized by the CO_2 from the reactant mixture. The Pt surface becomes and remains largely covered by carbon during the reaction but the catalyst retains its activity due to the sites at the metal-support interface remaining free of significant quantities of adsorbed carbon. (It should be noted that the reaction scheme depicted in Fig. 11.11 is a variant of the Mars–van Krevelen mechanism discussed in Chapter 7: The Kinetics and Mechanisms of Catalytic Reactions. Similar arguments concerning the importance of the metal-support interface are to be found in papers describing other metal-support systems for dry reforming and also for other similar reactions.)

11.3 Other Reactions of Natural Gas

As was shown in Fig. 11.1, there are a number of other potential conversion processes involving methane. We will now discuss two such processes, both of which are based on oxidation reactions. Thermodynamically, once oxygen is involved in a reaction of a hydrocarbon, the favored products will be CO_2 and water. Hence, in the processes that will be discussed, we are dealing with kinetic control of the reactions involved such that an intermediate is formed and not further oxidized. In the first of these processes, methane is converted by "oxidative coupling" to give a mixture of ethylene and ethane (Step 11, Fig. 11.1) while in the second partial oxidation of the methane occurs to give formaldehyde (Step 13, Fig. 11.1). As we will discuss further later, neither of these processes is as yet considered to be economically viable. Nevertheless, there was a very significant amount of research concerning these reactions in a period starting from about 1984, this continuing to a much lesser extent up until the present.

11.3.1 Methane Coupling

The oxidative coupling of methane can be represented by the following equations:

$$2\,CH_4 + \tfrac{1}{2}\,O_2 \rightarrow C_2H_4 + H_2O \tag{11.7}$$

and

$$2\,CH_4 + O_2 \rightarrow C_2H_6 + 2\,H_2O \tag{11.8}$$

It can be said rather generally that the ethylene formed in the first reaction is the primary product and that ethane is a secondary product. Further reaction of the ethylene and ethane then gives the total oxidation products, CO and CO_2, plus further water. (A number of papers on the subject also noted that some hydrogen was also formed and this led on later to work on catalysts for partial oxidation to give mixtures of CO and H_2 as discussed earlier.)

The first paper that described methane coupling was by Keller and Bhasin of Union Carbide, published in 1982.[14] They showed that methane could be coupled (or oxidatively dehydrogenated) to give a mixture of ethane and ethylene when using a range of metal oxides supported on alpha-alumina as catalysts. The work reported in this paper was carried out predominantly using a cyclic mode of operation in which successive pulses of air and then methane were fed to the catalyst bed, these being separated by short pulses of nitrogen. The catalyst bed temperatures were generally in the range of 600–1000°C. These authors reported that cobalt, manganese, and cadmium oxides all gave some C_2-forming activity. They also reported that when the reagents were cofed, the yields of C_2 products were much lower.

The work of Keller and Bhasin caused a great deal of interest and its publication was rapidly followed by two important papers by the groups of Baerns and Lunsford. Hinsen, Bytyn and Baerns showed in 1984 that it was possible to work in the cofeed mode and they showed that a PbO–alumina catalyst gave good yields of C_2 products.[15] The following year, Ito, Wang, Lin, and Lunsford showed that Li-doped MgO catalysts would also work under cofeed conditions and that the reaction involved the formation of gas-phase methyl radicals which then combined in the gas phase to give ethane and ethylene.[16] Work from the current author's laboratory then showed that the lead-containing catalysts steadily lost metallic lead during operation and that Li was lost from the Li/MgO samples as a result of the formation of volatile LiOH species.[17] This volatility, caused largely by the high temperature of operation, could be reduced by doping the catalyst with SnO_2 (with the formation of a $Li_2Mg_3SnO_6$ phase in the catalyst), this bringing about an improvement in activity and a consequent decrease in operation temperature. Further work showed that other more stable catalysts could be made, these including, for example, the rare earth oxides, particularly Sm_2O_3, a CaO/BaO mixture, and a K/Ca/Ni oxide material. A significant amount of research work was devoted to ascertaining the sequence of gas-phase radical reactions leading to the observed products when operating at high temperatures and it became clear that the conversions and selectivities obtained with most catalysts operating at high temperatures are limited by gas-phase radical reactions rather than by surface processes. Another important limitation is dictated by the explosion limits for the methane–oxygen system meaning that safe operation could only be achieved if the methane/oxygen ratio was at least 5.0. As maximum selectivities of about 80% to mixtures

---

[14] G.E. Keller, M.M. Bhasin, J. Catal. 73 (1982) 9.

[15] W. Hinsen, W. Bytyn, M. Baems, Proceedings of the 8th International Congress on Catalysis, Berlin, 3 (1984) 581.

[16] T. Ito, J.-X. Wang, C.-H. Lin, J.H. Lunsford, J. Am. Chem. Soc. 109 (1985) 5062.

[17] This work is reviewed in the paper: J.R.H. Ross, A.N.J. van Keulen, M.E.S. Hegarty, K. Seshan, The conversion of natural gas to useful products, Catal. Today 30 (1996) 193–199. More detail is to be found in: S.J. Korf, J.A. Roos, J.R.H. Ross, "Development of Doped Li/MgO Catalyst Systems for the Oxidative Coupling of Methane" in "Methane Conversion by Oxidative Processes", Van Nostrand Reinhold Catalysis Series, E.E. Wolf (Ed.), Van Nostrand (1992) 168–199.

of C_2 products are obtained and a significant proportion of the products are ethane, conversions above about 25% (total yields of about 20%) have seldom been reported. The consequence of these various observations, confirmed in a great deal of additional work, is that the process can only be operated commercially if a selective catalyst operating at much lower temperatures could be found that would limit the importance of gas-phase radical reactions or if the process could be engineered in such a way as to make use of the exothermic heat of the reaction and to use any unreacted methane for another purpose. It appears that these objectives may have been achieved by the company Siluria Technologies (www.siluria.com//) who claims to have developed a catalyst operating some 200°C lower than those reported in the earlier literature on methane coupling. Siluria also claims to have improved the engineering aspects of the process so that the price of ethylene formed is competitive with that of ethylene from hydrocarbon cracking. (See also Task 11.4.)

Task 11.4 Catalysts for Methane Coupling

Use some of the references in footnotes 14–17 as starting points for a citation search on methane coupling. Tabulate what information you can find on the yields obtained with the best catalysts reported. Using Google, examine the Siluria patents, US 9,040,762, US 9,446,387, and US 9,469,577. You will discover that these patents describe a very large range of possible compositions so do not attempt to get more than a general view of the type of material that Siluria has developed. Some of the background material given in the patents may be of some value to aid your understanding of the subject of methane coupling.

11.3.2 Methane to Other Products

While it would be highly desirable to be able to convert methane into methanol by a catalytic process, it appears that the reaction:

$$CH_4 + \frac{1}{2} O_2 \rightarrow CH_3OH \qquad (11.9)$$

does not give any significant yields with a catalyst even though it will occur to a limited extent as a purely homogeneous reaction. The homogeneous reaction requires long residence times, temperatures of the order of 400°C and high pressures; it involves a variety of gas-phase radical species.[18]

In contrast, it has been shown that the reaction to produce formaldehyde does occur with measurable but low yields when N_2O is used as oxidant:

$$CH_4 + N_2O \rightarrow CH_2O + N_2 \qquad (11.10)$$

Work has also been carried out using air as the oxidant but the yields appear to be much lower than N_2O.[19] (See also Task 11.5.)

[18] H.D. Gesser, N.D. Hunter, A review of C-1 conversion chemistry, Catal. Today 42 (1998) 183–189.

[19] M. Kennedy, A. Sexton, B. Kartheuser, E. Mac Giolla Coda, J.B. McMonagle, B.K. Hodnett, Selective oxidation of methane to formaldehyde: Comparison of the role of promoters on the hydrocarbon rich and lean conditions, Catal. Today 13 (1992) 447–454.

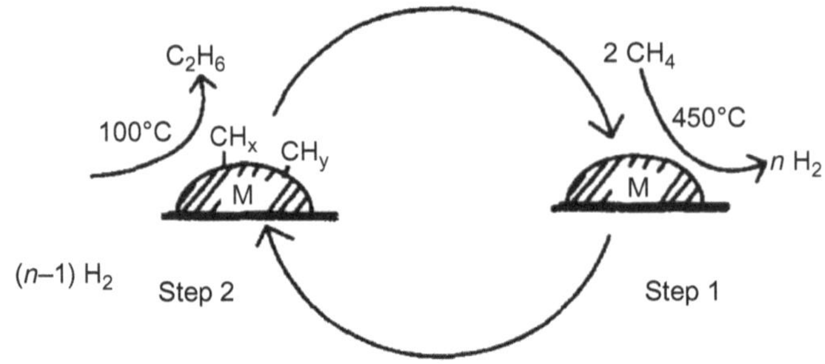

Figure 11.15
Schematic representation of the two-step reaction from methane to higher hydrocarbons.
Reproduced from reference of footnote 20, with kind permission of Elsevier.

Task 11.5 Conversion of Methane to Methanol and Formaldehyde

Use the references given in footnotes 18, 19, and 20 to find more recent papers on the conversion of methane to oxygen-containing products and higher hydrocarbons. What are the catalysts and conditions that have been used? Has there been any attempt to commercialize any such process?

Van Santen and his colleagues have reported the use of a two-step process for the production of hydrocarbons from methane.[20] In this procedure, the methane is first reacted on a Group VIII metal (e.g., Rh) at temperatures between about 200°C and 500°C to produce a surface carbide species; this then reacts with hydrogen at a lower temperature of between 30°C and 130°C to give ethane and several other lower hydrocarbons (C_3-C_5). The temperature gap is an essential feature for this procedure and "cheats" the thermodynamic limitations for the direct coupling of two methane molecules to give ethane. The reaction occurs in two distinct steps [Eqs. (11.11) and (11.12), the latter being given for ethane; similar ones would apply for higher hydrocarbons] and the combined reaction, Eq. (11.13), has a value of ΔG at 1000K that is highly unfavorable, namely 71.0 kJ/mol. Van Santen and his co-authors point out that the surface steps giving reaction 11.12 are equivalent to those thought to be occurring in the Fischer–Tropsch reaction (see Chapter 12). The reaction scheme is shown in Fig. 11.15).

$$2CH_4 + 6M \rightarrow 2M_3C + 4H_2 \tag{11.11}$$

$$2M_3C + 3H_2 \rightarrow C_2H_6 + 6M \tag{11.12}$$

$$2CH_4 \rightarrow C_2H_6 + H_2 \tag{11.13}$$

[20] T. Koerts, M.J.A.G Deelen, R.A. van Santen, Hydrocarbon formation from methane by a low-temperature two-step sequence, J. Catal. 138 (1992) 101–114.

CHAPTER 12

Catalytic Reactions Involving Syngas, Hydrogen, or Carbon Monoxide for the Production of Intermediates and Chemicals

Chapter Outline

12.1 Introduction

Syngas or either pure hydrogen or pure CO produced from methane or other hydrocarbons by the various routes described in Chapter 11 can be used to produce some important chemicals by catalytic routes. Fig. 12.1 summarizes some of these, showing a number of primary reactions such as ammonia and methanol synthesis and some of the secondary reactions to use these important platform chemicals. Some of the primary reactions and the subsequent conversions will now be discussed.

12.2 Ammonia Production and Uses

The production of ammonia by the Haber–Bosch process has already been considered in some detail in Chapter 1 while typical reactors used in industrial plants are described in Chapter 8 and the arrangement of an ammonia plant, including the steam reformer, etc., are shown in Fig. 11.2. Similarly, the use of ammonia to produce fertilizers has been discussed in Chapter 1. Further, the topic of the use of ammonia as a reductant in stack gas purification is covered in Chapter 13. Hence, no further discussion is given here;

Contemporary Catalysis.
DOI: https://doi.org/10.1016/B978-0-444-63474-0.00012-6

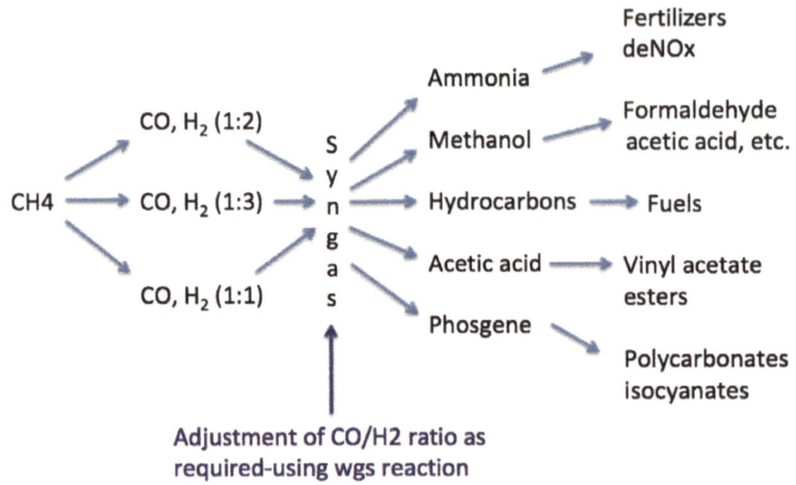

Figure 12.1
Various direct and indirect routes for the production of useful chemicals from natural gas showing some of the routes already discussed and the various products that can be obtained from syngas, hydrogen, or carbon monoxide.

however, the reader should recap on the material covered earlier, particularly that given in Chapter 1.

12.3 Synthesis and Uses of Methanol and Higher Alcohols

A brief introduction to the history of methanol synthesis was given in Chapter 1. The high-pressure reactors used for the reaction are similar to those used for ammonia synthesis as described in Chapter 8. The catalysts used for the reaction in modern plants are almost invariably Cu-based, generally with a formulation $Cu-ZnO-Al_2O_3$ that is similar to that used for the low-temperature water-gas shift reaction.

The history of the development of the commercial synthesis of methanol (Fig. 12.1) is in many ways similar to that of ammonia. Early work by Sabatier and Senderens (Ann. Chim. Phys. 4 (1905) 418), aimed at developing a catalyst for methanol synthesis, showed that methanol decomposition could be brought about by a Cu catalyst but that the reverse (synthesis) reaction was less effectively catalyzed. Thereafter, a patent by Patard (French Patent 540,343 (Aug. 1921)) claimed that a variety of materials, including ones containing Cu, were effective catalysts for methanol synthesis. However, despite this, the high-pressure process that was subsequently developed by BASF in 1923 (see Chapter 1: An Introduction to Heterogeneous Catalysis—Chemistry in Two Dimensions) made use of a "zinc chromite" catalyst. This process, which operated at a pressure of 200 atmospheres pressure and 350°C, was only superseded in the 1960s by the low-pressure process pioneered by ICI (now Johnson Matthey). For a more detailed review of this work, see the extensive review by Kamil Klier entitled "Methanol Synthesis" that was published in *Advances in Catalysis* (31 (1982) 243–313). (See also Task 12.1, Task 12.2.)

Task 12.1 The Active Site in Methanol Synthesis Catalysts

Use the paper by Klier noted earlier (Adv. Catal. (31 (1982) 243–313) as a source for a search using citation methods to find recent papers discussing the effect of CO_2 on methanol synthesis and also to learn more on the most up to date views on methanol synthesis catalysts. You should also examine the paper by G.C. Chinchen, K.C. Waugh, D.A. Whan, The activity and state of the copper surface in methanol synthesis catalysts, Appl. Catal. 25 (1986) 101–107, and more recent citations to that paper in relation to the nature of the active sites on Cu catalysts. These citations to this will in turn lead to work by A.T. Bell and his group on Cu/ZrO_2 catalysts which address, among other things, the importance of hydrogen spillover from Cu to zirconia during methanol synthesis.

BOX 12.1 Mega-Methanol Production

Lurgi has introduced a new large-scale technology for the production of methanol from natural gas that utilizes autoreforming to produce syngas and then a new highly active Cu–ZnO catalyst in the methanol synthesis step. The arrangement of the plant is shown schematically in the following figure. (J. Haid, U. Koss, "Lurgi's mega-methanol technology opens the door for a new era in downstream applications," in Natural Gas Conversion VI, in: E. Iglesia, J.J. Spivey, T.H. Fleisch (Eds.), Studies in Surface Science and Catalysis, Elsevier, 136 (2001) 399–404.)

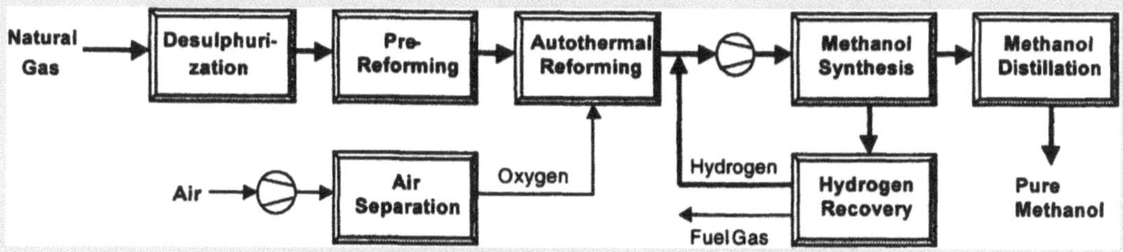

(*Continued*)

BOX 12.1 Mega-Methanol Production (Continued)

The methanol produced by the Lurgi Mega-Methanol plant is claimed to be significantly cheaper than that produced by plants using methane steam reforming as shown by the figures in the following table.

		Steam Reforming	Mega-Methanol Concept
Capacity	mtpd	2500	5000
Natural gas demand based on LHV	MMBtu/mt	30	28.5
Total fixed cost (EPC)	Million US $	225	300
Production cost	US$/mt	113	79

Basis

- Figures include O_2-production, power generation, and off-sites
- Natural gas price is set to US$ 0.5/MMBtu
- Depreciation is set to 10% of total fixed cost
- Return of investment is set to 20% of total fixed cost
- Operating cost for operator staff, plant overhead, maintenance labor, and material included

Figure and table reproduced by kind permission of Elsevier from the reference given earlier.

Task 12.2 Developments in Methanol Synthesis

Using the reference cited in Box 12.1, carry out a citation search on methanol synthesis and learn what you can about new developments in this area. Include in your study a web search for companies using Mega-Methanol Technology.

The catalysts developed by ICI were based on the $Cu/ZnO/Al_2O_3$ system. These catalysts bring about the effective conversion of syngas (also containing a proportion of CO_2) at 50–100 atm and 220–250°C:

$$CO + 2\,H_2 \rightarrow CH_3OH \qquad \Delta H^{o} = -\,100.5\ kJ \tag{12.1}$$

and

$$CO_2 + 3\,H_2 \rightarrow CH_3OH + H_2O \qquad \Delta H^{o} = -\,61.6\ kJ \tag{12.2}$$

Both these reactions occur reversibly but, as discussed in Chapter 8, acceptable yields can be obtained using staged reactors with interbed cooling or similar concepts. There has been much discussion in the literature over the nature of the active catalyst and its surface under reaction conditions and also concerning the importance of the catalyst composition. It is generally accepted that the CO_2 is necessary for good activities (the conversion of pure CO is lower than that of mixtures of CO and CO_2) and it has been suggested that the CO_2 causes the surface of the catalyst to be partially oxidized under operation conditions. More recent work has shown that the conclusion that surface oxide species are present under reaction conditions was based on misleading evidence and that it is most likely that the active surface is metallic copper. Nevertheless, the CO_2 addition is necessary and it is possible that this is because its presence is related to the participation of formate species in the reaction mechanism, these species being more easily formed from the CO_2 than from CO.

Task 12.3 Synthesis of Higher Alcohol Mixtures

Under certain circumstances and with specially formulated catalysts (e.g., promoted Cu–ZnO–alumina materials), instead of methanol, syngas can give a mixture of alcohols by what is effectively a variant of Fischer–Tropsch (FT) synthesis: a chain-growth mechanism. There was an upsurge of interest in this process in the 1980s as a result of a search for new additives for petroleum that would increase the octane and also the oxygen content. A review of the literature until 1987 by X. Xiaoding, E.B.M. Doesburg, and J.J.F. Scholten (Synthesis of higher alcohols from syngas—recently patented catalysts and tentative ideas on the mechanism, Catal. Today 2 (1987) 125–170) gives an excellent lead into the subject. You should examine that publication and also the papers citing it to obtain further knowledge of the subject.

Methanol is an important chemical feedstock with an estimated yearly worldwide demand for 2010 of 45 million metric tons. A very valuable source of information on methanol use is the website of the Methanol Institute (http://www.methanol.org/).

Two important uses of methanol are in the production of formaldehyde (as discussed in connection with unsupported metal catalysts, in this case silver, in Chapter 3: How Does a Catalyst Work?) and of acetic acid by the Monsanto process (a homogeneous Rh catalyst) or the Cativa process (a homogeneous Ir catalyst); as the latter processes are based on homogeneous catalysis, they will not be discussed further here but the interested reader should see the text associated with Task 12.5 below. Another use that for some time commanded much interest academically was the production of methyl tertiary butyl ether (MTBE); however, this activity has recently ceased since it was recognized in the early 2000s that MTBE can cause severe environmental problems. (See also Box 12.2.)

A rapidly developing use of methanol is in the so-called methanol to olefins process (MTO), the conversion of methanol to higher hydrocarbons (Fig. 8.1, reaction 9) over Mobil's ZSM-5 zeolite.[1] The process was originally developed following the energy crisis caused by the Arab oil embargo; the only plant until relatively recently was built in New Zealand and this ran for some time after its start-up in 1985. A full description of this plant

[1] The first publication on the subject dates from 1977: C.D. Chang, A.J. Silvestri, J. Catal. 47 (1977) 249.

BOX 12.2 MTBE and Higher Alcohol Mixtures

It is worth noting that interest in the potential use of higher alcohols as oxygen additives dwindled in the late 1980s because of the introduction of MTBE. This molecule was used as an oxygenated additive for petroleum (aiding in the reduction of CO emissions) for some time until it was recognized that its use presented serious environmental problems and it has now been phased out. The production of MTBE, using methanol as a reactant, is also a catalytic process carried out using an acid catalyst in a "reactive distillation" system. The literature on this subject can be easily accessed by doing a Google search on the subject of MTBE.

and its operation are given in a paper entitled "MTG Revisited" by C.D. Chang (Natural gas conversion, in: A. Holmen, K.-J. Jens, S. Kolboe (Eds.), Studies in Surface Science and Catalysis, Elsevier, 61 (1991) 393–404) who also gives his views on the mechanism of the process. In essence, the reaction occurs in two steps: the conversion of a mixture of methanol and dimethyl ether to olefins (with a predominance of propylene) and the subsequent conversion of the olefins to higher hydrocarbons, including aromatics. Essential to the process is the acidity of the zeolite (brought about by the presence of acidic Si–OH^+–Al groupings on the surface) that allows the formation of surface carbene (CH2 =) species that can in turn insert into C–H bonds:

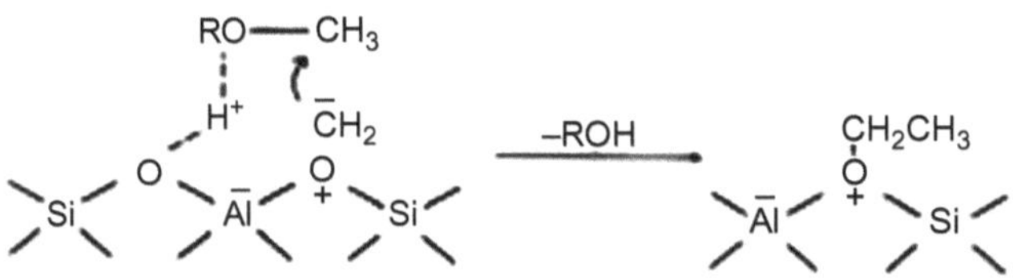

(12.3)

or undergo methylation:

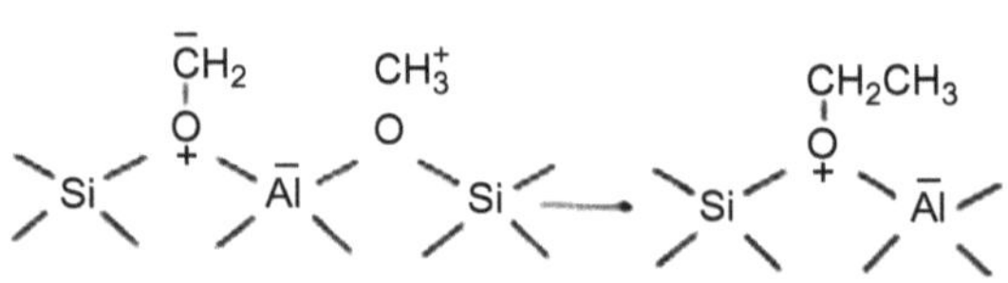

(12.4)

Once the first C–C bond is formed, the other products are formed by further insertion reactions.[2] The economics of the process were such that there has been little incentive to construct similar plants elsewhere; indeed, it is reported that the New Zealand plant is now used only to manufacture methanol. It is interesting to note, however, that a new plant has recently been started in Liangungang City, Jiansu Province, China, this producing a mixture

[2] For further details on the mechanism, the reader should refer to the paper by Chang referred to above and also some of the papers cited in that reference. Essential to the mechanism is also the open pore structure of ZSM-5.

of olefins: propylene for the manufacture of acrylonitrile, ethylene for ethylene/vinyl acetate copolymers, and C_4 olefins for butadiene. A variant of the MTO process, this producing predominantly propylene (methanol to propylene), is now offered by Lurgi.[3] In this Lurgi process, crude methanol is partially converted to dimethyl ether (DME) in a "DME prereactor" and the mixture is then converted over a "selective tailor-made zeolite" in the presence of steam at 400–450°C. Due to the increasingly high cost of propylene, an important chemical intermediate, the process economics are considered to be favorable if cheap methanol is available, for example, from the Lurgi Mega-Methanol Process. (See also Box 12.1.)

Formic acid production from methanol. A molecule that is rapidly gaining much attention in the research community is formic acid. As was discussed in Chapter 1, formic acid decomposition was long used as a model reaction to distinguish catalytic materials favoring dehydrogenation (Eq. 12.5) from those favoring dehydration (Eq. 12.6):

$$HCOOH \rightarrow CO_2 + H_2 \qquad \text{Dehydrogenation} \tag{12.5}$$

$$HCOOH \rightarrow CO + H_2O \qquad \text{Dehydrogenation} \tag{12.6}$$

However, it is now well recognized that the formic acid molecule is one of the most effective potential hydrogen carriers and storage materials: if a molecule of formic acid can decompose readily to give only CO_2 and hydrogen according to Eq. (12.5), the formic acid contains 4.3 wt% of usable hydrogen, one of the highest values compared with other hydrogen storage materials. Formic acid has other significant advantages, including the important property that it is environmentally acceptable as it is readily biodegradable. Although formic acid can be produced directly from biomass (and indeed, was for some time considered to be an unwanted by-product of biomass hydrolysis; see Chapter 15: Catalysis in Biomass Conversion), it is generally produced commercially from methanol by the BASF process. In this, CO is reacted with methanol in the presence of a strong base (Na methoxide) to give methyl formate with a selectivity of 96%:

$$CH_3OH + CO \xrightarrow{\text{base}} HCOOCH_3 \tag{12.7}$$

This is followed by a second step involving hydrolysis to produce formic acid:

$$HCOOCH_3 + H_2O \rightarrow HCOOH + CH_3OH \tag{12.8}$$

The methanol obtained in the second step is recycled and the methanol thus acts as a catalyst for the allover reaction (12.9):

$$CO + H_2O \rightarrow HCOOH \tag{12.9}$$

[3] See reference in Box 12.1.

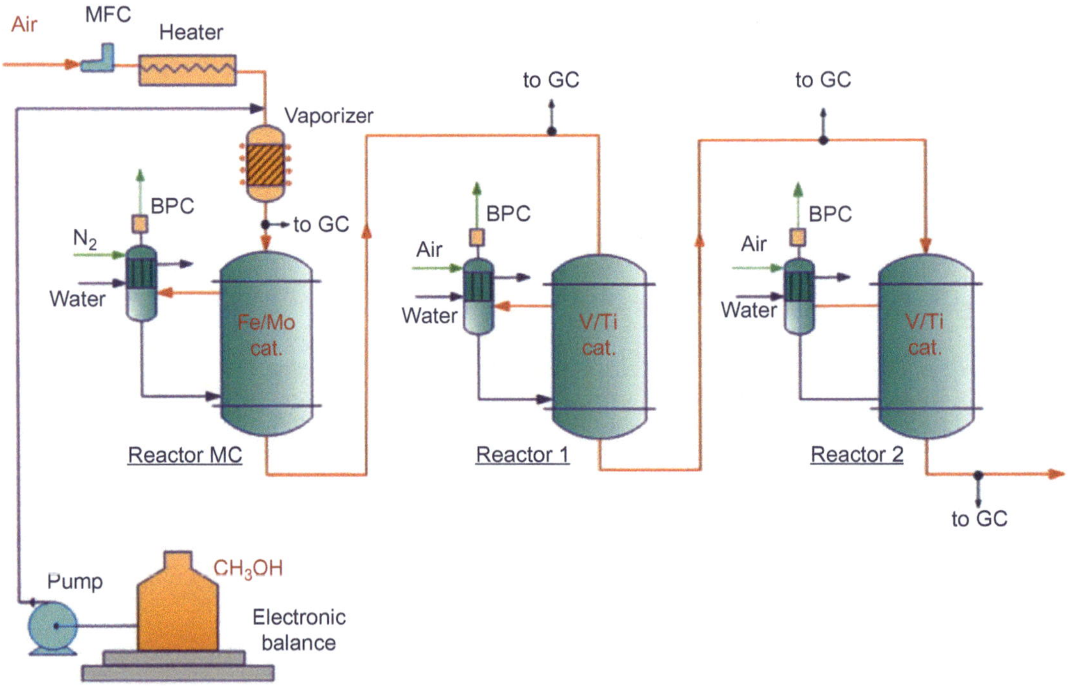

Figure 12.2
The Boreskov Institute of Catalysis process for the preparation of formic acid. *Reproduced with kind permission of Elsevier from reference given in footnote 4.*

It should be noted that neither reaction (12.9) nor the equivalent reaction of hydrogen with CO_2 to produce formic acid (12.10):

$$CO_2 + H_2 \rightarrow HCOOH \qquad (12.10)$$

is thermodynamically feasible as a single step and so the only way in which either can be achieved is to separate it into two separate thermodynamically feasible steps as in the BASF process described earlier: forming a thermodynamically permissible product in the first step and decomposing it (or converting it into formic acid plus another product) in the second. In line with this general approach, another process for formic acid preparation has been pioneered by the Boreskov Institute of Catalysis in Novosibirsk, Russia; see Fig. 12.2.[4] In this process, which depends on a ready supply of methanol, methanol is first oxidized to formaldehyde over an iron–molybdenum catalyst (Eq. 12.11):

$$CH_3OH + \frac{1}{2}\,O_2 \rightarrow HCHO + H_2O \qquad (12.11)$$

[4] E.V. Danilevich, G.Y. Popova, T.V. Andrushkevich, V.V. Kaichev, I.G. Danilova, Y.A. Chesalov, V.A. Rogov, V.I. Bukhtiyarov, V.N. Parmon, Appl. Catal., A 475 (2014) 98–108.

The formaldehyde is then oxidized to formic acid using a vanadia/titania monolayer catalyst (Eq. 12.12):

$$HCHO + \frac{1}{2} O_2 \rightarrow HCOOH \tag{12.12}$$

The yield reported for the first step is 94% and that in the second step is 87%–88%. If the methanol is formed from syngas, the allover reaction is again Eq. (12.9). (If the methanol were to be formed from hydrogen and CO_2, the allover reaction would—at least formally—be depicted by Eq. (12.10).

BOX 12.3 Synthesis of Formic Acid from CO_2 Using Homogeneous Catalysts

A very considerable amount of work has been carried out on formic acid formation from CO_2 using homogeneous catalysts. The most favorable route appears to be one in which an amine is first formed and this is subsequently converted into formic acid. The formation of ammonium salts is also possible. A very large number of homogeneous catalysts have been studied for these reactions since the pioneering work of Inoue and his coworkers.[5] Much of the work in this area has been discussed in an important review by Beller and his colleagues.[6] Complexes of Ru or Ir appear to give the best results. However, it must be remembered that homogeneous catalysts are difficult to isolate and recycle and that for this reason the homogeneous approach is very unlikely to be used for formic acid production despite the very high turnover frequencies reported for these complexes.

[5] Y. Inoue, Y. Sasaki, H. Hashimoto, Synthesis of formats from alcohols, carbon dioxide and hydrogen catalyzed by a combination of Group VIII transition metal complexes and tertiary amines, J. Chem. Soc., Chem. Commun. (1975) 718.

[6] K. Sordakis, C. Tang, L.K. Vogt, H. Junge, P.J. Dyson, M. Beller, G. Laurenczy, Homogeneous catalysis for sustainable hydrogen storage in formic acid and alcohols, Chem. Rev. 118 (2018) 372–433.

Use of methanol as a feed for fuel cell operation. Another use of methanol of current interest but with as yet relatively limited application is the methanol-fed fuel cell. As there has been a very intense period of research on that subject, we will now consider this topic briefly.

Fuel cells have high efficiencies for the production of electricity and are being developed for use in a variety of situations. One such application is in automobile power trains in which their use would have both environmental and energy saving advantages. The most effective fuel for use in such applications would be pure hydrogen but its use is still not effective because of the difficulties in distribution and storage, the latter particularly in relation to the storage in the automobile itself because of the need for either high pressures or for a suitable hydrogen storage material with sufficient capacity for storing the hydrogen. Methanol is considered to be a promising molecule that would be used as a source of pure

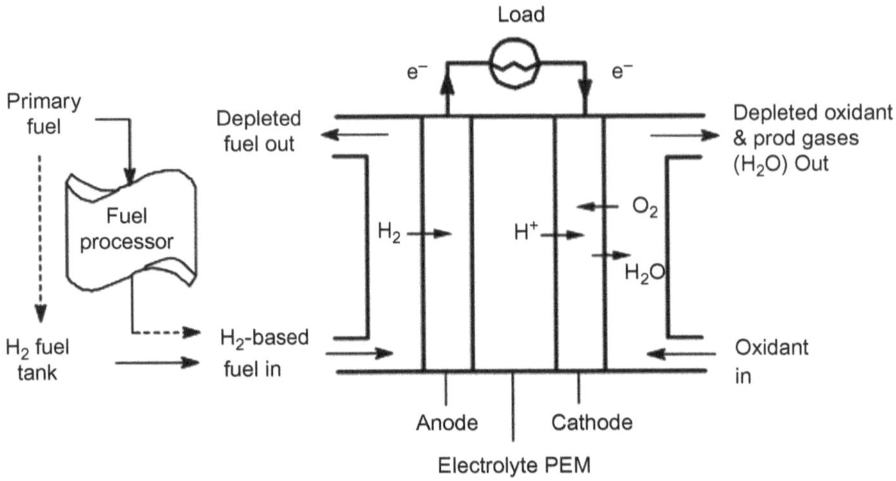

Anode (fuel) reaction:	$H_2 = 2\,H^+ + 2\,e^-$
Cathode (oxidant) reaction:	$\frac{1}{2}\,O_2 + 2\,H^+ + 2\,e^- = H_2O$
Total reaction:	$H_2 + \frac{1}{2}\,O_2 = H_2O$

Figure 12.3
Schematic representation of the arrangement of a PEMFC. *Reproduced from C. Song, Fuel processing for low-temperature and high-temperature fuel cells—challenges and opportunities for sustainable development in the 21st century, Catal. Today 77 (2002) 17–49 with kind permission of Elsevier.*

hydrogen or even directly in the fuel cell ("Direct Methanol Fuel Cell," DMFC).[7] The fuel cells used can be either low-temperature or high-temperature designs (see, e.g., C. Song, Fuel processing for low-temperature and high-temperature fuel cells—challenges and opportunities for sustainable development in the 21st century, Catal. Today 77 (2002) 17–49). One of the most-highly developed fuel cells for such an application is the "proton exchange membrane fuel cell" (PEMFC) illustrated schematically in Fig. 12.3. The membrane through which the protons diffuse is a fluorinated sulphonic acid polymer such as Nafion (DuPont) that is very selective for the transport of only H + ions. Hydrogen is generally produced in the fuel processor (rather than being supplied from external sources as hydrogen gas) and passes to the anode of the fuel cell where it is converted into protons. These then pass through the membrane to the cathode where they react with oxygen from the air to produce water. Fig. 12.4 shows one application of a direct methanol fuel cell (DMFC).

[7] Formic acid, discussed in the previous section, would also be a suitable hydrogen carrier. Discussion of the production of hydrogen from formic acid is included in Chapter 15, Catalysis in Biomass Conversion.

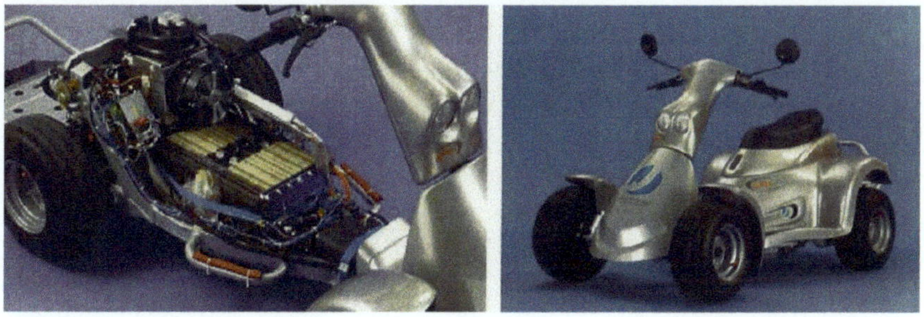

Figure 12.4
A direct methanol fuel-cell vehicle (JuMOVe) developed at the Research Centre Jülich. *Taken with permission of Forschungscentrum Julich, Zentralbibliothek from P. Biedermann, T. Grube, B. Höhlein, Methanol as an energy carrier, Schriften des Forschungscentrum Jülich/Energy Technol., 55 (2006), 33.*

Although as mentioned earlier, various other molecules have been tested as hydrogen carriers, the main work appears to have been done using hydrogen produced by methanol steam reforming:

$$CH_3OH + H_2O \rightarrow CO_2 + 3\ H_2 \quad (12.13)$$

The catalysts that have received most attention for this reaction (see Task 8.5) are those based on formulations containing copper, similar to those used for the methanol synthesis. As an example of such work on the methanol reforming reaction, we will now consider some of the results obtained in a study carried out by J.P. Breen and J.R.H. Ross as part of a project aimed at developing a ship powered by a fuel-cell as this illustrates some of the important parameters which need to be considered in developing such catalysts. (Methanol reforming for fuel-cell applications: development of zirconia-containing Cu−Zn−Al catalysts, Catal. Today 51 (1999) 521−533)). In this work, we demonstrate the importance of carrying out the relevant thermodynamic calculations when carrying out such experiments.

Fig. 12.5 shows typical data obtained for a sequentially precipitated $Cu{-}ZnO{-}Al_2O_3{-}ZrO_2$ catalyst as a function of reaction temperature and compares them with thermodynamically calculated concentrations (Box 12.4) for a situation in which CO formation was either not allowed (lines labeled m) or was allowed (lines labeled n). It can be seen that methanol conversion begins to occur at temperatures as low as 130°C and that it is complete at 340°C. The maximum possible percentage of hydrogen formed if no CO can be formed is just below 70% and the corresponding percentage of CO_2 is about 23% (balance H_2O, not shown). The experimentally determined value of the hydrogen and CO_2 contents approached the thermodynamically calculated values (see Box 12.4) at the temperature at which the methanol conversion was 100% (at about 540°C) and then dropped off slightly at a higher temperature; the concentrations of both products were well above those permitted

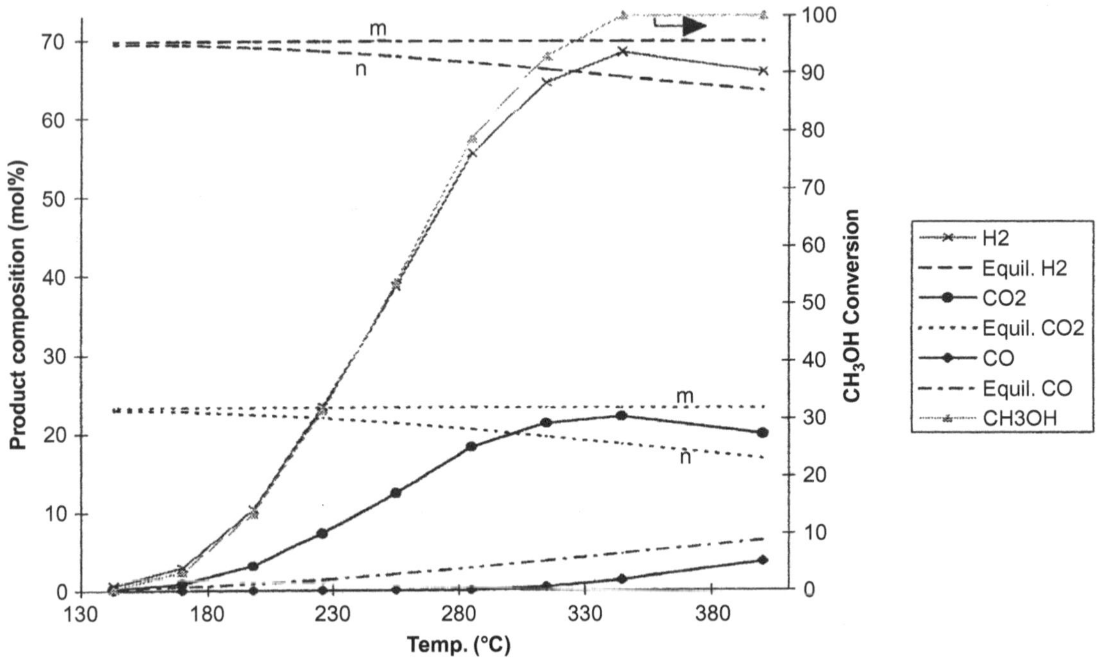

Figure 12.5
Steam reforming of methanol over sequentially precipitated Cu catalysts. The influence of temperature on the product compositions for a $Cu-ZnO-ZrO_2-Al_2O_3$ catalyst prepared by sequential precipitation as a function of reaction temperature. The data are compared with the product mix which would have been expected thermodynamically if CO was not (lines m) or was (lines n) allowed as a product in the reaction. (Reaction conditions: $H_2O/CH_3OH = 1.3$, $P = 101$ kPa, $W/F = 0.00259$ g min^{-1} cm^{-3}.) *Reproduced, with kind permission of Elsevier, from J.P. Breen, J.R.H. Ross, Catal. Today 51 (1999) 521–33.*

BOX 12.4 Thermodynamic Calculations of Gas Composition for Methanol Reforming

It is often desirable to know whether or not a possible reaction in a catalytic reaction is or is not in equilibrium. In the present case, it is desirable to know whether the reverse water-gas shift reaction (Eq. 12.8) is at equilibrium or not during the steam reforming of methanol but a similar approach can be used to determine whether or not the same reaction is in equilibrium during the CO reforming of methane or whether or not carbon deposition is likely to occur on a catalyst. Let us consider the case shown in Fig. 12.5. Two approaches are possible. One is to calculate the apparent equilibrium constant for the reaction in question using the measured partial pressures the reactants and products for each set of experimental conditions and to compare this value with the thermodynamically calculated equilibrium constant. However, the approach adopted here is to calculate for each conversion of

(Continued)

BOX 12.4 Thermodynamic Calculations of Gas Composition for Methanol Reforming (Continued)

methanol the equilibrium compositions that would be expected if the water-gas shift equilibrium held and to compare these compositions with what was obtained experimentally. The calculations were carried out using a simple thermodynamic computer program HSC Chemistry (www.outotec.com) which works out the equilibrium composition for any set of input pressures and reaction temperature. In this case, the input concentrations were calculated from the conversion of methanol and the partial pressures of water, hydrogen and CO_2 that would have resulted if no CO was formed. Similar calculations have been carried out to show the stable bulk phase of a catalyst that might be expected at any stage of a reaction, for example, whether or not a carbide might remain stable in the methane + CO_2 reaction.

if CO was formed in equilibrium. It was therefore concluded that the primary reaction in the steam reforming of methanol is that shown in Eq. (12.13) and that the reverse water–gas shift reaction:

$$CO_2 + H_2 \leftrightarrows CO + H_2O \tag{12.14}$$

occurs only once the conversion of methanol approaches 100%.[8] (See also Task 12.4.)

Task 12.4 Methanol Reforming Catalysts

Catalysts used for the steam reforming of methanol for fuel cell use have been reviewed by C. Song (Fuel processing for low-temperature and high-temperature fuel cells—challenges and opportunities for sustainable development in the 21st century, Catal. Today 77 (2002) 17–49) and by A.F. Ghenciu (Review of fuel processing catalysts for hydrogen production in PEM fuel cell systems, Curr. Opin. Solid State Mater. Sci. 6 (2002) 389–399). Use these reviews as a starting point to examine the literature on methanol reforming catalysts, paying particular attention to any significant advances in catalyst formulations and to current views on reaction mechanism.

Methanol conversion to other products. The oxidation of methanol to formaldehyde (Fig. 12.1) was discussed in Chapter 3, How Does a Catalyst Work? and hence it will not be discussed further here. The principal routes for the synthesis of acetic acid, the Monsanto

[8] It is interesting to note that contrasting conclusions have been reached in a more recent paper on the use of very similar catalysts but prepared by coprecipitation rather than by sequential precipitation) for the oxidative steam reforming of methanol, that is, the same reaction but with an addition of oxygen to the feed.
S. Velu, K. Suzuki, M. Okazaki, M.P. Kapoor, T. Osaki, F. Ohashi, Oxidative steam reforming of methanol over CuZnAl(Zr)-oxide catalysts for the selective production of hydrogen for fuel cells: catalyst characterization and performance evaluation, J. Catal. 194 (2000) 377–384.

or Cativa processes, involve the use of homogeneous Rh or Ir catalysts and so will also not be discussed further here. A new route to acetic acid involving the oxidation of ethane will be described in Chapter 14. Acetic acid is also produced (as vinegar) by fermentation of ethanol as well as being formed as a by-product in a number of other reactions. (See also Task 12.5.)

Task 12.5 Acetic Acid Synthesis

Carry out a web search for acetic acid synthesis and find out about the Monsanto and Cativa processes mentioned in the main text. Can you find any other processes that are, or have been, used for acetic acid manufacture? Did any of these use heterogeneous catalysis in any of the steps involved? What are the factors determining the preferred use of the two homogeneously catalyzed processes?

12.4 Hydrocarbon Formation by the Fischer–Tropsch Process

The FT synthesis of higher hydrocarbons

$$n\,CO + (n + m/2)\,H_2 \rightarrow C_nH_m + n\,H_2O \quad \Delta H^o < 0 \tag{12.15}$$

was discussed briefly in Chapter 1 where some aspects of the historical importance of the process were discussed. The reaction uses the appropriate mix of CO and hydrogen, utilizing the water–gas shift reaction to alter the composition as required. (As the predominant products are most frequently long–chain saturated paraffins, $m = 2\,n + 2$, and so the required ratio is close to 2.0.) Strictly speaking, the reaction is reversible and so the single arrow could be replaced by a reversible one. (Indeed, the reverse reaction is the steam reforming of a hydrocarbon (other than methane) and the thermodynamics of the FT and steam reforming processes are therefore directly interrelated.) High yields in the FT process will result if the temperature is low and the pressure is high (Le Chatelier's Principle) while the opposite is the case for the steam reforming of hydrocarbons. In practice, as with other reversible processes such as ammonia synthesis (discussed in Chapter 8: Mass and Heat Transfer Limitations and Other Aspects of the Use of Large Scale Catalytic Reactors), the forward FT reaction has to be carried out at sufficiently high temperatures to ensure reasonable conversions without restricting the conversion too much. Another important limitation of the process is that the product formed is almost always a mixture of molecules of very different molecular weights, ranging from methane to waxes, this being largely determined by a so-called Schultz–Flory (SF) distribution. This arises from the mechanism of the reaction which involves a step-wise polymerization chain process in which the probability of chain growth versus termination is given by the

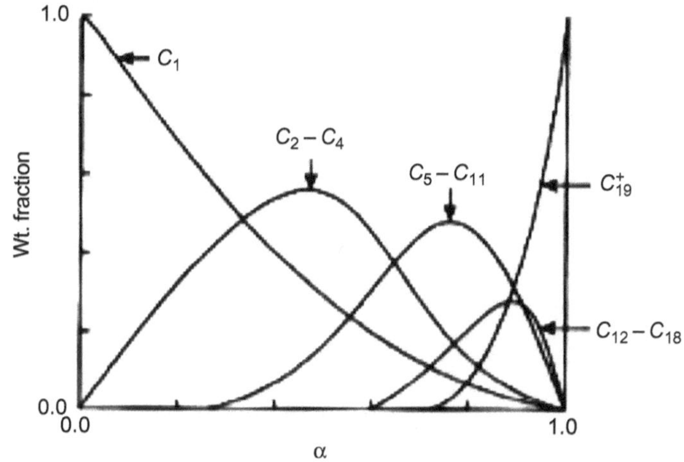

Figure 12.6
Weight fraction of FT products as a function of α, the chain-growth probability. *Reproduced from C.H. Bartholomew and R.J. Farrauto, Fundamentals of Industrial Catalytic Processes, 2nd Edition, 2006 with kind permission of Wiley Interscience.*

symbol α.[9] Fig. 12.6 shows a simplified plot of the weight fraction of any particular product range to be expected for any value of α. It can be seen that with low α, methane is the expected product while at high α (approaching unity), only high molecular weight products are expected. In general, three different metallic components of catalysts have received attention: Ni, which has a low value of α and tends to give methane as a predominant product; Fe, which has an intermediate value of α and gives a relatively wide range of products; and Co which has a high value of α and tends to give high molecular products such as waxes.[10] Iron and cobalt are therefore the metals of choice. (See also Box 12.5, Task 12.6.)

A particularly interesting review of the subject of the FT process and its catalysts, including some fascinating historical detail only recently unearthed, has been published by Casci, Lok, and Shannon.[11] These authors analyzed the patent activity over the period 1980–2007

[9] There are examples where the distribution deviates from the strict mathematical expression of the SF distribution. For an advanced treatment of this and other aspects of the Fischer–Tropsch process, the reader should refer to more advanced texts such as "Fundamentals of Industrial Catalytic Processes," C.H. Bartholomew, R.J. Farrauto, Wiley-Interscience, New Jersey, 2006.

[10] The only other metal that has received any attention is ruthenium; however, it is very scarce and is therefore not usable as a catalysts component: any large-scale demand for its use would immediately give rise to a significant increase in price. Copper gives alcohol production, predominantly methanol, as is discussed in the next section.

[11] J.L. Casci, C.M. Lok, M.D. Shannon, Fischer–Tropsch cataysis: the basis for an emerging industry with origins in the early 20th century, Catal. Today 145 (2009) 38–44. This article was used as the basis for a Task in Chapter 1, An Introduction to Heterogeneous Catalysis—Chemistry in Two Dimensions. See also (Task 1.5.)

BOX 12.5 Mechanism of the FT Process

A detailed description of much of the early German work on the subject, as well as a possible mechanism for the FT process, was given by H.H. Storch, N. Gulumbic, R.B. Anderson entitled "The Fischer–Tropsch and Related Syntheses" (John Wiley & Sons, New York, 1951, 610 p). Their mechanism involved chain growth using oxygenated C_1 species (CHOH(ads)). Strong support for this chain-growth mechanism was given by Emmett and coworkers in their papers describing work carried out using labeled molecules. The more recent, and more generally accepted, approach involves CH_2 entities as the chain-growth species. However, it seems that oxygenated species are still the predominant species involved in the surface processes if the main products of the reaction are alcohols.

Much has been written about the history of the development of the FT process. A very comprehensive compilation of the literature on the process has been assembled on a special website: http://www.fischer-tropsch.org/. For the present purposes, it is sufficient to say that there have been four periods in the development of the process to its current position:

- Development of the original process and its use in Germany up until the end of the World War II, when the majority of effort was devoted to Fe-based catalysts.
- The use of the process in South Africa during the period when that country was unable to import oil as a result of the trade embargoes in force and when SASOL operated a full-scale plant for the production of petroleum and diesel.
- Research in the United States and elsewhere, particularly spurred on by the energy crisis of the early 1970s, when attention began to focus once more on Co catalysts.
- The modern development of GTL for use in situations where there is an excess of natural gas, this generally being associated with oil reserves and hitherto flared off; there have also recently been developments in relation to biomass conversion to liquids and to suitable microchannel reactors.

Task 12.6 Catalysts for FT Synthesis

M.E. Dry was a senior scientist with Sasol during the period when the only significant work on FT synthesis occurred in South Africa. During the period of political isolation, the FT process was the only source of automobile fuel in South Africa. Some of the work at Sasol is described in the influential article: "The Fischer Tropsch Process: 1950–2000," by M.E. Dry, Catal. Today 71(2002) 227–241. Analysis with Scopus or other citation tools shows that this article is still very highly cited. Examine the citation behavior, paying particular attention to work on new developments such as gas to liquid plants (GTL) and biomass conversion. Examine also the types of catalysts that have been studied and the effect of catalyst composition, reaction conditions, etc., on product distribution. You should also encounter reference to SF distributions of products.

Include in your review of the literature an explanation of such distributions and their importance.

and demonstrated very clearly a very significant upsurge of interest from 1996 onward, particularly by Sasol, Shell, and Mossgas (now PetroSA); see Fig. 12.7. They show that the majority of the then current effort concerned the development of Co-containing materials, often supported on preformed supports such as titania, silica, or alumina using cobalt nitrate for the impregnation step; these catalysts often contain oxide and noble metal promoters; see Fig. 12.8. As indicated earlier, most of the current activity in the area of FT synthesis is directed toward the production of fuels in GTL plants. Several of these have been built and

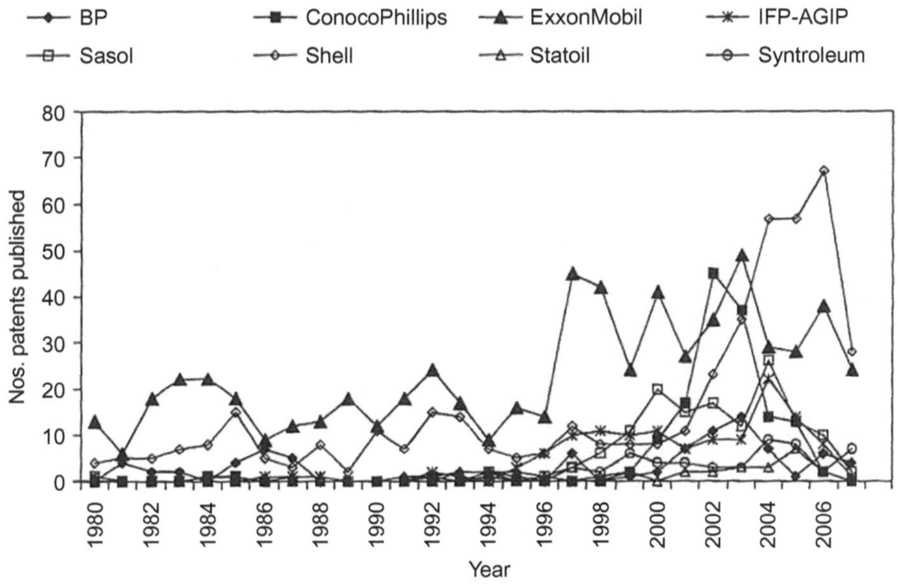

Figure 12.7

The numbers of patents on FT technology as a function of time from 1980 to 2007. *Reproduced from Casci et al.*[11] *with kind permission of Elsevier.*

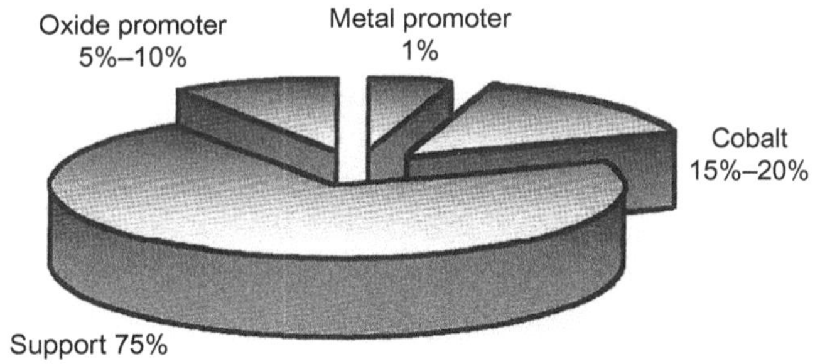

Figure 12.8

Typical compositions of Co-based FT catalysts. *Reproduced from Casci et al.*[11] *with kind permission of Elsevier.*

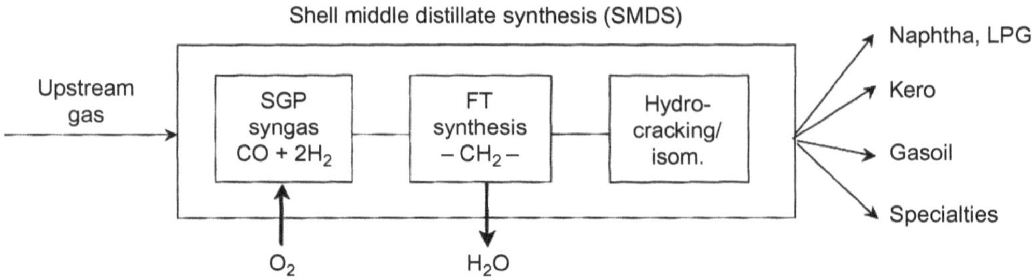

Figure 12.9
The shell middle distillates process as carried out at Bintulu (Malaysia). *Reproduced from A. Hoek, L.B.J.M. Kersten, Stud. Surf. Sci. Catal. 147 (2004) 25–30, with kind permission of Elsevier.*

commissioned in the Gulf States although the first plant using the modern technology was built by Shell at Bintulu (Sarawak, Malaysia). Fig. 12.9 shows a schematic representation of this plant. In essence, it first converts natural gas to syngas by a noncatalytic partial oxidation process. The syngas is then converted into paraffins by a FT process involving multitubular fixed bed reactors. Finally, the FT product is hydrocracked/hydro-isomerized to give a mixture of naphtha, middle distillates and a waxy residue. Further details of these developments and also some of the science behind them are available in a range of proceedings of the Natural Gas Conversion Symposia; see, for example, Studies in Surface Science and Catalysis, Vol. 147 (2004), edited by X. Bao and Y. Xu.

Task 12.7 FT Developments

Use the article by Casci et al. cited in footnote 11 as a basis for a citation search to find out about the most recent developments in the field of FT technology.

CHAPTER 13

Environmental Catalysis

Chapter Outline

13.1 Introduction

The use of catalysis in environmental control has become very important in everyday life and for most people in the street the term "catalysis" is synonymous with car exhaust emission control: destruction of traces of unburnt hydrocarbon fuel as well as of CO and NO_x formed during the combustion process. However, there are many other applications of catalysis in environmental control, these including the removal of SO_2 and NO_x (the latter also formed in a combustion process) from the stack-gas emitted from power stations, the destruction of volatile organic compounds (VOCs) and CO, and the destruction of pollutants in water by photocatalysis.

In the early days of research on emission control using the methods of environmental catalysis, publications on the subject were dotted throughout the literature. However, journals that concentrate on such subjects later appeared, notably Applied Catalysis B: Environmental, this having been first published in 1992. A number of companies have been at the forefront in research and development work in emission control, for example, Engelhard (now part of BASF http://www.catalysts.basf.com/main/) and Johnson Matthey (http://www.matthey.com/). The latter company has for many years now published their own journal, now available only electronically: *Platinum Metals Review*;[1] this is an

[1] The full searchable archive for *Platinum Metals Review* is available at http://www.platinummetalsreview.com/dynamic/volume/archive/. This is a very valuable source of information on all aspects of the use of the noble metals, particularly on their use in catalysts, and includes a large number of valuable review articles on the subject. As there is no link between the various articles, you will have to return to the search page to find each.

Contemporary Catalysis.
DOI: https://doi.org/10.1016/B978-0-444-63474-0.00013-8

excellent source of short review articles that summarized the state of the art at any stage over the last four or five decades in various aspects of the use of noble metals, most specifically in catalysis but occasionally in other areas as well. Although this is unashamedly a "house journal," it is still an excellent source of background material. As an example, an article by H. Windawi and M. Wyatt on Catalytic destruction of halogenated volatile organic compounds: mechanism of platinum catalyst systems (Platinum Met. Rev. 37 (1993) 186–193) gives an excellent background introduction to that area of activity as well as giving a listing of earlier contributions to the same periodical on other aspects of VOC control. Similarly, a number of companies active in the field have very informative websites that provide valuable information on (and often good animated diagrams of the operation of) their technology.

This chapter has separate sections, each devoted to a different area of catalytic environmental control. Each discusses the uses of the technology in practice and then gives some examples of relevant research publications and their main conclusions. Each section includes some related tasks aimed at encouraging literature study as well as several boxes containing more detailed information on some aspects of the topic concerned.

13.2 Selective Catalytic Reduction of NO_x

Power stations create electricity by the use of turbines that depend on heat generated by the combustion of a fuel that may be coal, oil, or natural gas; in some cases, biomass combustion is also used. Because of the high temperatures involved in the combustion process, some of the nitrogen and oxygen of the air combine to form what is termed as NO_x: a mixture of NO, N_2O, and NO_2, there being a predominance of NO. In addition, the flue gas from the power station may include traces of unburnt fuel (or molecules derived from the fuel) as well as SO_2 and SO_3 (SO_x) derived from sulfur-containing molecules present in the fuel as well as nitrogen; in addition, there will be some unreacted oxygen remaining from the air used from combustion. It is well established that the emission of NO_x and SO_x is responsible for the creation of acid rain, known to cause severe environmental damage. The catalytic control of NO_x emissions from power stations (as well as from nitric acid plants and incinerators) by "selective catalytic reduction" (SCR) using ammonia as reductant has therefore been a subject of interest for many decades. The selective reduction reaction, written here for the reduction of NO (generally 90%–95% of the NO_x in the flue gas), occurs according to the following reaction:

$$4NO + 4NH_3 + O_2 \rightarrow 4N_2 + 6H_2O \tag{13.1}$$

If the reactant is NO_2 (5%–10% of the NO_x), then oxygen is not necessary and the reaction is:

$$6NO_2 + 8NH_3 \rightarrow 7N_2 + 12H_2O \tag{13.2}$$

Other reductants such as CO, H_2, CH_4, and higher hydrocarbons have also been tested for use but these molecules are not selective in their action and hence some oxidation of each of them also occurs using the oxygen present in the reacting mixture arising from the air not used during the combustion process.

A number of commercial SCR processes are available; see Box 13.1. The catalyst used for this processes is generally composed of V_2O_5 supported on TiO_2 as it has been found that

BOX 13.1 SCR Catalyst Manufacturers

An example of a website giving information on a commercial catalytic system using selective reduction with ammonia from CRI Catalyst Company (part of the Shell Group) is given at http://www.cricatalyst.com/ where you will find a short video describing the process in simple terms. Websites for some other companies manufacturing ammonia de-NO_x systems are: www.cormetech.com/; www.ducon.com/; www.etecinc.net/; www.tenviro.com/; www.ceram-ibiden.com/. Information on catalysts made by BASF is to be found at www.catalysts.basf.com/. Additional links can be found by carrying out a search on SCR plants or similar search terms. Look out for a U-tube presentation on SCR technology.

this catalyzes the reduction of NO_x without catalyzing unwanted oxidation reactions. The catalytically active material is wash-coated onto ceramic honeycomb matrices manufactured in the form of blocks (Fig. 13.1). These blocks are then mounted in a larger structure of the type depicted in Figs. 13.2 and 13.3. Ammonia is dosed to the system in stoichiometric

Figure 13.1
The upper surface of the blocks of a typical SCR de-NO_x catalyst assembly. *Reproduced with kind permission from www.ceram-ibiden.com*

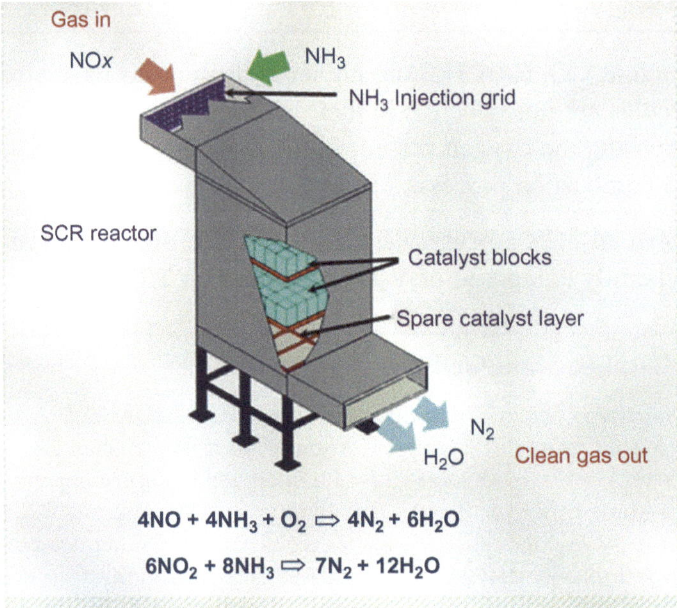

Figure 13.2
A cutaway view of a typical SCR plant. *Reproduced with kind permission from www.siriusengineers.com*

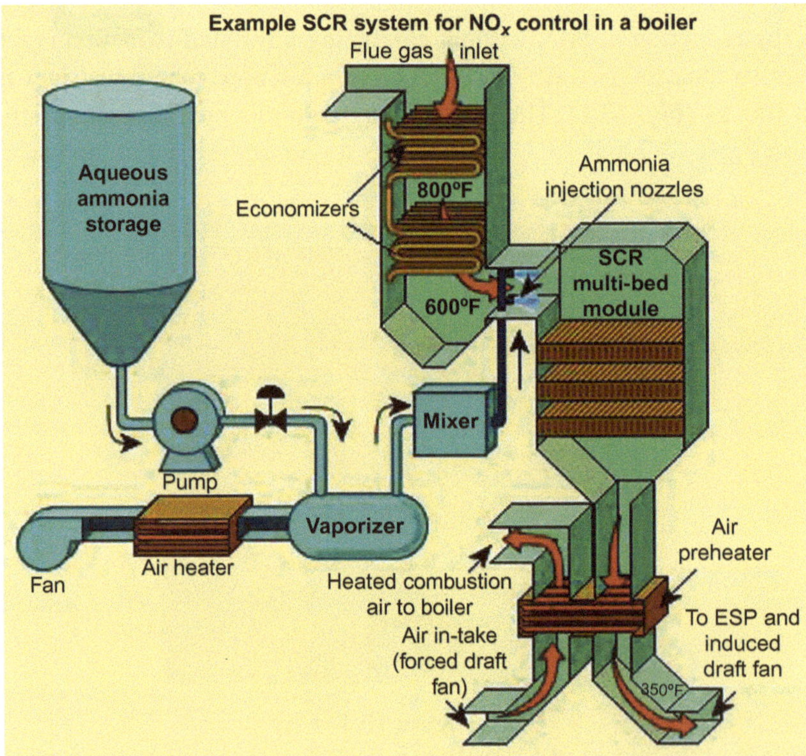

Figure 13.3
The SCR unit and its surrounding equipment (DOE).

quantities, the rate of introduction depending on the NO_x content of the gas being treated. The conversions obtained approach 100% with 100% selectivity to nitrogen. The main problem with such plants is ammonia "slip" (emission of unreacted ammonia) and hence the control of the ammonia addition is the most critical factor in the operation of such a plant.

Janssen and his colleagues have studied the mechanism of the selective reduction of NO_x over a number of typical supported vanadia catalysts and have carried out a series of experiments using isotopically labeled $^{18}O_2$ and $^{15}NH_3$. They have suggested that the mechanism of the reaction is as shown in Fig. 13.4. In this mechanism, emphasis is placed

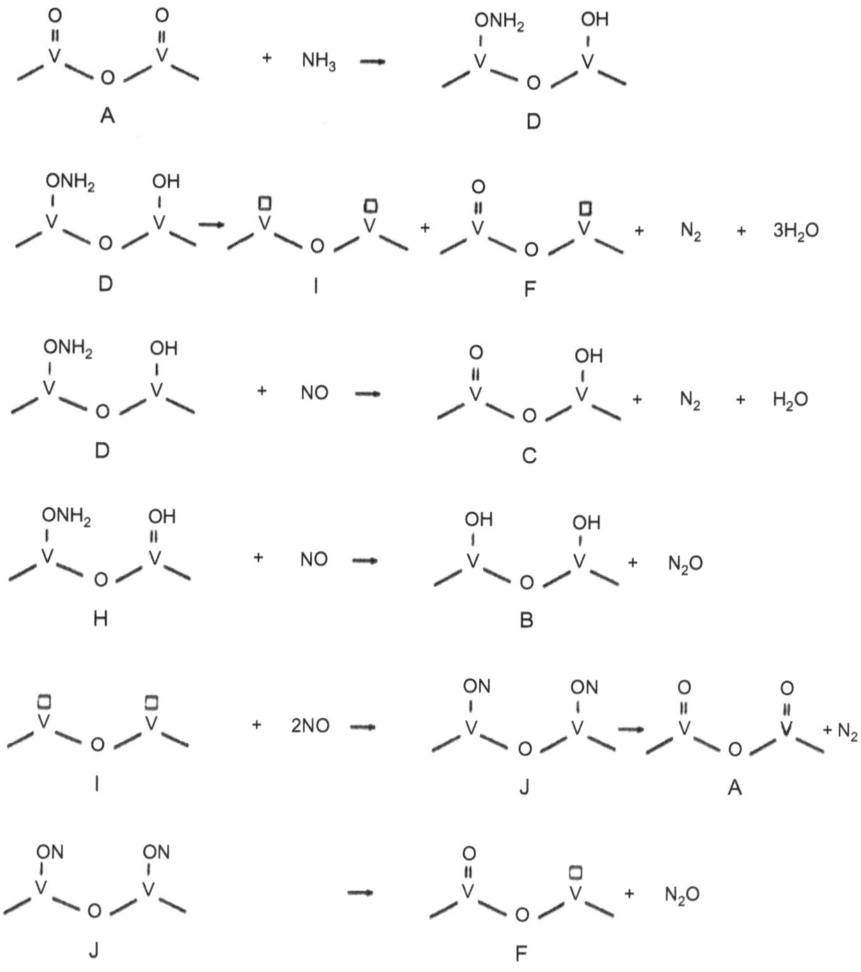

Figure 13.4
A mechanism for the selective catalytic reduction of NO over a vanadia catalyst. *Taken from F.J.J.G. Janssen, F.M.G. van den Kerkhof, H. Bosch, J.R.H. Ross, J. Phys. Chem. 91 (1987) 5931 and 6633; F.J.J.G. Janssen, Ph.D. Thesis, University of Twente, 1987. Reproduced with kind permission of American Chemical Society.*

on the reactivity of the V = O species at the catalyst surface, suggesting that a Mars–van Krevelen mechanism (oxidation–reduction, see Chapter 7: The Kinetics and Mechanisms of Catalytic Reactions) applies to the reaction.

The review by Busca and his colleagues (see reference given in Box 13.2) favors a rather different mechanism, this being shown in Fig. 13.5. Their mechanism places emphasis on

BOX 13.2 Literature on Selective Catalytic Reduction

One of the earliest and most extensive reviews on the subject of NO/NH_3 SCR appeared as a monograph written by H. Bosch and F.J.J.G. Janssen, Catalytic reduction of nitric oxides: a review, Catal. Today, 2 (1988) No.4.[a] You can also find other reviews of the subject by following some of the citations of Bosch and Janssen review or use as starting point the reviews by P. Forzatti, Present status and perspectives in de-NO_x SCR catalysts, Appl. Catal., A 222 (2001) 221–236, and another by G. Busca, L. Letti, G. Ramis, and F. Berti, Chemical and mechanistic aspects of the selective catalytic reduction of NO_x by ammonia over oxide catalysts: a review, Appl. Catal. B 18 (1998) 1–36.

An example of a highly cited paper on the subject of SCR catalysts is by M.D. Amarides, I.E. Wachs, G. Deo, J.-M. Jehng, and D.S. Kim, Reactivity of V_2O_5 catalysts for the selective catalytic reduction of NO by NH_3: influence of vanadia loading, H_2O and SO_2, J. Catal. 161 (1996) 247–253; other papers by the same group have also been highly cited. Amarides and colleagues have also examined the use of such catalysts for the oxidation of chlorinated hydrocarbons: S. Krishnamoorty, J.P. Baker, and M.D. Amiridis, Catalytic oxidation of 1,2-dichlorobenzene over V_2O_5/TiO_2 catalysts, Catal. Today 40 (1998) 39–46. A very complete review of NO_x removal is given in the following article: V.I. Pârvulescu, P. Grange, and B. Delmon, Catalytic removal of NO, Catal. Today 46 (1998) 628.

[a]In order to find this monograph on Science Direct, one should search for *Catalysis Today*, Volume 2, Issue 4. You should omit the authors' names since these do not appear as search terms.

the prior interaction of the ammonia with Lewis acid type sites on the vanadium species (not the V = O groups, these remaining intact) and that the NO interacts with the resultant adsorbed NH_2 groups.

Topsøe and his colleagues have proposed yet another mechanism, depicted in Fig. 13.6, in which both acid–base and redox functions of the catalyst are brought into play. This scheme has been rather heavily criticized by Busca and his colleagues in their review of the mechanisms that have been proposed for the reaction. Busca et al. conclude their review by suggesting a number of points that need further investigation, these including the role of polymeric V species on the catalyst surface and the possible role of the exposed TiO_2 surface. Further reading on this subject is recommended. (See also Task 13.1.)

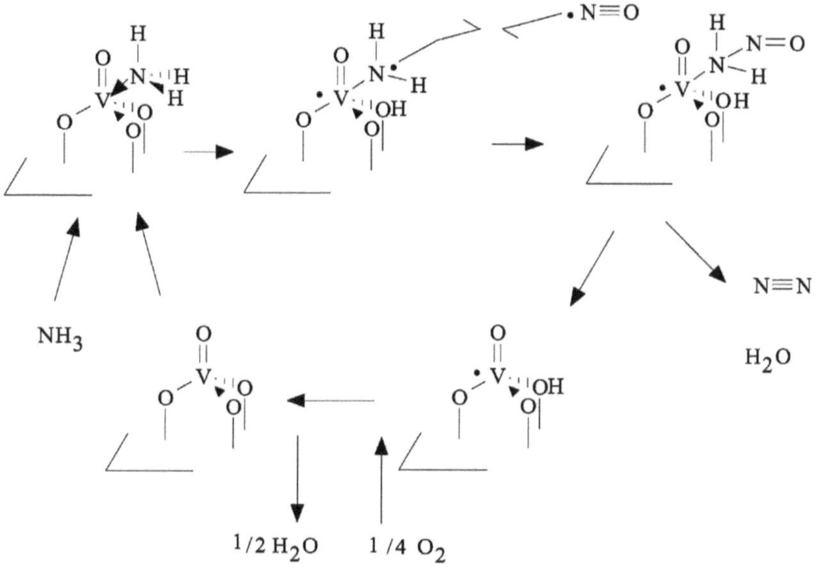

Figure 13.5
Mechanism proposed by Busca et al. for the selective reduction of NO_x over supported vanadia catalysts. *Reproduced from G. Ramis, G. Busca, F. Bregani, P. Forzatti, Appl. Catal. 64 (1990) 259, with kind permission of Elsevier.*

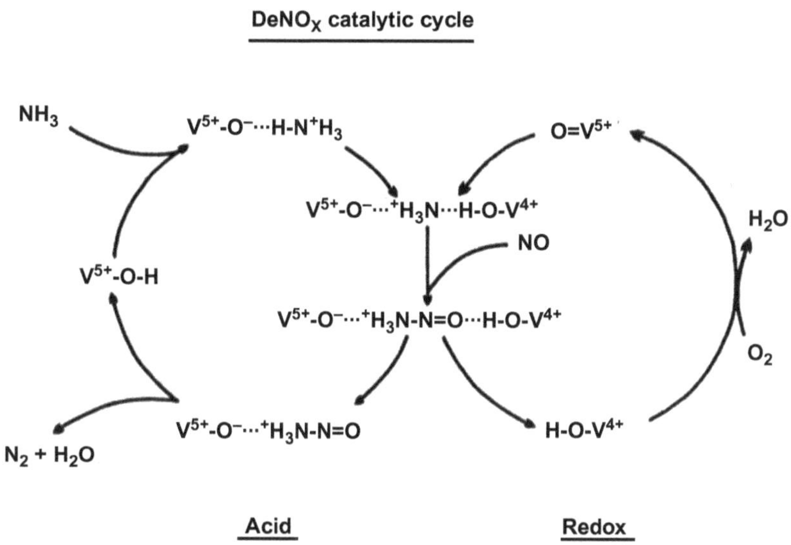

Figure 13.6
Mechanism for the selective catalytic reduction of NO_x proposed by Topsoe and her colleagues. *Reproduced from N.Y. Topsoe, H. Topsoe, J.H. Dumesic, J. Catal. 151 (1995) 241–252, with kind permission of Elsevier.*

Task 13.1 Mechanism of Selective Catalytic Reduction of NO_x in Flue Gases

Paying particular attention to the techniques used to study the mechanistic aspects of the SCR reaction, use the references given with Figs. 13.4–13.6 to examine the literature on the mechanism.

SCR catalysts have also been used for the control of emissions from incinerators and also other combustion systems and the catalysts used are generally the same as those discussed earlier for power station applications. One of the very considerable problems of the use of incinerators is the possible additional emission of dioxins, these being chlorinated polyaromatic compounds formed during the incineration process from precursors such as chlorinated phenols and polychlorinated biphenyls. There has been some evidence in the literature that the vanadia–titania catalysts of the type considered earlier would also bring about the destruction of such species. Some work was therefore carried out in the author's laboratory on the possible destruction of some model chlorinated hydrocarbons, ethyl chloride and chlorobenzene, over a range of different vanadia-containing catalysts while also carrying out the SCR reaction[2]. Fig. 13.7 shows the conversions of NO and ethyl chloride over a commercial vanadia–titania catalyst that also contained WO_3; it can be seen that there was a very narrow window of temperature, centered at 573 K, in which the conversions of both molecules reached almost 100%. However, at that temperature, the conversion of NO_x had already started to decrease (due to the formation of N_2O by

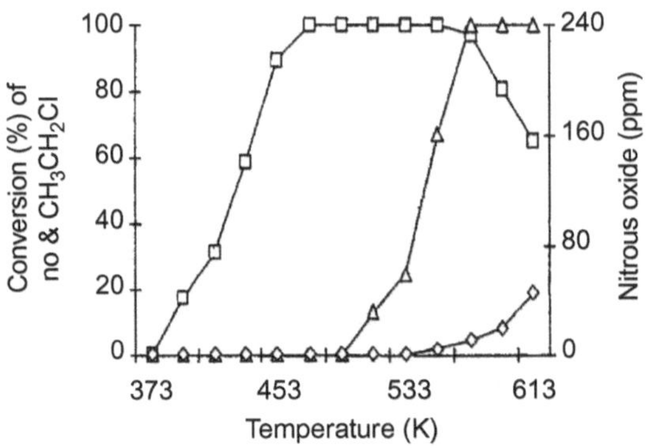

Figure 13.7
Combined reduction of nitric oxide and oxidation of ethyl chloride as a function of temperature of reaction over a commercial V_2O_5–WO_3–TiO_2 catalyst. □, NO conversion; Δ, ethyl chloride conversion; ◊, N_2O formation. *Reproduced from J. Jones, J.R.H. Ross, Catal. Today 35 (1997) 97–105, with kind permission of Elsevier.*

[2] J. Jones, J.R.H. Ross, Catal. Today 35 (1997) 97–105.

oxidation of the ammonia) before the oxidation of the chloro-compound became appreciable. It was found that the use of zirconia as a support gave a catalyst that gave 100% conversion of NO over a wide range of temperature before oxidation of the ammonia commenced; titania and alumina were found to be less effective supports. As shown in Fig. 13.8, promotion of the zirconia with alumina was particularly beneficial. The alumina-promoted zirconia-supported material also gave a reduction of the temperature at which complete oxidation of the ethyl chloride occurred (Fig. 13.9) so that both total NO_x

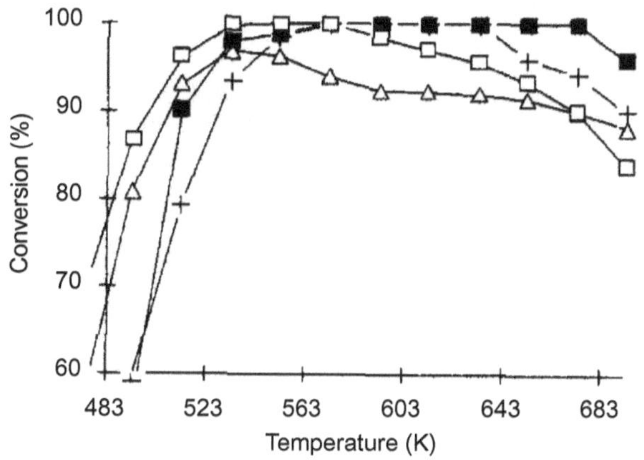

Figure 13.8

Conversion of NO over a series of zirconia-supported catalysts promoted with different oxides (all containing 4 wt% vanadia) as a function of temperature of reaction. □, unpromoted; +, SiO_2; ■, Al_2O_3; △, La_2O_3. *Reproduced from J. Jones, J.R.H. Ross, Catal. Today, 35 (1997) 97–105, with kind permission of Elsevier.*

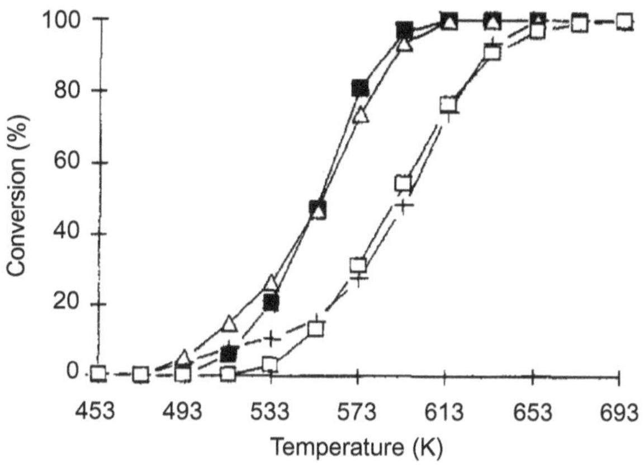

Figure 13.9

Oxidation of ethyl chloride over vanadia on zirconia with various additives as a function of temperature. □, unpromoted; +, SiO_2; ■, Al_2O_3; △, La_2O_3. *Reproduced from J. Jones, J.R.H. Ross, Catal. Today 35 (1997) 97–105, with kind permission of Elsevier.*

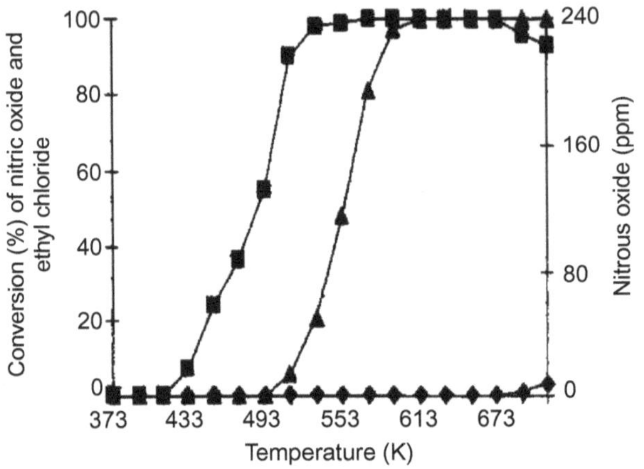

Figure 13.10
The combined de-NO_x and oxidation reactions over the vanadia on alumina–zirconia sample; the formation of N_2O at higher temperatures is also shown. ■, NO_x conversion; ▲, ethyl chloride conversion; ♦, N_2O formation. *Reproduced from J. Jones, J.R.H. Ross, Catal. Today 35 (1997) 97–105, with kind permission of Elsevier.*

conversion and ethyl chloride oxidation could be carried out over a relatively wide range of temperatures (Fig. 13.10). The zirconia–alumina-supported catalyst also converted chlorobenzene effectively over the same temperature range as for ethyl benzene but the catalyst developed was not tested for the conversion of any more complicated molecules. (For further information, complete Task 13.2.)

Task 13.2 Destruction of Dioxins

Starting with a literature citation search based on the reference of footnote 3 as well as a more general web-based search, find out what you can about the destruction of chlorinated molecules produced in incineration systems.

13.3 Automobile Exhaust Catalysts

13.3.1 Normal Engine Operation

The use of catalysts in car exhaust purification ("the catalytic muffler"), one of the most important applications of catalysis in the environmental area, is probably one of the best-known applications of catalysis of which almost everyone will have heard. The aim of the catalytic exhaust treatment is to minimize the emission of NO_x formed in the combustion chamber through combination of nitrogen and oxygen and to remove at the same time any

Figure 13.11
A metal foil catalyst (left) and an extruded cordierite material (right) used as supports in car exhaust catalysts. *Reproduced from M.V. Twigg, Catal. Today 163 (2011) 33–41, with kind permission of Elsevier.*

CO formed as well as any unburnt hydrocarbon fuel. It is interesting to follow the gradual development of the requirements for three-way catalysts needed to control these three components over the years, from the earliest stages following the introduction of legislation in the United States demanding emission control (1970 Clean Air Amendments Act) to the present day. These steady improvements were necessitated by increasingly strict environmental control requirements in both the United States and Europe related to automobile emissions. Further information on these changes can be gained by reading some of the articles in *Platinum Metals Review* (see footnote 1); Box 13.2 lists some of the most appropriate of these items. Studying them will also give the reader a good idea of the technology applied to the control of exhaust emissions and the effect of the stoichiometry of the combustion process on the effectiveness of the catalysts used.

Fig. 13.11 shows end views of two typical forms of catalyst used in car exhaust systems: (1) metal monolith and (2) ceramic honeycomb structures. The metal monoliths are made (often from Fecralloy, an alloy of Fe, Cr, Al, and Y, originally patented by the UK Atomic Energy Authority) by a crimping process while the ceramic monolith is formed by extrusion methods using ceramic technology.[3] In both cases, the matrix is then coated with a "washcoat," this being deposited on the surface of the matrix from a slurry containing both the support and the active components (see later), followed by drying and calcination. Fig. 13.12 shows expanded images of two wash-coated ceramic substrates in which the thickness of the coating is clearly visible. As can be seen, the washcoating technique tends

[3] Many of the substrates used are made by Corning (http://www.corning.com/environmentaltechnologies/index.aspx) who also supply ceramic blocks of the type used in power stations (see Section 13.3) and diesel filters (see later).

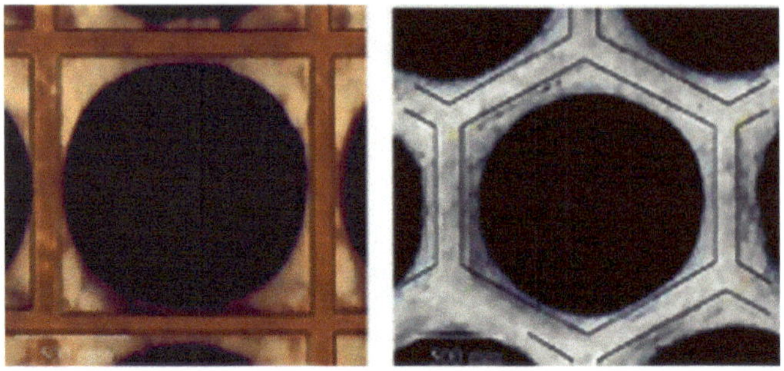

Figure 13.12

Enlarged diagrams of typical wash-coated monolith structures showing clearly that the washcoat tends to round off the corners of the channels. *Reproduced from S. Matsumoto, Catal. Today 77 (2003) 419–447, with kind permission of Elsevier.*

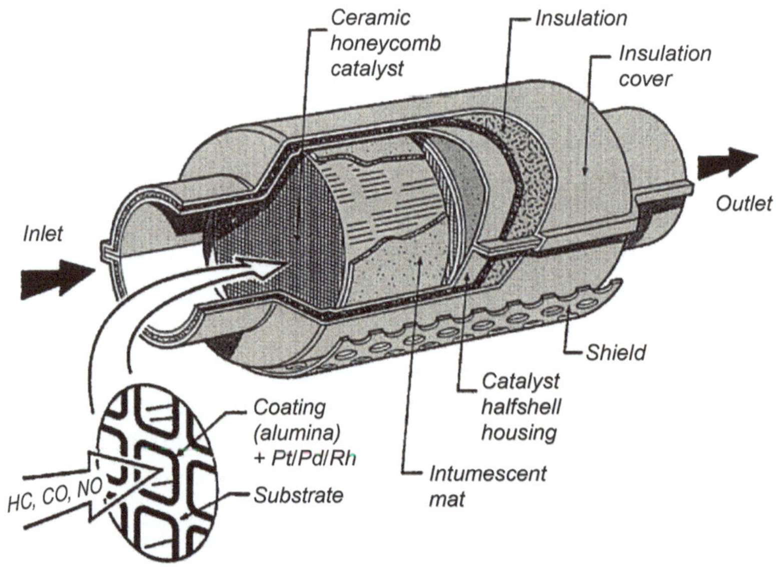

Figure 13.13

A typical catalytic converter using a ceramic honeycomb material (usually cordierite) into which is incorporated a washcoat containing the various catalytic components. *Reproduced from R.M. Heck, R.J. Farrauto, Appl. Catal., A 221 (2001) 443–457, with kind permission of Elsevier.*

to fill the corners of the structures so that the channel is almost cylindrical. Fig. 13.13 shows the way in which a typical ceramic honeycomb is then mounted in a container that is, in most cases, placed between the engine and the conventional silencer module as shown in Fig. 13.14. The whole system is controlled very carefully using sensors placed before and after the catalyst module in order that the fuel mix fed to the engine ensures optimal behavior of the catalyst. (Some relevant articles are listed in Box 13.3.)

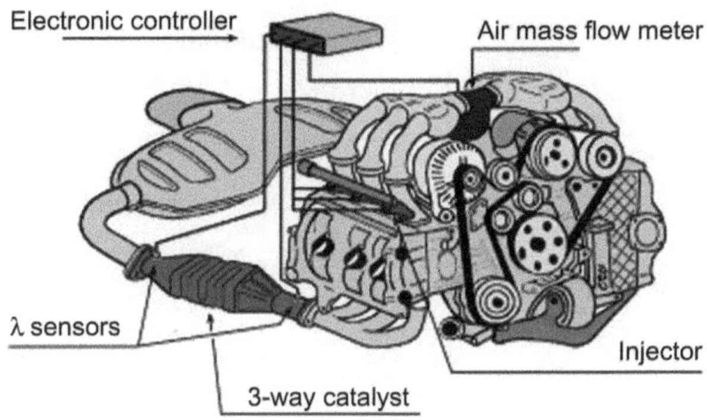

Figure 13.14
Typical positioning of a three-way exhaust catalyst relative to the engine. *Reproduced from J. Kaspar et al., Catal. Today 77 (2003) 419–449, with kind permission of Elsevier.*

BOX 13.3 Some Articles on Car Exhaust Treatment From *Platinum Metals Review*

Some examples of the many relevant articles in PMR on exhaust catalysis are:

- B. Harrison, B.J. Cooper, A.J.J. Wilkins, Development of Rh/Pt three-way catalyst systems, Platinum Met. Rev. 25 (1981) 14–22.
- B.J. Cooper, Challenges in emission control catalysts for the next decade, Platinum Met. Rev. 38 (1994) 2–10.
- M.V. Twigg, Twenty-five years of autocatalysis, Platinum Met. Rev. 43 (1999) 168–171.

An article by G.J.K. Acres in the more general literature also provides useful background knowledge, The development of emission control technology for motor vehicles, Stud. Environ. Sci. 44 (1991) 376–396.

We will now consider very briefly the chemistry of the reactions involved and the nature of the catalysts used. Again, the reader should consult the articles cited for more details. Critical to the operation of a catalytic exhaust control system is the air–fuel ratio of the feed to the engine. Fig. 13.15 depicts schematically the engine power and also the concentrations of carbon monoxide, NO_x, and unburnt hydrocarbons as a function of the air–fuel ratio; the position of the stoichiometric mix (i.e., the composition required to give 100% combustion of the fuel) is also shown by a *vertical line*. If there is an excess of fuel in the feed (left-hand side of the diagram), then the concentration of unburnt hydrocarbon increases and as does the amount of carbon monoxide formed. With an excess of air, the carbon monoxide concentration falls to zero and the unburnt hydrocarbon concentration also decreases. (The latter increases once more once the mix enters the lean-burn region and

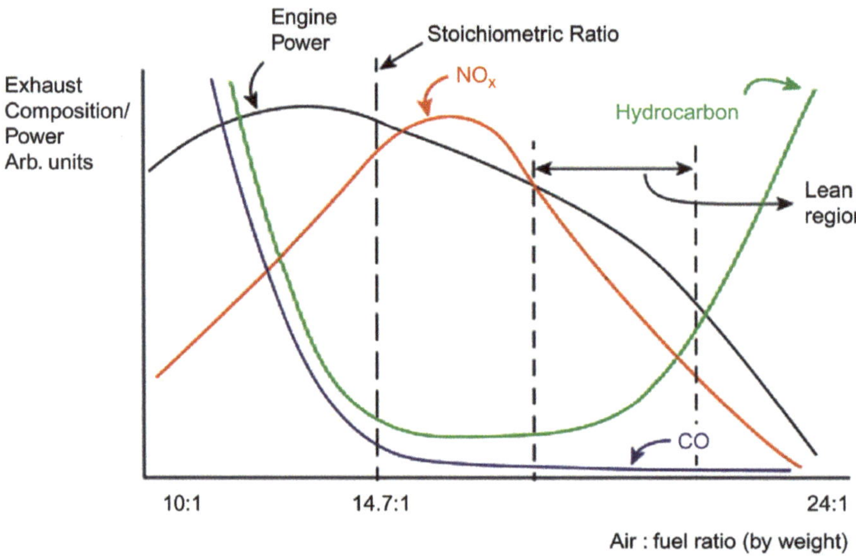

Figure 13.15
The effect of air–fuel ratio on engine operation.

gets worse still in the misfire region.) The formation of NO_x is relatively low in the fuel-rich region as a result of a lower temperature[4] and also of the relatively low concentration of oxygen present. However, the concentration of the NO_x increases and goes through a maximum as the gas composition becomes close to the stoichiometric mixture and then drops once more when the mixture corresponds to the lean-burn region. The so-called three-way catalyst used for an engine, operating close to the stoichiometric gas composition, has therefore to lower the concentrations of NO_x, CO, and unburnt hydrocarbon in an atmosphere containing largely N_2, CO_2, and water. Under these conditions, the concentration of oxygen is close to zero. Hence, the reactions occurring[5] are:

$$2CO + 2NO \rightarrow N_2 + 2CO_2 \tag{13.3}$$

$$HC + NO \rightarrow CO_2 + H_2O + N_2 \tag{13.4}$$

$$2H_2 + 2NO \rightarrow 2H_2O + N_2 \tag{13.5}$$

The hydrogen needed for reaction (13.5) is produced by steam reforming of the unburnt hydrocarbon and also by the water–gas shift reaction. The earliest three-way catalysts

---

[4] NO_x is formed by a homogeneous reaction between oxygen and nitrogen in the engine cylinder. The reaction is very temperature dependent.

[5] The reactions involving the hydrocarbon are not balanced.

contained largely Pt or a Pt/Rh mixture supported on alumina but more recently Pd has also been used and an oxygen-storage component (a CeO_2-ZrO_2 mixed oxide) is also added. For more information on these catalyst compositions, the reader should see some of the recent reviews on the subject listed in Box 13.4.

BOX 13.4 Recent Reviews on Exhaust Catalysts and Reactions

The following reviews on the subject of exhaust gas cleanup provide useful entry points to the literature on this subject:

- V.I. Pârvulescu, P. Grange, B. Delmon, Catalytic removal of NO, Catal. Today 46 (1998) 628.
- J. Kašpar, P. Fornasiero, N. Hickey, Automotive catalytic converters: current status and some perspectives, Catal. Today 77 (2003) 419–449.
- R. Burch, J.P. Breen, F.C. Meunier, A review of the selective reduction of NO_x with hydrocarbons under lean-burn conditions with non-zeolitic and platinum group metal catalysts, Appl. Catal., B 39 (2002) 283–303.
- S. Matsumoto, Recent advances in automobile exhaust cleanup, Catal. Today 90 (2004) 183–190.
- W.S. Epling, L.E. Campbell, A. Yezerets, N.W. Currier, J.E. Parks II, Overview of the fundamental reactions and degradation mechanisms in NO_x-storage/reduction catalysts, Catal. Rev. Sci. Eng. 46 (2004) 163–245.
- M.V. Twigg, Progress and future challenges in controlling automotive exhaust gas emissions, Appl. Catal., B 70 (2007) 2–15.
- R.M. Heck, R.J. Farrauto, Automobile exhaust catalysts, Appl. Catal., A 221 (2001) 443–457.
- M.V. Twigg, Catalytic control, of emissions from cars, Catal. Today 163 (2011) 33–41.

13.3.2 Lean-Burn Engines

With the introduction of more stringent requirements for fuel efficiency in the 1990s, automobile manufacturers started to introduce so-called lean-burn engines. These work with an excess of oxygen above the stoichiometric amount required for combustion of the fuel (see Fig. 13.15). The three-way catalysts discussed in the last section do not work effectively under these conditions. This led to a new area for research on selective NO_x reduction catalysts, notably in Japan, led by developments in Toyota Motor Corporation, first on selective de-NO_x using hydrocarbon reductants and later on so-called NO_x-storage catalysts. Some of the background is covered in the article by S. Matsamoto shown in Box 13.4 that also cites a number of the important related academic publications up until that time. (Follow up by carrying out Task 13.3.)

Task 13.3 Control of Lean-burn Engine Emissions

Using the reference to the work of Matsumoto listed in Box 13.4 as a starting point for a citation-based search, examine the literature for work on lean-burn engines and the control of their NO_x emissions.

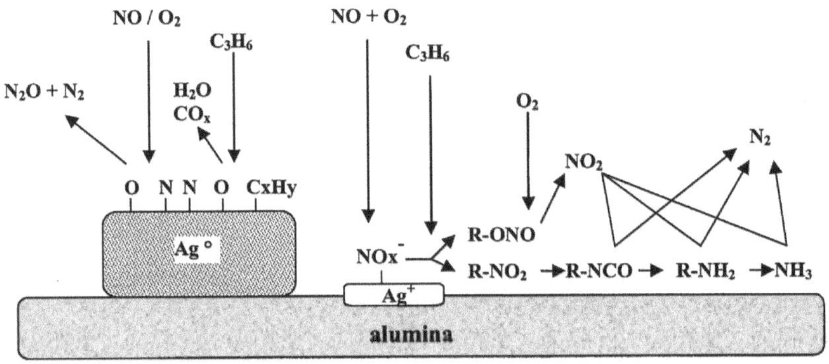

Figure 13.16

Schematic representation of some of the steps occurring on a silver/alumina catalyst under lean-burn conditions. *Reproduced from Meunier et al., footnote 7, with kind permission of Elsevier.*

One of the most effective catalysts developed for use under lean-burn conditions is composed of silver supported on alumina. Work by Meunier et al.[6] showed that the reaction occurs on both silver crystallites and on the alumina surface and that Ag^+ also participates, as shown schematically in Fig. 13.16. These workers also found that a variety of organic species containing nitrogen and oxygen were formed under their reaction conditions.

Unfortunately, although these de-NO_x catalysts give high conversions with high selectivities to nitrogen, they are particularly susceptible to impurities (particularly sulfur-containing molecules) in the fuels and so do not have the lifetimes required for practical use. As a results of this type of work, Toyota developed the so-called "NO_x-storage catalysts," these consisting of Pt and BaO supported on alumina. These catalysts are used under consecutively oxidizing and reducing conditions, adsorbing NO_x during the first oxidizing period and then reducing it during the reducing cycle. Fig. 13.17 shows the model put forward by the Toyota researchers for the catalyst surface. It shows NO being oxidized on the Pt sites and being adsorbed as nitrate on Ba sites when the air–fuel ratio is lean; then,

[6] F.C. Meunier, J.P. Breen, V. Zuzaniuk, M. Olsson, J.R.H. Ross, Mechanistic aspects of the selective reduction of NO by propene over cobalt and silver promoted alumina catalysts: kinetic and in-situ DRIFTS study, J. Catal. 187 (1999) 493–505.

when the air–fuel ratio is changed to the stoichiometric value (reducing conditions), the nitrate is reduced by HC, CO, and H_2O. A slight variant on this picture is shown in the review by Kaspar et al. (Box 13.4) as shown in Fig. 13.18. In this model, it is recognized that the Ba present in the catalyst will form $BaCO_3$ and so the scheme shows its participation in the reaction.

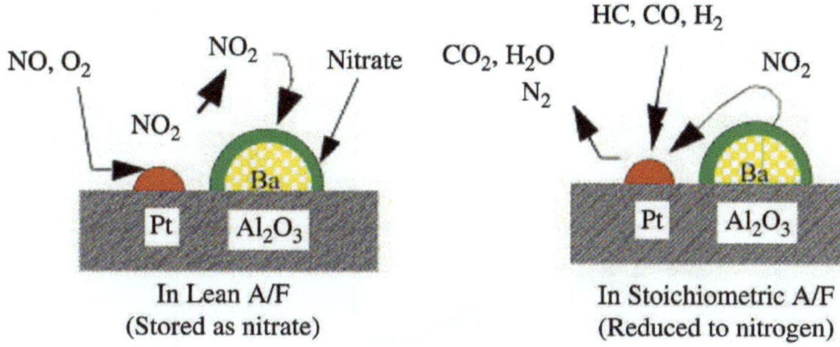

Figure 13.17
The Toyota model for the NO_x-storage catalyst. *Reproduced from S. Matsumoto, reference in Box 13.3, with kind permission of Elsevier.*

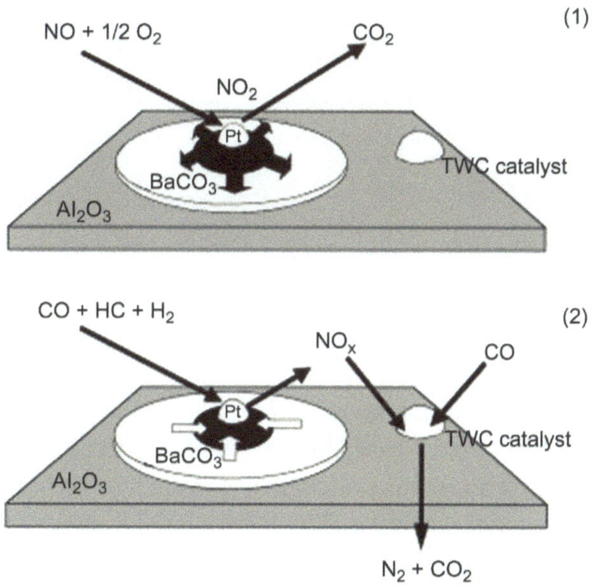

Figure 13.18
Model for NO_x storage, including $BaCO_3$ formation. *Reproduced from Kaspar et al., reference in Box 13.4, with kind permission of Elsevier.*

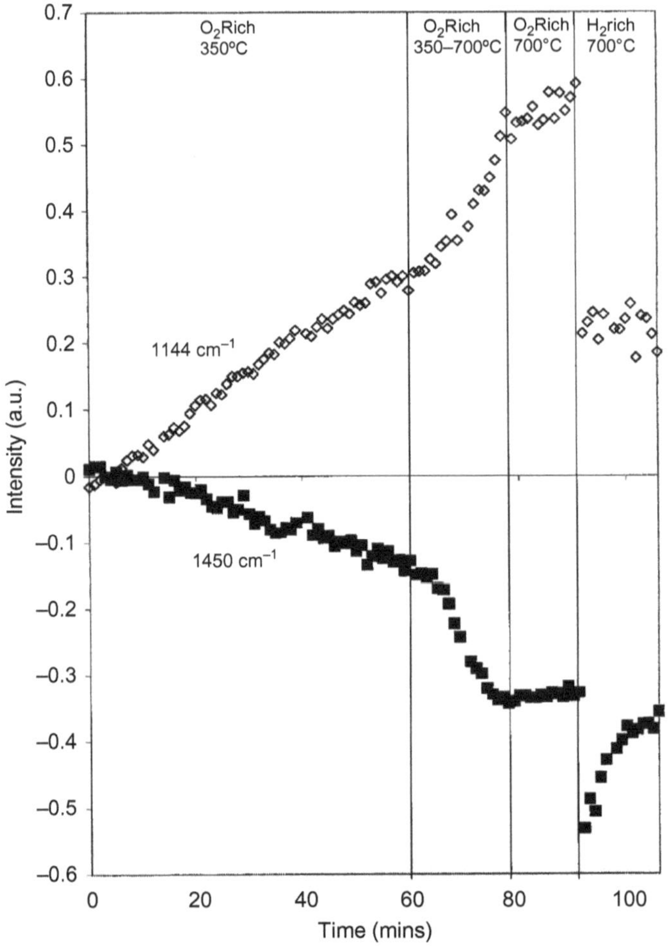

Figure 13.19
Concentration of bulk carbonate and bulk sulfate species over time during an in-situ DRIFTS experiment carried out at different temperatures and feed compositions. First 90 min: feed 50 ppm SO_2, 5% O_2, and 10% CO_2; followed by 15 min without O_2 in feed: 50 ppm SO_2, 5% H_2, and 10% CO_2. *Taken from reference of footnote 7, with kind permission of Springer.*

Neither of these models allows for the possible reaction of the NO_x-storage traps with SO_2 should that be present in the exhaust gases; as this is highly likely, the effect of added SO_2 needs to be investigated. Breen and his colleagues[7] have examined the interaction of SO_2 with a typical NO_x trap material and have shown by DRIFTS measurements that the $BaCO_3$ (1450 cm^{-1}) of the trap rapidly interacts to form $BaSO_4$ (1144 cm^{-1}). However, the latter can be partially converted back to the carbonate by exposure to H_2 and CO_2 at higher

7 J.P. Breen, M. Marella, C. Pistarina, J.R.H. Ross, Sulfur-tolerant NO_x storage traps: an infrared and thermodynamic study of the reactions of alkali and alkaline-earth metal sulfates, Catal. Lett. 80 (2002) 123–128.

Table 13.1: Typical composition of diesel exhaust gases

Gas	O_2/%	H_2O/%	CO_2/%	CO/ppm	NO_x/ppm	SO_2/ppm	HC/ppm	N_2
	5–15	10	≤10	100–400	500–1500	50	1000	Balance

temperatures; see the results shown in Fig. 13.19. They show thermodynamically that $BaSO_4$ is the stable product over a large range of temperatures but that can be converted back to $BaCO_3$ by treatment in CO_2/H_2 mixtures at temperatures above about 600°C.

13.3.3 Control of Diesel Emissions

Another form of lean-burn system is the diesel engine. A diesel engine operates at relatively very low temperatures (generally 120–150°C) with an excess of oxygen ($\lambda > 1$) and so the concentrations of CO and unburnt hydrocarbon in the exhaust gas (Table 13.1) is relatively very low; the levels of these gases are comparable to those found in the exhaust of a normal gasoline powered (Otto) engine after a three-way exhaust catalyst has been used. The other main contaminants that are found are NO_x, SO_2 and soot particles. The NO_x levels, although low compared with the Otto engine, still require to be controlled. This is now done by applying methods similar to those used for the SRC systems for stack-gas control described in Section 13.2 using ammonia as the reductant. (The ammonia required is produced in situ by hydrolysis of urea.) The SO_2 levels are dependent on the sulfur content of the fuel used but this is now generally very low as a consequence of the introduction of stricter norms for the processing of diesel fuels. However, the emission of diesel soot particulates is a very severe problem, causing potential health problems, and so these particles have to be removed from the exhaust gases.[8]

It has been shown by Cooper and Roth[9] that a Pt catalyst can be used to oxidize NO to NO_2 and that the NO_2 formed can be used to remove particulate C by the reaction of this molecule with the carbon. The reactions occurring in such a system are as follows:

$$NO + \tfrac{1}{2}O_2 \rightarrow NO_2 \qquad (13.6)$$

$$NO_2 + C \rightarrow NO + CO \qquad (13.7)$$

[8] A very comprehensive review of the whole area of the catalytic control of emissions from cars is given by M.V. Twigg, Catalytic control of emissions from cars, Catal. Today 163 (2011) 33–41. This review gives a particularly useful summary of the current situation regarding diesel emissions.

[9] B.J. Cooper, S.A. Roth, Flow-through catalyst for diesel engine emission control: Pt-coated monoliths reduce particulates, Platinum Met. Rev. 35 (1991) 178–187.

$$NO_2 + CO \rightarrow NO + CO_2 \qquad (13.8)$$

Together, these give the allover reaction:

$$C + O_2 \rightarrow CO_2 \qquad (13.9)$$

The NO thus acts as a catalyst for the reaction and is itself not consumed.

Hence, as long as there is little or no SO_2 present or the catalyst does not encourage the formation of sulfates on the catalyst surface by allowing the following reaction to occur:

$$SO_2 + \frac{1}{2}O_2 \rightarrow SO_3 \qquad (13.10)$$

it should be possible to reduce the amount of particulate carbon collected on a so-called "catalytic filter." Fig. 13.20 shows the structure of a typical filter element in which alternative channels are blocked at one end or the other, ensuring that the exhaust gas must pass through the walls of the channels.

Fig. 13.21 shows typical configurations of the catalytic elements relative to the filters that are currently in use. The first generation of filter systems operated using an oxidation catalyst (used to oxidize unburnt hydrocarbons as well as to produce NO_2 as discussed earlier). In these systems, the filter performs just the function of removing the soot from the exhaust in order that the reaction between NO_2 and C can take place. One of the main problems with this approach is that the reaction between C and NO_2 requires a relatively high temperature that is only achieved in the diesel engine when operated for relatively long times and does not work properly immediately after start-up. It is used nowadays only for heavy-duty vehicles for which the operating temperatures are high. If this system is to

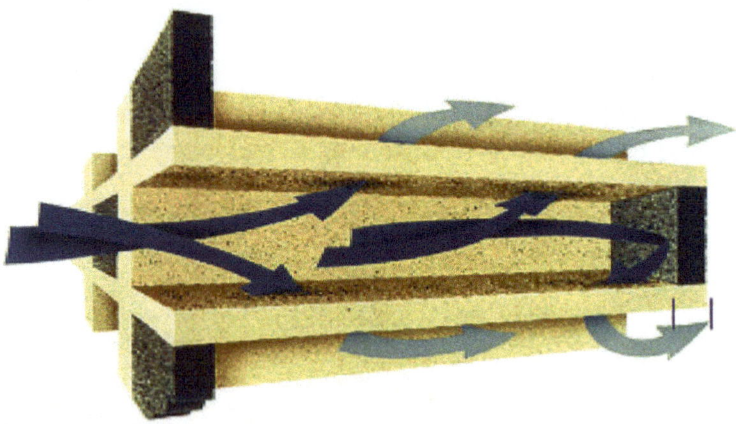

Figure 13.20
Schematic representation of the structure of a ceramic filter used for the removal of particulate matter from a diesel exhaust stream. *Reproduced from reference of footnote 9; with kind permission of Elsevier.*

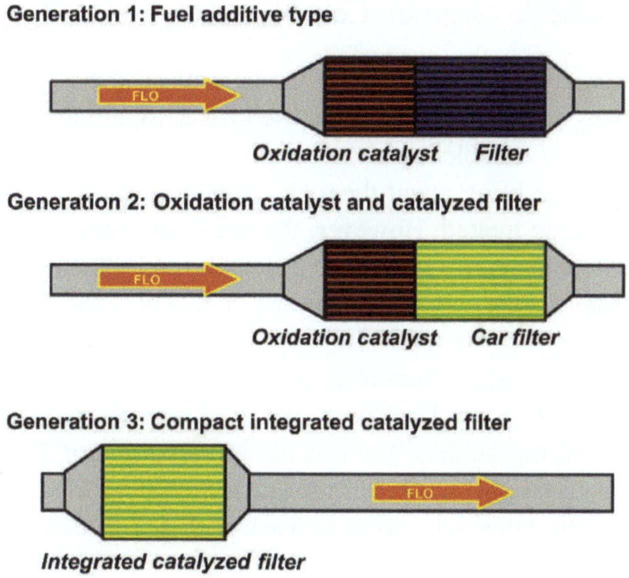

Figure 13.21
Three different particulate filter systems used for the removal of particulates from diesel exhausts. *Reproduced from reference of footnote 9; with kind permission of Elsevier.*

be applied to cars, an active form of regeneration of the filter has to be incorporated by which the exhaust gas temperature is increased periodically (every 400–2000 km, depending on the driving conditions) to oxidize the carbon. This approach had a further drawback: the need to introduce additives into the fuel system to lower the temperature required for the oxidation of the particulate matter. In Generation 2 (Fig. 13.20), the filter has incorporated in it a catalyst (containing Pt or a Pt/Pd alloy) to bring about the oxidation of the particulate C while, in Generation 3, all the constituent catalysts required are incorporated into a single unit.[9] (Carry out Task 13.4.)

Task 13.4 Control of Diesel Particulates

Examine the control of diesel particulates by catalytic means along the lines shown in the main text, particularly in relation to hindering SO_2 oxidation while encouraging NO_2 formation. Use as a possible starting point for a citation search an article by E. Xue, K. Seshan, and J.R.H. Ross, Roles of supports, Pt loading, and Pt dispersion in the oxidation of NO to NO_2 and of SO_2 to SO_3, Appl. Catal., B 11 (1996) 65–79, where it was shown that the relative rates of the two reactions is very dependent on the catalyst support used.

13.4 Control of Volatile Organic Compounds

Another area in which catalysis is applied is in the control of VOCs. The catalysts used for VOC control are in many ways similar to those used for automobile exhaust cleanup as they often require the use of a noble metal (generally Pt or Pd) wash-coated onto a ceramic support. These catalysts then bring about the total oxidation of any hydrocarbon species in the effluent gas stream being treated. However, the situation is generally less complex since the reaction conditions are generally less restrictive than those encountered in a car exhaust catalyst. The temperature is generally rather lower (25–400°C) and the concentration of organic molecules is also quite low (100–1000 ppm). (The technology used is also similar to that used for the catalytic combustion of natural gas discussed in Chapter 9: CO_2 and Energy.) The subject of VOC control has been reviewed by many authors and there are also several books dealing with this subject. For this reason, the topic of VOC control by catalytic combustion methods will not be discussed further here. The interested reader is advised to follow up on the references given in Task 13.5 and in Boxes 13.5 and 13.6.

Task 13.5 VOC Control

A review by J.J. Spivey, Complete catalytic oxidation of volatile organics, Ind. Eng. Chem. Res. 26 (1987) 2165–2180, has been highly cited. Use the citations of this article to examine the more recent literature on the subject of VOC control.

It would also be instructive to carry out a web search using such search terms as "VOC control," "air pollution control equipment," and "odor abatement," refining your search to find information on the types of technology available commercially.

BOX 13.5 Some Less Usual Applications of VOC Control

Coffee Roasting

This author first encountered emission control technology when, as a young academic, he was asked to try to find out a way in which a catalytic gauze material that had been used for odor abatement purposes could be regenerated after having become deactivated. This sample came from a company that was roasting coffee beans and had been having problems with neighbors who were finding that the aromas that were emitted to the surroundings were overpowering. The company had therefore installed a catalytic oxidation system fitted with a catalyst in the form of a compact netting of tape-like wires that was intended to work at a temperature close to that of the exit of the coffee roaster. The catalyst had worked well for some time but was now allowing the coffee aromas to be detected once more; more complaints were now being made and pressure was being applied to the company by the local authority to resolve the

(*Continued*)

BOX 13.5 Some Less Usual Applications of VOC Control (Continued)

problem. They had been advised by the suppliers of the technology to "wash the catalyst" but, needless to say, simple washing in water had no effect. The author had available a simple experimental rig that was used in undergraduate teaching to follow the oxidation of CO over a Pt/C catalyst and he was able to modify it to test samples of the catalytic tape for their catalytic activity for CO oxidation. The sample as supplied was inactive and no method of washing had any effect; however, when the sample was heated in air at about 200°C and the experiment was repeated, it exhibited good activity. The company was therefore advised to "roast" their catalyst under similar conditions to those used for the coffee beans. There were, to the best of my knowledge, no further complaints.

Ozone Control

Ozone, when formed at ground level by reactions between NO_x and VOCs in the presence of sunlight, contributes to various problems such as the creation of smog. It can be controlled by so-called passive catalytic technologies, as illustrated in the accompanying figures. In the case illustrated, the catalyst is incorporated into a structure that is placed either in the radiator of a car or in the entrance vents of an air conditioning system in such a way that the reaction:

$$2O_3 \rightarrow 3O_2$$

occurs. The catalysts used contain noble metals, predominantly Pt or Pd. No special control of these systems is required and virtually 100% of the ozone in the contaminated air entering the devices is converted. (See Fig. 13.22.)

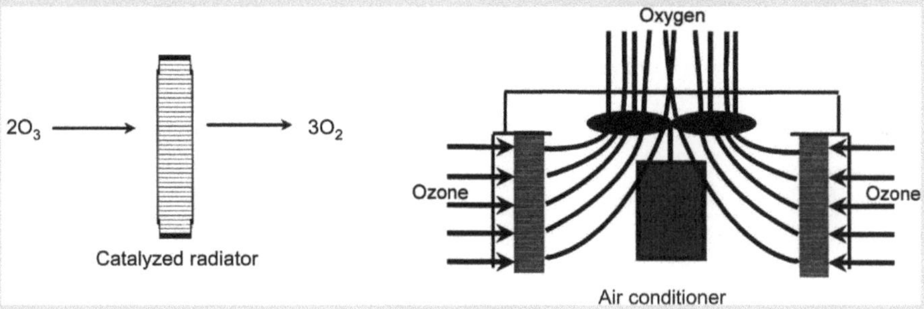

Figure 13.22

Passive devices for the control of ozone on car radiators and in air conditioning systems. *Taken from R.J. Farrauto, R.M. Heck, Environmental catalysis into the 21st century, Catal. Today 55 (2000) 179–187; reproduced with permission of Elsevier.*

Self-cleaning Glass

Self-cleaning glass is an interesting application of passive technology. The glass is coated with a thin film of titania, a photocatalyst (see Box 13.5), deposited onto the glass surface by coating it with a solution containing a suitable Ti complex, followed by calcination to form the anatase phase. The photocatalyst is then used to decompose, under the action of solar radiation, any organic compounds that become attached to the glass. Similar technologies are applied for the protection against fingerprints of the screens of tablets and smart phones.

BOX 13.6 Photocatalysis in Water Purification

Photocatalysis is an important field of research and there are many practical applications including the use self-cleaning glass mentioned in Box 13.5. The topic of photocatalysis as applied to the removal of trace contaminant from water has been comprehensively reviewed by S. Malato, P. Fernández-Ibáñez, M.I. Maldonado, J. Blanco, and W. Gemjak, Decontamination and disinfection of water by solar photocatalysis: recent overview and trends, Catal. Today 147 (2009) 1–59. Although other photocatalysts are available, the most commonly used one is titania (TiO_2) and this is available as a fine powder that is added to the water to be treated in a suitable vessel that can be exposed to sunlight. The accompanying figure (Fig. 13.23), taken from the reference cited earlier, shows for a typical demonstration plant, on the left, the handling system required to separate out the TiO_2 added to the water being treated and, on the right, the solar array through which the water containing the catalyst slurry is passed.

Figure 13.23
Photographs of a solar detoxification demonstration plant constructed at HIDROCEN (Madrid, Spain). Left: TiO_2 separation system. Right: compound parabolic collector.
From reference given earlier; reproduced with permission of Elsevier.

CHAPTER 14

Conversion of Hydrocarbons to Intermediates by Catalytic Hydrogenation and Oxidation Routes

Chapter Outline

14.1 Introduction

The fields of petrochemicals and industrial organic chemistry, including pharmaceutical applications, are immense and will only be given relatively brief treatment in this text, the emphasis being on selective hydrogenation and selective oxidation reactions. In the past, many important organic chemical reactions were carried out industrially using stoichiometric reaction systems and little attention was given to factors such as environmental concerns. For example, the oxidation of a variety of molecules was commonly carried out using stoichiometric quantities of reagents such as potassium permanganate or dichromate and no catalyst was involved; these reactions were therefore accompanied by the production of by-products that presented serious disposal problems. Hence, there is currently a great deal of interest in substituting these potentially polluting technologies by "green" heterogeneous catalytic processes exhibiting high selectivities to the desired products and resulting in minimal quantities of undesirable side-products. Most of the various processes based on the conversion of hydrocarbon species can be divided into two distinct categories: those carried out under reducing conditions and those carried out under oxidizing conditions. In the following sections, we will concentrate firstly on a number of hydrogenation reactions and then on some heterogeneous oxidation processes.

Contemporary Catalysis.
DOI: https://doi.org/10.1016/B978-0-444-63474-0.00014-X

As we will see, hydrogenation is achieved almost exclusively using metallic catalysts while selective oxidation generally involves oxidic materials, often combinations of two or more oxides. We are now dealing with reactions exhibiting high selectivity since the products are seldom thermodynamically the most favored and hence their production is kinetically controlled.

14.2 Catalytic Hydrogenation Reactions

Almost all hydrogenation reactions that have been examined involve selective processes since the products could in almost all cases be converted further to less desirable products. One of the simplest examples of such a selective process is the hydrogenation of ethylene (ethene) to give ethane:

$$C_2H_4 + H_2 \rightarrow C_2H_6 \quad (14.1)$$

a reaction that can be catalyzed by almost all the transition metals. In that case, further reaction to give two molecules of methane is thermodynamically permissible[1]:

$$C_2H_6 + H_2 \rightarrow 2\,CH_4 \quad (14.2)$$

In this simple case, the desired selectivity is achieved by operating at a temperature that is high enough to permit step (14.1) to occur but not sufficiently high that step (14.2) can take place, at least not to any significant degree. Reactions (14.1) and (14.2) have no practical application in industrial practice since ethylene (generally produced by thermal cracking of naphtha (Chapter 10)) is more valuable that is ethane or methane. However, a closely related reaction, the selective hydrogenation of acetylene (ethyne) in the presence of ethylene is of industrial significance:

$$C_2H_2 + H_2 \rightarrow C_2H_4 \quad (14.3)$$

Acetylene is produced in relatively low concentrations during the thermal cracking process and it would be a poison for any subsequent catalytic process for the conversion of the ethylene, being more strongly adsorbed on the catalysts used than is ethylene. It is therefore desirable to be able to hydrogenate one of the C–C bonds in the acetylene to give ethylene by reaction (14.3) without further hydrogenation to ethane by reaction (14.1), this hydrogenation being carried out in the presence of excess ethylene. In the absence of added ethylene, a variety of transition metals are effective for the selective production of ethylene

[1] Yet another reaction could also have occurred, the decomposition of the methane formed to give C and H_2, but this is excluded in the current discussion. If it occurred, it would most probably bring about poisoning of the primary reaction.

from acetylene by reaction (14.1) although the best results are obtained with the noble metals. Bond[2] shows that the activities are in the order:

$$Pd \gg Rh,\ Pt > Ru > Ir > Os$$

Pd is also much more selective than are the other metals. In practical applications, the Pd catalysts used contain very low concentrations of the metal, typically 0.04 wt% Pd, and the support is most frequently alumina. The catalysts are frequently promoted by other metals (e.g., silver, copper, or gold), these giving significant improvements in the selectivities to ethylene, while selective poisons, typically CO, can also be added to the reaction mixture in low concentrations, this again giving much improved selectivities. This added CO apparently encourages the formation and subsequent desorption of di-adsorbed ethylene species rather than further hydrogenation. Work has also been carried out using unpromoted gold catalysts and it is reported that these can also have very high selectivities for reaction (14.3); in this case, the results obtained are very dependent on the way in which the catalyst is prepared.

There are many similar applications of selective hydrogenation of hydrocarbons, including the selective hydrogenation of butadiene to form a mixture of butenes (without or with isomerization). This reaction again requires that the reaction goes no further than the various butenes following the hydrogenation of the first bond:

$$C_4H_6 \rightarrow C_4H_8 \nrightarrow C_4H_{10} \tag{14.4}$$

Much attention has been focused on the nature of the C_4H_8 formed as this molecule can be *either* but-1-ene ($CH_2=CH-CH_2-CH_3$), formed by selective reduction of one of the double bonds of butadiene without any rearrangement, *or* the isomer formed by rearrangement of the remaining double bond to give but-2-ene ($CH_3-CH=CH-CH_3$). Further, the but-2-ene can be formed in either the *cis* or *trans* form, the actual isomeric distribution depending strongly on the catalysts and conditions used:

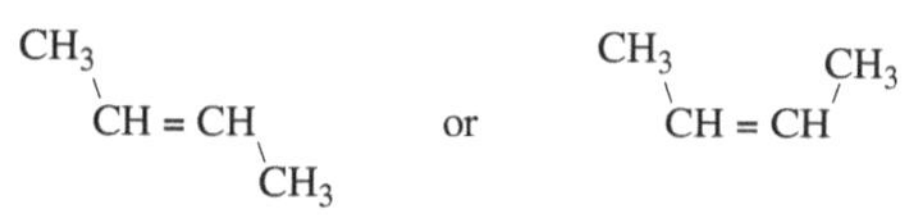

trans but-2-ene ***cis but-2-ene*** (14.5)

With more complex molecules, the possible range of products can become even more complex. We take as an example the selective hydrogenation of 1,5-hexadiene found as an impurity in a hex-1-ene feedstock, the aim being to convert the diene selectively into the 1-alkene, as shown in Fig. 14.1. It is undesirable to allow any isomerization to occur as that would result in a significant number of other possible by-products, both dienes and internal

[2] G.C. Bond, Metal-Catalysed Reactions of Hydrocarbons, Fundamental and Applied Catalysis series, Springer, NY, 2005.

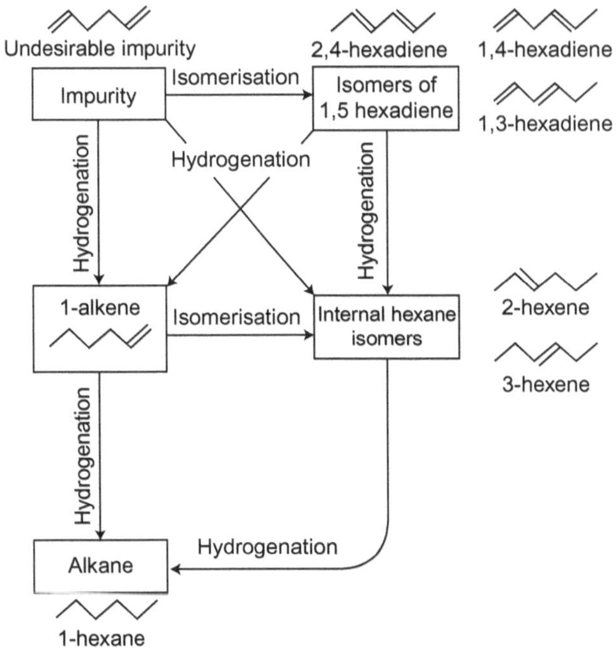

Figure 14.1
Reaction scheme for 1,5-hexadiene hydrogenation. *Taken from McEwan et al. (reference in Box 14.1), reproduced with kind permission of Springer.*

alkenes. It has been found that gold is particularly good at catalyzing this type of reaction very selectively as it gives total selectivity to the partially hydrogenated alkene; this desorbs and, in the presence of the multiply unsaturated species, is then not readsorbed to initiate the next step.

Much work has been carried out to examine the hydrogenation and isomerization steps in this and similar reactions with a wide variety of catalysts including Pd and Au. Much depends on the catalysts and supports used, the addition of any promoters and the actual reaction conditions applied. Further discussion of these parameters is beyond the scope of this text and the reader is referred for further information to some of the references given in Box 14.1 and hence, if desired, to the extensive and detailed research literature.

Selective hydrogenation has also been used for much more complex molecules for use in a whole range of processes. A number of such processes have been summarized by Chen et al. in a comprehensive review of the subject, listed in Box 14.1; some of these are shown in Table 14.1. The catalysts needed for many of these processes are available from suppliers such as Evonic or Johnson Matthey. The latter company offers a useful App entitled "Catalytic Reaction Guide" giving information regarding the uses of the various catalysts they supply.

BOX 14.1 Some Important Papers on Hydrogenation Reactions

A large number of industrial organic processes involve selective reduction reactions. These can use either homogeneous or heterogeneous catalysis. The applications of heterogeneous catalysis are many and varied, including the production of margarine and hydrogenation alkynes to alkenes. Examples of source references for these applications are listed as follows.

A. Borodziński and G.C. Bond, Selective hydrogenation of ethyne in ethene-rich streams on palladium catalysts. Part 1: Effect of changes to the catalyst during reaction, Catal. Rev. Sci. Eng. 48 (2006) 91–144. There are also many other papers describing the important work carried out over a long period by G.C. Bond and his colleagues. Some of this work is treated in the book by Bond detailed in footnote 1.

B. Nohair, C. Especel, G. Lafaye, P. Marécot, L.C. Hoang, J. Barbier, Palladium supported catalysts for selective hydrogenation of a sunflower oil, J. Mol. Catal. A (Chemical) 229 (2005) 117–126.

A. Molnar, A. Sarkany, M. Varga, Hydrogenation of carbon–carbon multiple bonds: chemo-, regio- and stereo-selectivity, J. Mol. Catal. 173 (2001) 185–221. As well as reviewing many aspects of the mechanistic features of the various different types of hydrogenation processes that have received academic attention, this review gives details of the purification of industrial feedstocks containing multiply unsaturated species.

An excellent general review of industrial processes (that also includes details of some homogeneous catalytic processes) that discusses many examples of selective hydrogenation in the production of drugs and fine chemicals has been written by a team of authors from Degussa: B. Chen, U. Dingerdissen, J.G.E. Krauter, H.G.J. Lansink Rotgerink, K. Mobus, D.J. Ostgard, P. Panster, T.H. Riermeier, S. Seebald, T. Tacke, H. Trauthwein, New developments in hydrogenation catalysis, particularly in the synthesis of fine and intermediate chemicals, Appl. Catal. A (General) 280 (2005) 17–46.

A very readable review on the use of gold catalysts is by L. McEwan, M. Julius, S. Roberts, J.C.Q. Fletcher, A review of the use of gold catalysts in selective hydrogenation reactions, Gold Bulletin 43 (2010) 298–306. (It should be noted that this journal is "open access.")

Two reviews on the use of gold catalysts: G.J. Hutchings, Supported gold and palladium catalysts for selective chemical synthesis, Catal. Today 138 (2008) 9–14; A. Hashmi, G.J. Hutchings, Gold catalysis, Angew. Chem. Int. Ed. 45 (2006) 7896–7936.

Hydrogenation reactions of oxygen-containing molecules are also applied in industry. A typical reaction is the conversion of acrolein by selective hydrogenation of the C = O group as shown in Fig. 14.2. A parallel route involving hydrogenation of the C = C double bond is less economically desirable since the product, propionaldehyde, can be produced more cheaply by other routes. Clearly, it is necessary that the reaction does not proceed beyond the first step in either case since propanol is a much less valuable product. It has been found that a supported gold catalyst is particularly selective for the reaction even though the path (b) is favored both kinetically and thermodynamically. This appears to be because gold selectively hydrogenates CO bonds, probably as a result of the ability of gold to adsorb

Table 14.1: Some examples of selective hydrogenation reactions carried out in the fine chemicals and intermediate chemicals industries

Product	Process	Application	Catalyst
Butane-1,4-diol	Hydrogenation of butyne-1,4-diol (single step process or two-step process via butene-1,4-diol)	Monomers	Supported Ni, ABMC Pd—Ag/alumina, Pd—Pb, and Pd—Zn
Hexen-1-ols	Selective hydrogenation of sorbic acid, hexyn-1-ols or hexadienols	Fruit and vegetable fragrances and flavors	Lindlar-catalysts, supported Pd colloids, and ABMC
Methyl-2-hexenoates	Hydrogenation of methyl sorbate	Flavors	Pd/C
Linalool, citral	Selective hydrogenation of dehydrolinalool	Terpenic fragrant substances for scent compositions and cosmetic use	Pd, Pd—Mn, Pd—Pb, Pd—Bi, Pd—polymer, supported Pd colloids, and Pd/alumina
C14, C16, C18 monoene and diene alcohols or acetates	Selective hydrogenation of monoyne alcohols or enynols	Pheromones	Lindlar-catalysts, Pd/BaSO4, and Pd—Pb alloy
Pyrethrins	Partial hydrogenation of cis-propenyl pyrethroids	Insecticides	Lindlar-catalysts
Edible oils	Partial hydrogenation of natural oils (fatty acid hydrogenation)	Food additives	Supported Ni, ABMC, and Pd/C
β-Carotene	Selective hydrogenation of 15-b-dihydrocarotene or dehydrolinalool	Food additives (vitamin A)	Lindlar-catalysts
Biotin	Selective C = C double bond hydrogenation of an intermediate	Food additives (vitamin H)	ABMC
Menadiol diacetate	Selective C = C double bond hydrogenation of menadione	Food additives (vitamin K3 derivative)	Pd/C
Sertraline	Selective C = C double bond hydrogenation of an intermediate	Pharmaceuticals (antidepressant)	Pd/C
Betamethasone	Selective C = C double bond hydrogenation of an intermediate	Pharmaceuticals (glucocorticcides)	Pd/C, Pd/CaCO3
Dihydroergotamine	Ergotamine	Pharmaceuticals (antimigraine agent)	Pd/BaSO4

Source: *Reproduced with slight modification from Chen et al. (see Box. 14.1). ABMC: "activated base metal catalysts" such as Raney nickel; Lindlar catalysts: catalysts composed of Pd supported on calcium carbonate poisoned by lead or sulfur.*

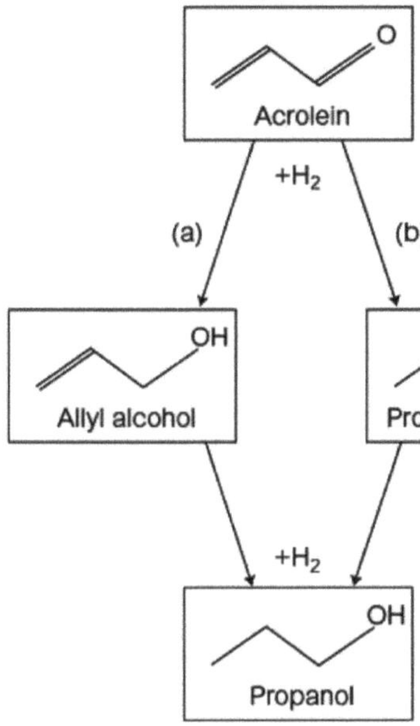

Figure 14.2
Reaction scheme for acrolein hydrogenation. *Taken from McEwan et al. (reference in Box 14.1), reproduced with kind permission of Springer.*

CO.[3] A similar reaction is the hydrogenation of crotonaldehyde (Fig. 14.3) using Au/ZrO_2 or Au/ZnO catalysts; this reaction using these gold catalysts gives high selectivities to crotyl alcohol (up to 80% at conversions of 5%–10%) by preferential hydrogenation of the carbonyl group.

Another important example of the selective hydrogenation of a molecule containing oxygen is the hydrogenation of maleic anhydride (MA), the production of which is discussed in Section 14.3. Fig. 14.4 shows a reaction scheme for the reaction; the primary hydrogenation product is succinic anhydride that can be converted to either γ-butyrolactone or 1,4-butanediol; the γ-butyrolactone can be converted further to tetrahydrofuran but it can also be converted reversibly to the diol. The final desired product, tetrahydrofuran, can also be formed by dehydration of the diol. The most commonly used catalyst for this reaction is either a Pd or a Raney Ni catalyst doped with Re. Computational methods have been applied by Neurock and his group (V. Pallasana, M. Neurock, G. Coulston, Towards understanding the mechanism for the selective hydrogenation of maleic anhydride to tetrahydrofuran over palladium, Catal. Today 50 (1999) 589–601). They have compared

[3] Full details of the experimental results associated with this scheme are to be found in a paper by P. Claus, Appl. Catal. A (General) 291 (2005) 222–229.

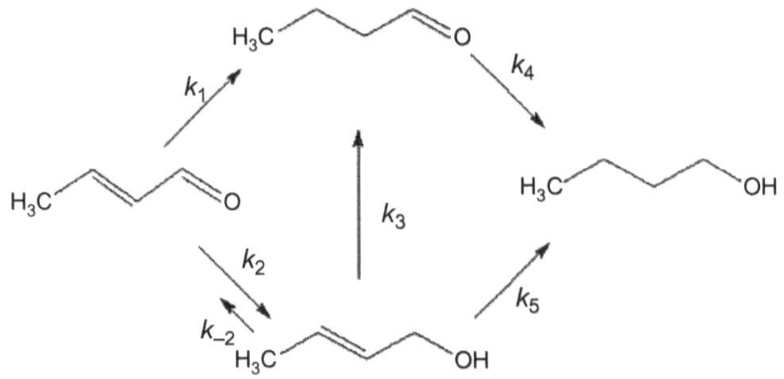

Figure 14.3
Reaction scheme for the hydrogenation of crotonaldehyde. *Reproduced from Hutchings (reference in Box 14.1), with kind permission of Elsevier.*

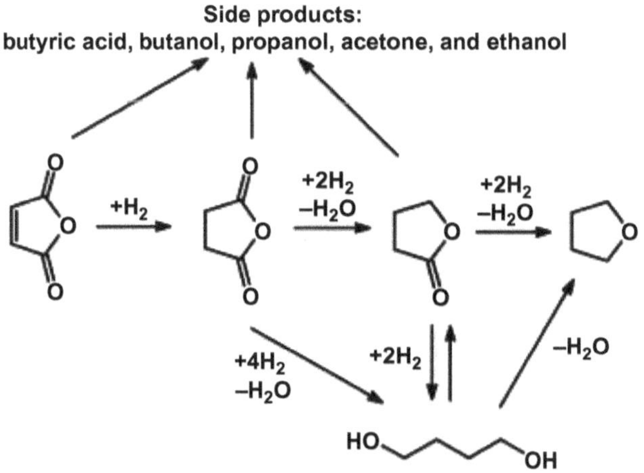

Figure 14.4
Scheme for the hydrogenation of maleic anhydride to produce tetrahydrofuran. *Reproduced from Chen et al. (reference in Box 14.1), with kind permission of Elsevier.*

experimental data with their calculated values for reaction using a Pd (111) cluster model for the active site and have obtained excellent agreement between the two.

In fine chemical applications, it is often desirable that the reactions occur enantioselectively to produce one of two optical isomers. It has been found that the addition of a reaction "modifier," generally itself optically active, will affect bring about the hydrogenation of a reactant (often termed the "substrate") over a suitable catalyst to give a single stereoisomer selectively. Fig. 14.5 shows as an example two routes for the hydrogenation of the substrate ethylpyruvate (etpy). The modifiers used are shown in the lower half of the diagram: cinchonidine (Cd, $R = CH = CH_2$) or dihydro-Cd (HCd, $R = C_2H_5$). In both cases, the

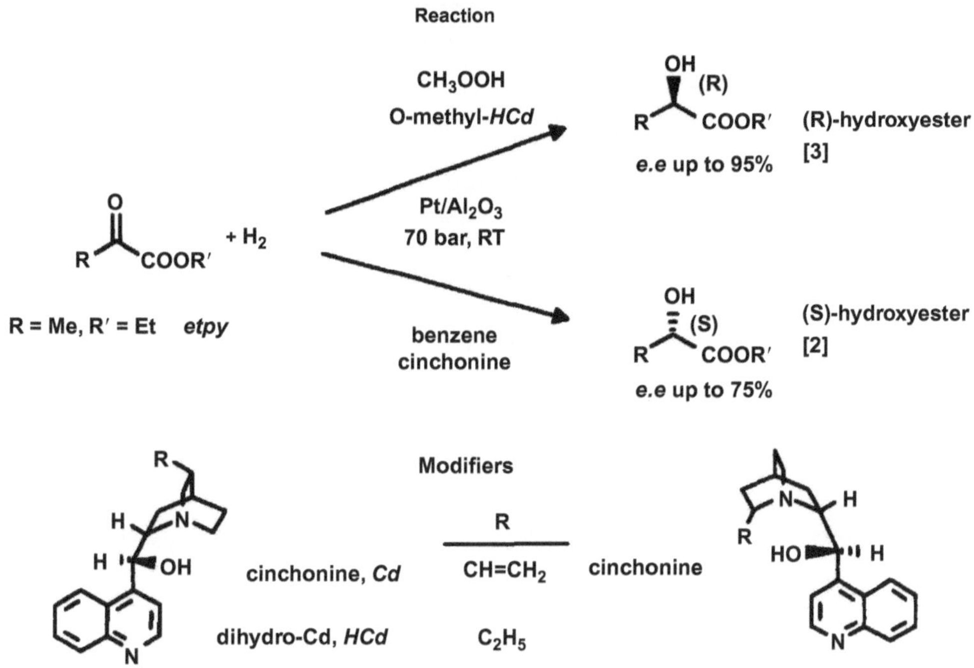

Figure 14.5

Hydrogenation of ketoesters catalyzed by cinchona-modified Pt catalysts. *Reproduced from Blaser et al. (reference second paper listed in Box 14.2), with kind permission of Elsevier.*

catalyst is Pt on alumina, the pressure is 70 bar and the temperature is 20°C. Using acetic acid as solvent and *o*-methyl substituted dihydrocinchonidine, the product of the hydrogenation of the carboxy group is the (R)-hydroxyester. On the other hand, if the solvent is benzene and the modifier is cinchoidine, the product is the (S)-hydroxyester. The review by Blaser and his colleagues from which this diagram was taken gives details of the experimental methods used and of effects of the choice of solvents, catalysts and modifiers. Other publications from Blaser's group and from other authors who have worked on this type of system are listed in Box 14.2. (See also Task 14.1.)

A fascinating application of selective reduction is that of molecular oxygen to give hydrogen peroxide as shown schematically in the Fig. 14.6. The reaction of hydrogen with oxygen generally produces only water but it has been shown by Hutchings and his colleagues (see references below) that supported Pd/Au alloys (support: silica, C, or titania) give good conversions and selectivities to hydrogen peroxide when the reaction is carried out in an autoclave. This conversion has to be carried out so that the gas composition is outside the explosion limits and this means that the H_2 concentration has to be below 4%. The reaction is carried out at ambient temperature and selectivities above 90% are obtained.

BOX 14.2 Papers and Reviews on Enantioselective Hydrogenation

H.U. Blaser, C. Malan, B. Pugin, F. Spindler, H. Steiner and M. Studer, Selective hydrogenation for fine chemicals: recent trends and developments, Adv. Synth. Catal. 345 (2003) 103–151.
H.U. Blaser, H.-P. Jalett, M. Müller and M. Studer, Enantioselective hydrogenation of α-ketones using cinchona modified Pt catalysts and related systems: a review, Catal. Today 37 (1997) 441–463.
H.U. Blaser, Heterogeneous catalysis for fine chemicals production, Catal. Today 60 (2000) 161–165.
H.U. Blaser, F. Spindler, A. Studer, Enantioselective catalysis in fine chemicals synthesis, Appl. Catal. A (General) 221 (2001) 119–143.
Another review from Blaser's group deals with not only selective hydrogenation but also hydrogenolysis, dehydrogenation and oxidation: H.U. Blaser, A. Indolese, A. Schnyder, H. Steiner, M. Studer, Supported palladium catalysts for fine chemicals synthesis, J. Mol. Catal. A (Chemical) 173 (2001) 3–18.
Baiker, Progress in asymmetric heterogeneous catalysis: design of novel chirally modified platinum metal catalysts, J. Mol. Catal. A (Chemical) 115 (1997) 473–493.
T. Mallat, E. Orglmeister, A. Baiker, Asymmetric catalysis at chiral metal surfaces, Chem. Rev. 107 (2007) 4863–4890.
P. McMorn, G.J. Hutchings, Heterogeneous enantioselective catalysts: strategies for the immobilisation of homogeneous catalysts, Chem. Soc. Rev. 33 (2004) 108–122.
G.C. Bond, P.A. Meheux, A. Ibbotson, P.B. Wells, Origin of enhanced rate in the platinum-catalysed enantioselective hydrogenation of methyl pyruvate, Catal. Today 10 (1991) 371–378.

Task 14.1 Enantioselective Catalysis

Using some of the references given in Box 14.2, use citation analysis to examine the more recent literature on enantioselective catalysis.

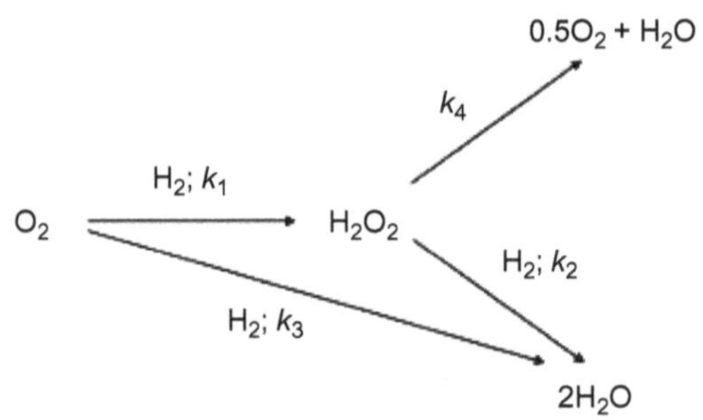

Figure 14.6
Reaction scheme for the selective oxidation of molecular oxygen to form hydrogen peroxide. *Reproduced from Hutchings (second reference to work by Hutchings in Box 14.1), with kind permission of Elsevier.*

(This reaction, although included in a section on selective reduction, could equally well have been included as an example of selective oxidation: the selective oxidation of hydrogen.) (See also Task 14.2.)

Task 14.2 Hydrogen Peroxide Formation Using Pd and Pd/Au Catalysts

Using the references to the work of Hutchings given in Box 14.1, examine some of the other work that has been carried out on the production of hydrogen peroxide from H_2/O_2 mixtures. Include in your study an examination of some of the papers published by J.H. Lunsford and his group on the reaction over supported Pd catalysts promoted by Cl^- or Br^- anions and by M. Haruta on the use of gold catalysts. Using citation analysis, examine any more recent work on this subject.

14.3 Selective Catalytic Oxidation

We have already encountered oxidation reactions in various situations, particularly in relation to the combustion of fuels (Chapter 9) and to the removal of volatile organic compounds (VOCs) (Chapter 13). We will now proceed to a short discussion of a vast field, that of the selective oxidation of various molecules, with particular attention being given to the selective oxidation of various different hydrocarbon species. Hydrocarbons from crude oil are the building blocks from which the vast majority of industrial chemicals in everyday use are produced and selective oxidation in one form or another is thus often the first step in the synthesis of these products. In general terms, any chemical process involves an increase of the value of the product relative to that of the starting material and oxidation processes are no exception to this rule: a molecule containing oxygen in the structure is almost invariably more valuable that a saturated hydrocarbon molecule. Similarly, if we can remove hydrogen oxidatively (or by straightforward dehydrogenation) to form the corresponding olefin, we have also ascended the value chain.[4] We will now first consider oxidative dehydrogenation reactions of some simple hydrocarbons and then proceed to discuss some important examples of the introduction of oxygen into hydrocarbon structures.

14.3.1 Dehydrogenation and Oxidative Dehydrogenation of Paraffins

Ethylene, propylene, and isobutene are important industrial intermediates, for example, in the production of polyethylene and polypropylene, and are normally made by noncatalytic steam cracking. As the demand for these olefins, particularly propylene, has increased, sources other than thermal cracking of naphtha or liquefied petroleum gas (LPG) (see Chapter 10) have been sought, these including oxidative routes to convert the ethane,

[4] This rule of thumb is not entirely general; in the case of biomass-derived materials (see Chapter 15: Catalysis in Biomass Conversion), the starting molecules (e.g., sugars or cellulose) often contain too much oxygen and so one must remove oxygen to go up the value chain.

Table 14.2: Some processes used for the dehydrogenation of light hydrocarbons

	Catofin	Oleflex	STAR	FDB-4
Reactor	Fixed bed Adiabatic	Moving bed Adiabatic	Fixed bed Isothermal	Fluidized bed Isothermal
Diluent	None	H_2	Steam	None
Pressure (atm)	0.33–0.66	>1	1–20	1.2–1.5
Regeneration frequency	7–15 min	Continuous	8 h	Continuous
Catalyst	Cr_2O_3/Al_2O_3	Pt/Al_2O_3	Pt/Al_2O_3	Cr_2O_3/Al_2O_3
Catalyst life (years)	2	>1	1–2	
Temperature (°C)	525–635		510–620	525–580
Conversion[a](%)	85–95	50	45–55	
Selectivity[a](%)	75–90	91	85–95	
Coke (%)	1.2	0.02		

[a]Conversions and selectivities are given for the reaction of isobutene to give isobutene.

propane, and butane fractions of LPG. Before proceeding to discuss oxidative dehydrogenation of these hydrocarbons, we will discuss briefly the processes that exist for the catalytic dehydrogenation of propane to produce propylene:[5]

$$C_3H_8 \leftrightarrows C_3H_6 + H_2 \tag{14.6}$$

This reaction is an endothermic reversible process ($\Delta H^\circ = +\ 117\ \text{kJ mol}^{-1}$) that requires relatively high temperatures (>600°C) and low pressures to obtain reasonable yields. The reaction is beset by the problem of coking so that relatively frequent regeneration of the catalysts used is generally necessary. Examples of the commercial processes used are Oleflex (Honeywell-UOP), Star (originally Phillips Petroleum, now supplied by ThyssenKrupp), Catofin (Clariant), and FDB-4 (Snam Progetti-Yarsintez).[6] Typical conversions are between 20% and 25% with selectivities between 90% and 95%. Table 14.2 gives some details of these processes, including the catalysts used.[7] (See also Task 14.3.)

Task 14.3 Processes for the Dehydrogenation of Propane and Isobutane

Do a web search on the Oleflex, Star, Catofin, and FDB-4 processes to learn more about the scope of these processes and the most up-to-date situation regarding their operation.

5 Strictly speaking, this is also an oxidation route and so, although it is not really within the context of this chapter, a short description is included for comparative purposes.

6 This process originated in the former Soviet Union and plants have been operating there for approximately 50 years. Yarsintez is a major company specializing in the production of synthetic rubber.

7 Data from R.H.H. Smits, PhD Thesis, University of Twente, 1994.

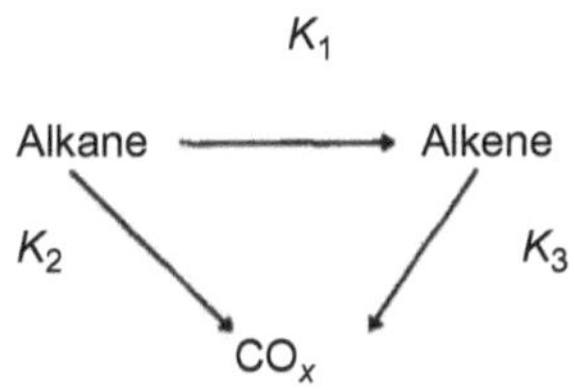

Figure 14.7
Scheme for the formation of an olefin from an alkane with parallel and sequential formation of CO_x. *Reproduced from Blasco and Nieto (T. Blasco, J.M.L. Nieto, Appl. Catal. A: General, 157 (1997) 117–142), with kind permission of Elsevier.*

To be competitive with the straightforward dehydrogenation routes, the oxidative dehydrogenative (ODH) reactions (14.7) and (14.8) that will now be discussed need to be able to achieve at least similar conversions and selectivities.

$$C_2H_6 + \frac{1}{2} O_2 \rightarrow C_2H_4 + H_2O \quad (14.7)$$

$$C_3H_8 + \frac{1}{2} O_2 \rightarrow C_3H_6 + H_2O \quad (14.8)$$

Both of these ODH reactions have received a significant amount of attention over the last couple of decades and much work was carried out in parallel to the work on the oxidative coupling of methane described in Chapter 11. As with methane coupling, the reactions occur in parallel to the production of CO_2 and H_2O and these two products can also be formed from the product olefin, as shown in Fig. 14.7.[8]

Assuming that the ODH reaction is first order, Blasco and colleagues have calculated the selectivities for the reaction as a function of conversion for two situations (Fig. 14.8), one in which the parallel oxidation reaction does not occur (A) and the other when the rate constants for both the ODH and total oxidation reactions are the same (B) for a series of different ratios of K_3 to K_1. It is clear that a wide variety of different selectivies will be obtained, depending strongly on the extent to which total oxidation occurs on the catalysts in question.

To illustrate this approach, Blasco and Nieto give the results shown in Fig. 14.9 for the ODH of ethane, propane, and butane over an unsupported V_2O_5 catalyst. The results for ethane behave in a way that corresponds to the data of Fig. 14.8A while those for propane and butane are closer in behavior to those of Fig. 14.8B. The implication of these results is that in the case of butane, step 1 of Fig. 14.7 for the formation of the olefin (with rate constant K_1) is fast compared with the parallel formation of oxidation products; (step 2) and that the majority of the CO_2 forms by further oxidation of the butene (step 3); in contrast, the rates of steps 1 and 3 are more similar for the cases of ethane and propane. It has been

[8] Strictly speaking, CO and H_2 are also possible additional products.

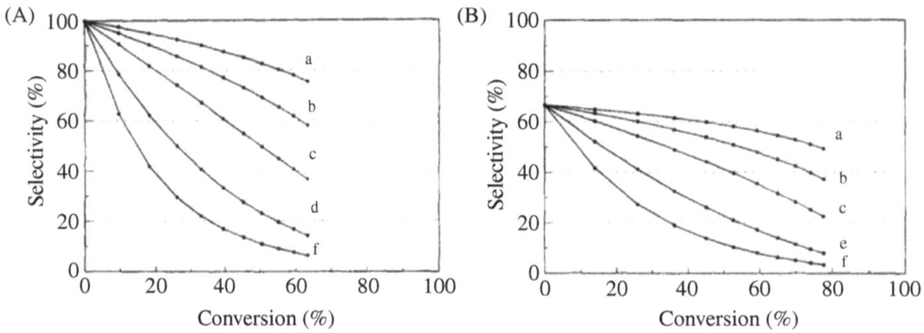

Figure 14.8

Variation of the selectivity to oxydehydrogenation products as a function of the conversion during the ODH of an alkane according to the scheme of Fig. 14.7. (A) $K_2/K_1 = 0.0$ and (B) $K_2/K_1 = 1.0$. The selectivities are calculated for a range of values of K_3/K_l: (a) 0.0; (b) 1.0; (c) 2.0; (d) 5.0; and (e) 10.0. *Reproduced from Blasco and Nieto (reference in Fig. 14.7), with kind permission of Elsevier.*

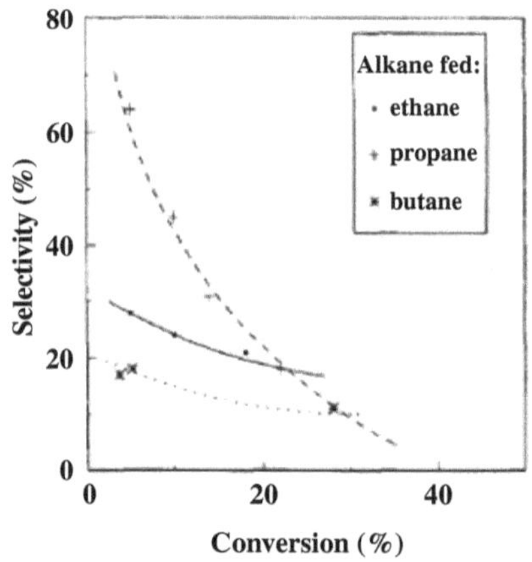

Figure 14.9

The selectivity as a function of conversion for the ODH of ethane, propane, and butane over unsupported V_2O_5. *Reproduced from Blasco and Nieto (reference Fig. 14.7), with kind permission of Elsevier.*

suggested that this difference in behavior may be related to the level of basicity of the olefins formed and so the acid/base behavior of the catalyst is imprtant in determining the selectivity of the reaction.

Fig. 14.10 shows some results demonstrating the effect of the support used for V_2O_5 catalysts for the ODH of ethane, propane, and *n*-butane. It can be seen that the more basic supports (20 wt% V_2O_5 on MgO or 27 wt% V_2O_5 on MgO/Al_2O_3) give better results than the more acidic ones (6.7 wt% V_2O_5 on MgO/SiO_2 (sepiolite) or 7 wt% V_2O_5 MgO/Al_2O_3) for the ODH of *n*-butane or propane but that the situation is reversed for the conversion of

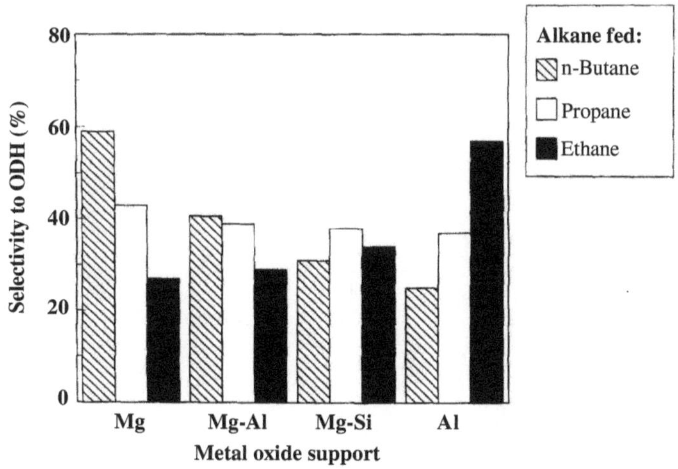

Figure 14.10

Selectivities to olefins for the ODH of ethane, propane, and *n*-butane over various supported V_2O_5 catalysts. Experimental conditions: $T = 550°C$ and an alkane/O_2/He molar ratio of 5/20/75 (*n*-butane) and 4/8/88 (ethane and propane); alkane conversion of 30%. *Reproduced from Blasco and Nieto (reference in Fig. 14.7), with kind permission of Elsevier.*

ethane. Blasco and his colleagues suggest that this is because ethylene is more acidic than are butene or propene and that the intermediate formed from ethane will thus be more easily desorbed from the more acidic catalysts. (See also Task 14.4.)

Task 14.4 Literature Study on Catalysts for the ODH Reaction

The following article, which discusses the reaction over V-based catalysts, can be used as a source on which a fuller literature study can be based:

T. Blasco, J.M. López Nieto, Oxidative dehydrogenation of short-chain alkanes on supported vanadium oxide catalysts, Appl. Catal. A (General) 157 (1997) 117–142.

Use this paper to examine the mechanism of the reaction and then use citation methods to examine more recent work on the subject.

14.3.2 Acetic Acid Production by Selective Oxidation of Ethane

One of the first papers reporting the oxidative dehydrogenation of ethane was that by Thorsteinson and his colleagues from Union Carbide, now Dow.[9] They showed that

[9] E.M. Thorsteinson, T.P. Wison, F.G. Young, P.H. Kasai, The oxidative dehydrogenation of ethane over catalysts containing mixed oxides of molybdenum and vanadium, J. Catal. 52 (1978) 116–132.

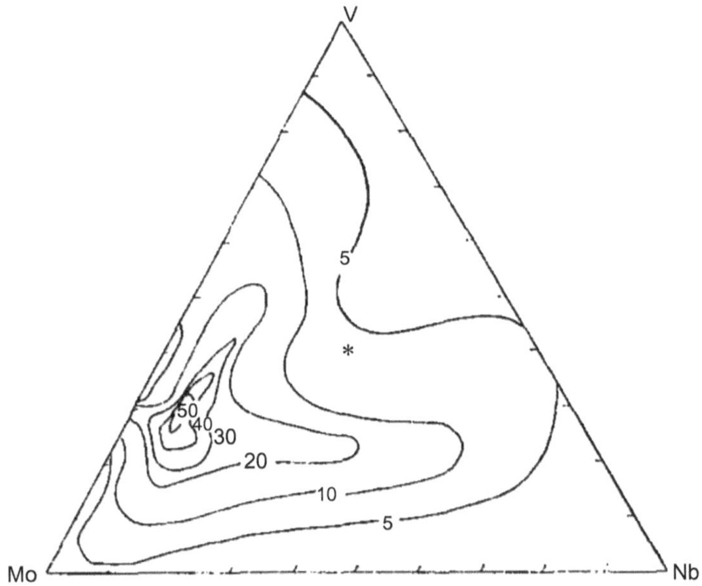

Figure 14.11
MoVNb composition diagram showing the contours for the conversion of ethane at 400°C.
Reproduced from reference of footnote 9, with kind permission of Elsevier.

catalysts containing molybdenum and vanadium with a third component such as niobium, titanium, or tantalum were particularly effective for the reaction. They examined the Mo, V, and Nb system in some detail and found (Fig. 14.11) that the optimal behavior occurred in a small molybdenum-rich region $Mo_{0.61-0.65}V_{0.31-0.27}Nb_{0.08}$ for samples supported on α-alumina; a somewhat broader set of compositional limits ($Mo_{0.61-0.77}V_{0.31-0.19}Nb_{0.08-0.04}$) was found for unsupported samples.

A particularly interesting result from the paper by Thorsteinson et al. was that significant proportions of acetic acid were formed by the subsequent oxidation of the ethylene formed, especially if the reaction was carried out at above atmospheric pressure. Fig. 14.12 shows a reaction scheme taken from a paper Rahman et al.[10] that includes acetic acid formation, showing that the acetic acid is formed from the ethylene by a subsequent oxidative reaction. These authors also showed that the acetic acid yield was significantly improved by the addition of water to the reaction mixture; while the maximum selectivity to acetic acid without water addition was about 50%, the selectivity

[10] F. Rahman, K.F. Loughlin, M.A. Al-Saleh, M.R. Saeed, N.M. Tukur, M.M. Hossain, K. Karim, A. Mamedov, Kinetics and mechanism of partial oxidation of ethane to ethylene and acetic acid over MoV type catalysts, Appl. Catal. A (General) 375 (2010) 17–25.

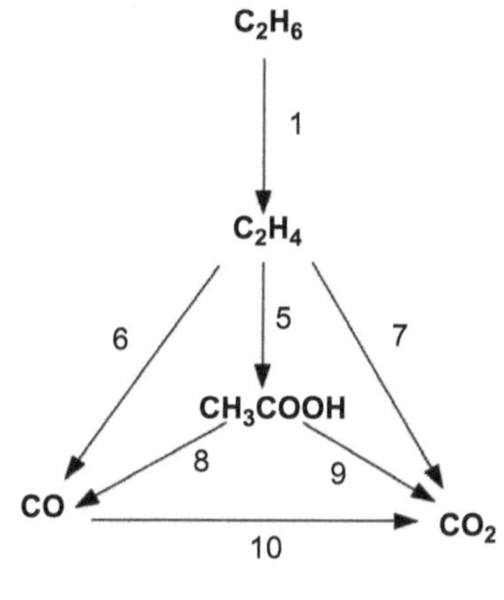

Figure 14.12
Possible reaction scheme for the formation of acetic acid in the ethane ODH system (with no H_2O in feed). *Reproduced from reference of footnote 10, with kind permission of Elsevier.*

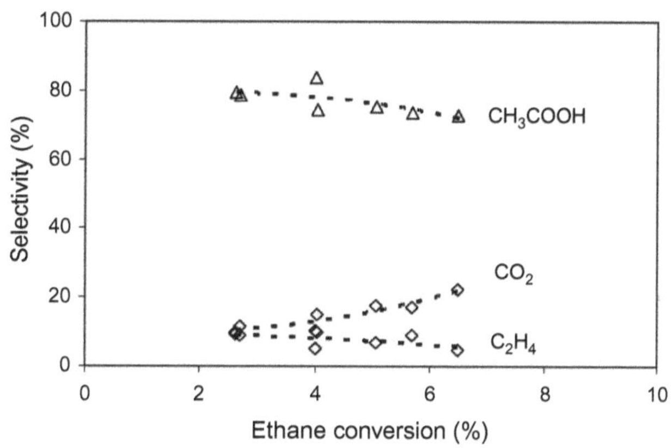

Figure 14.13
Product selectivity versus conversion of C_2H_6 with steam in feed at varying space velocity ($T = 553$ K, $P = 14$ bar, $C_2H_6 = 40$ mol%, $O_2 = 8$ mol%, $H_2O = 20$ mol%, and $N_2 = 32$ mol%). *Reproduced from reference of footnote 10, with kind permission of Elsevier.*

in the presence of water was up to 80% (see Fig. 14.13). It appears that the water promotes desorption of the adsorbed acetate species. A number of companies have now commercialized the production of acetic acid from ethane, these processes being in competition with the homogeneously catalyzed Monsanto and Cativa processes (involving CO insertion into methanol) described in Chapter 12. (See also Fig. 14.13; Task 14.5.)

Task 14.5 Acetic Acid Production From Ethane

Carry out a web search to find out about the current status of the "Sabox" process that has been introduced by Sabic for the production of acetic acid by the oxidation of ethane.[a] According to Ivars and López Nieto,[b] other companies that are involved in similar processes are Standard Oil Company, Mitsubishi Chemical Corporation, Celanese, and Hoechst Company (now a subsidiary of Sanofi). Include these companies in your search to find out how their processes differ from that used by Sabic.

[a] E. Mamedov, K. Karim, Selective oxidation at SABIC: innovative catalysts and technologies, in: D. Duprez, F. Cavani (Eds.), Handbook of Advanced Methods and Processes in Oxidation Catalysis—From Laboratory to Industry, Imperial College Press, 2014, pp. 291–301. This article gives details of the catalysts used in the Sabox process.

[b] F. Ivars, J.M. López Nieto, Light alkanes oxidation: targets reached and current challenges, in: D. Duprez, F. Cavani (Eds.), Handbook of Advanced Methods and Processes in Oxidation Catalysis—From Laboratory to Industry, Imperial College Press, 2014, pp. 767–834.

14.3.3 Oxidation and Ammoxidation of Propylene to Produce Acrolein and Acrylonitrile

Both acrolein and acrylonitrile are important monomers for the production of a variety of polymers as well as of synthetic rubber and a number of other chemicals, as discussed earlier in the chapter (see Fig. 14.2). One of the great success stories concerning research on the selective oxidation of alkanes comes from work carried out the Sohio labs, now part of the British Petroleum Company, that showed that it was possible to synthesize these molecules by selective oxidation or ammoxidation.[11] This research was initiated, in a very speculative manner, by Franklin Veatch in 1953 who realized that the incorporation of oxygen into a paraffin would have considerable economic value. Soon after, in 1955, Jim Callahan and Emily Ross found that they had produced acrolein by the oxidation of propylene over a bismuth phosphomolybdate catalyst:

$$CH_2 = CH - CH_3 + \tfrac{1}{2}\, O_2 \rightarrow CH_2 = CH - CHO \qquad (14.9)$$

It was then shown that the acrolein could be converted to acrylonitrile in a subsequent step:

$$CH_2 = CH - CHO + NH_3 + \tfrac{1}{2}\, O_2 \rightarrow CH_2 = CH - CN + 2\, H_2O \qquad (14.10)$$

Shortly after, in 1957, Idol and Evelyn Jonak produced acrylonitrile directly from propylene, with acetonitrile and hydrogen cyanide as coproducts, using the same bismuth phosphomolybdate catalyst. A very rapid commercialization of the ammoxidation process ensued and the first plant was ready for operation by mid-winter 1959–60. Work was then

[11] The development of the Sohio process is described in some detail in a commemorative booklet issued by the American Chemical Society in 1966. See https://www.acs.org/content/dam/acsorg/education/whatischemistry/landmarks/acrylonitrile/sohio-acrylonitrile-process-commemorative-booklet-1996.pdf.

carried out, using the same catalyst systems, on the production of acrolein and acrylic acid, both of which were also rapidly commercialized. Acrylonitrile production now rates as one of the highest volume processes worldwide and acrolein and acrylic acid production are also very important processes.

Following the initial work involved in developing and using the bismuth phosphomolybdate catalysts, a large number of new but related catalysts were discovered and there has since been a great deal of academic work aimed at gaining an understanding of the behavior of these catalysts. An example of such a modified composition, described in the type of shorthand used for these materials, is MoV(Nb)TeO. The catalyst contains two phases, designated as M1 and M2, the former being orthorhombic in structure and the latter being pseudo-hexagonal. The M1 phase performs the majority of the conversion to acrylic acid or acrylonitrile while the M2 phase carries out a "mop-up" function, bringing about reaction of any propylene formed as an intermediate. Fig. 14.14 shows a reaction scheme proposed by Grasselli and coworkers for the oxidation reaction (no ammonia added) over such catalysts

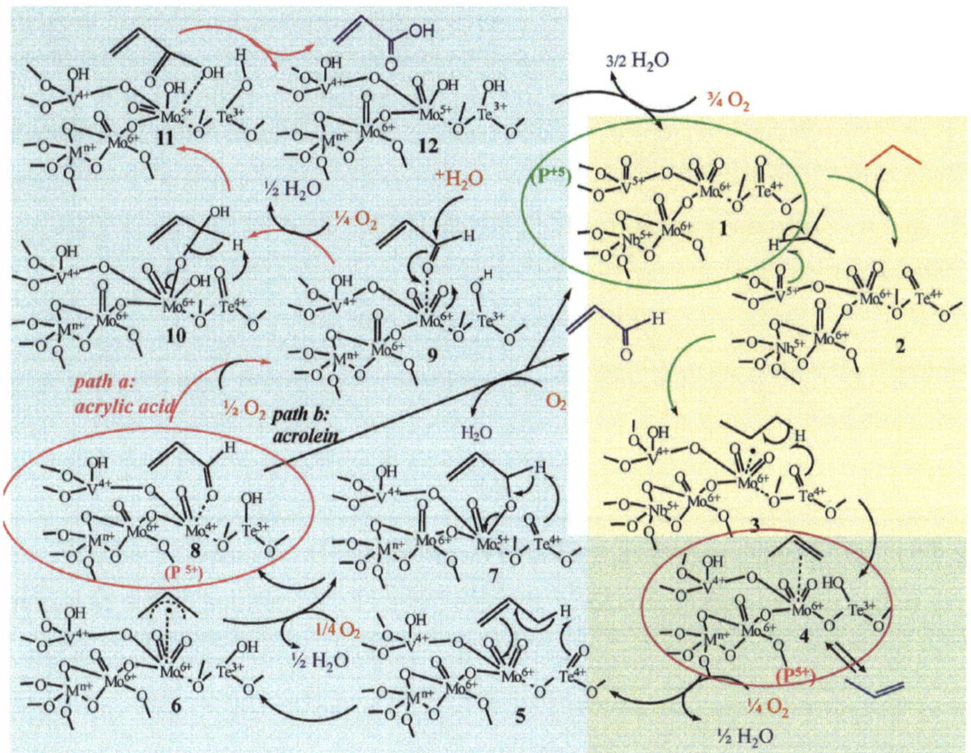

Figure 14.14

Proposed propane/propylene oxidation mechanism over MoV(Nb)TeO catalysts. *Reproduced from reference of footnote 14, with kind permission of Elsevier.*

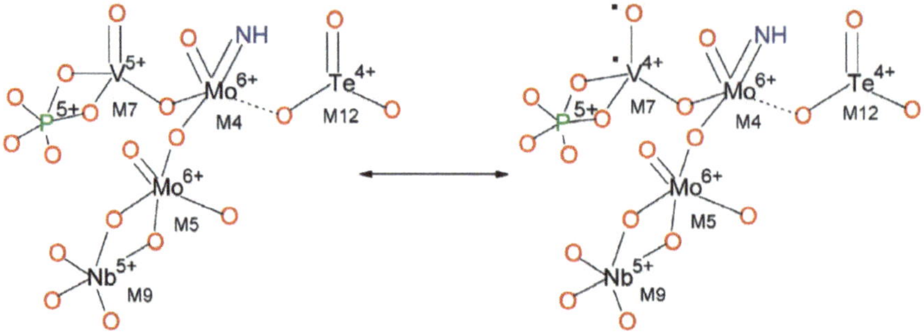

Figure 14.15

Schematic representation of the catalytically active center of $Mo_{7.8}V_{1.2}NbTe_{0.94}O_{28.9}$ *(M1 phase)*. *Reproduced from reference of footnote 14 with kind permission of Elsevier.*

doped with phosphorus.[12] (The phosphorus was added as it is known to possess substantial acidic properties and the consequence was that it increased the production of acrylic acid from the acrolein formed.) It is not the intention to examine the minute detail of the mechanism shown in Fig. 14.14 (although interested readers should follow up the detail by reading this and other papers by Grasselli and his colleagues). The important features to observe are the complex arrangements of the Mo, V, Nb, and Te entities in the structure and how oxidation/reduction steps occur on these. In other words, a Mars-Van Krevelen mechanism applies; see Chapter 7.

Grasselli and his colleagues argue that the phosphorus has an effect on the V sites, encouraging the reaction:

$$V^{5+} = O \leftrightarrow {}^{\bullet}V^{4+} - O^{\bullet} \tag{14.11}$$

to proceed to the right-hand side, this increasing the activity of the catalyst; see Fig. 14.15. (See also Task 14.6.)

Task 14.6 Literature on Acrylic Acid and Acrylonitrile Production

Use the reference given in footnote 13 to examine some of the literature on the development of catalysts for the selective oxidation and ammoxidation reactions of propylene.

12 R.K. Grasselli, C.G. Lugmair, A.F. Volpe Jr., A. Andersson, J.D. Burrington, Enhancement of acrylic acid yields in propane and propylene oxidation by selective P doping of MoV(Nb)TeO-based M1 and M2 catalysts, Catal. Today 157 (2010) 33–38.

14.3.4 Maleic Anhydride Production

MA is another important monomer used in polymer synthesis as well as an intermediate used in the preparation of many other important chemicals; an example was the production of tetrahydrofuran from MA by selective reduction as discussed earlier in this chapter (Fig. 14.4). The selective oxidation of butane to MA occurs as follows using a vanadium-phosphorus oxide (VPO) catalyst:

$$2\,C_4H_{10} + 7\,O_2 \longrightarrow$$ 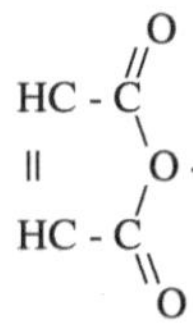 $$+\,8\,H_2O \qquad (14.12)$$

Before this process was introduced, the production of MA carried out by the selective oxidation of benzene:

$$C_6H_6 + 3\,½\,O_2 \rightarrow C_4H_2O_3 + 2\,CO_2 \qquad (14.13)$$

However, this reaction wastefully uses two carbon atoms, liberating carbon dioxide, and so the process involving the much cheaper molecule butane is now generally favored. As with the selective oxidation of propylene discussed in Section 14.3.3, there has been a great deal of academic work on the nature of the VPO catalyst and it is now generally accepted that the most active and selective catalyst is derived from a $VOHPO_4 \bullet 0.5H_2O$ precursor. This precursor can have a variety of structures and crystalline appearances (Fig. 14.16). Many aspects of the method of preparation and pretreatment of the catalyst are critical in determining the behavior of the catalyst, much also depending on the type of reactor system used.

As with other catalytic oxidation processes, it is important to operate outside explosion limits and this generally means that butane conversion to MA is carried out using an excess of oxygen. An interesting approach to avoid this problem is one introduced by DuPont[13] in which the catalyst is circulated between a reactor in which the butane interacts with the oxygen-containing catalyst to form MA and is then regenerated in another reactor, the oxygen and butane being well separated from one another (Fig. 14.17). The selectivity obtained with such a reactor (Fig. 14.18) is higher than that obtained with a more normal reactor. The catalyst used is produced by a granulation technique using silica as a binder

[13] R.M. Contractor, H.E. Bergna, H.S. Horowitz, C.M. Blackstone, B. Malone, C.C. Torardi, B. Griffiths, U. Chowdry, A.W. Sleight, Butane oxidation to maleic anhydride over vanadium phosphate catalysts, Catal. Today 1 (1987) 49–58.

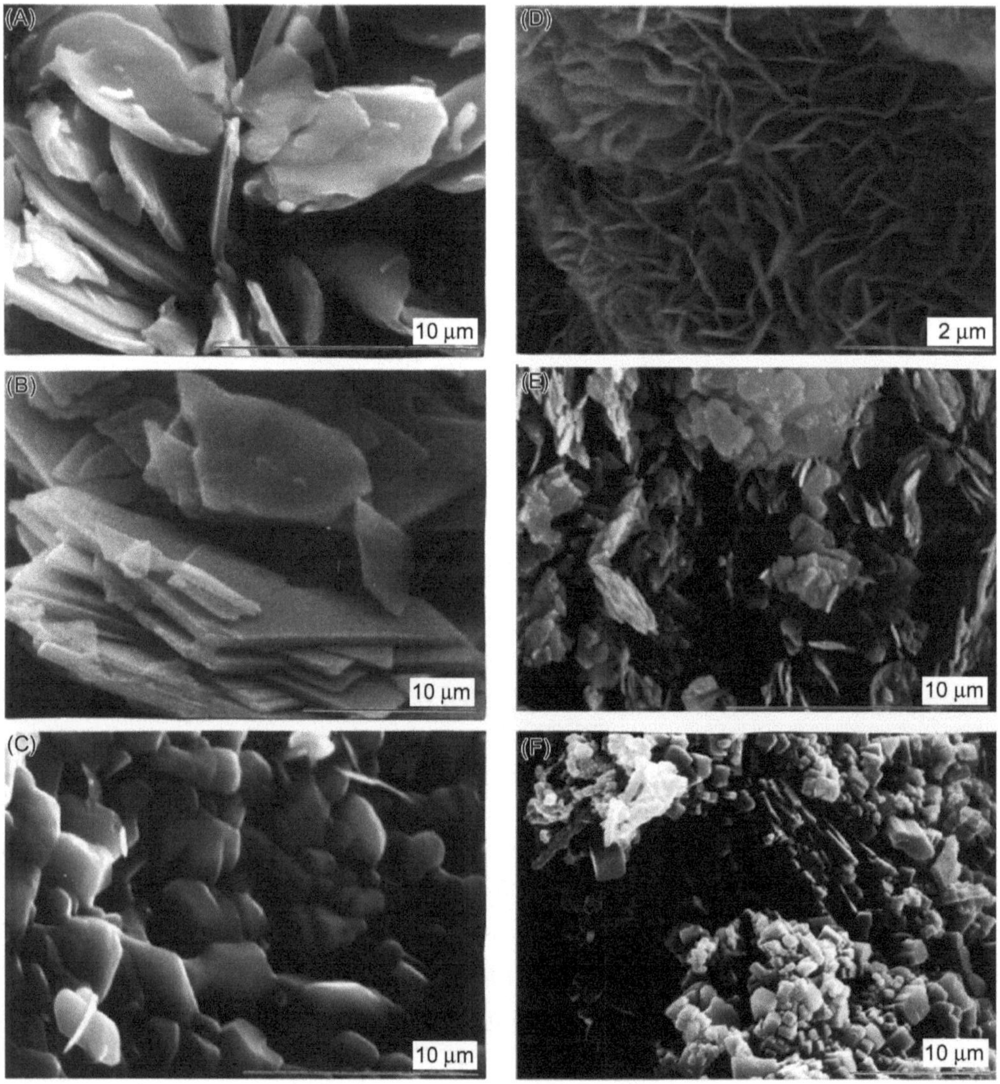

Figure 14.16
Electron micrographs of different phases of $VOPO_4$ *Reproduced from Kiely et al. (reference in Task 14.7), with kind permission of Elsevier.*

(Fig. 14.19). The advantage of the use of the material prepared using polysilicic acid is that the small amount of silica used does not cause any degradation of the catalyst properties. The authors point out that the VPO material used in this application is not really a catalyst and might be better referred to as an "inorganic carrier." (See also Task 14.7.)

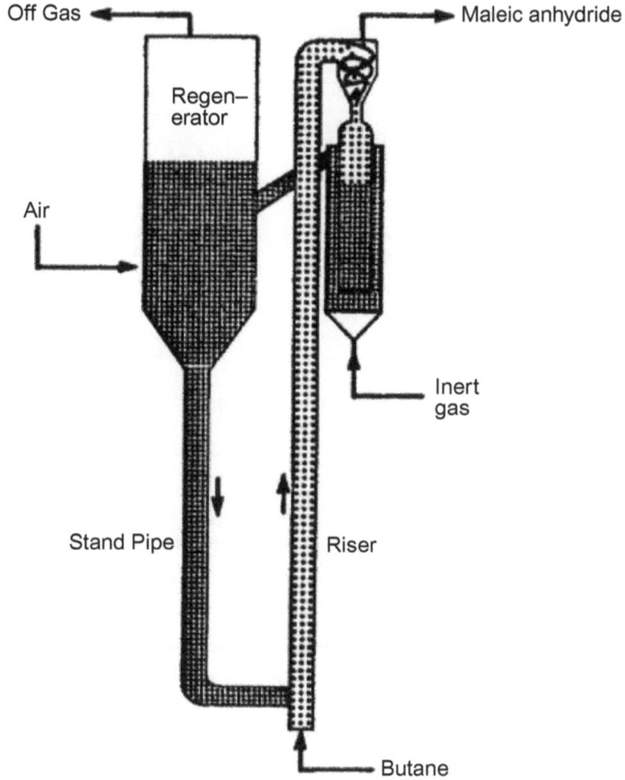

Figure 14.17

Schematic diagram of the DuPont recirculating solids reactor in which butane is oxidized in the riser section and the VPO catalyst is reoxidized in the regenerator section. *Reproduced from reference of footnote 15, with kind permission of Elsevier.*

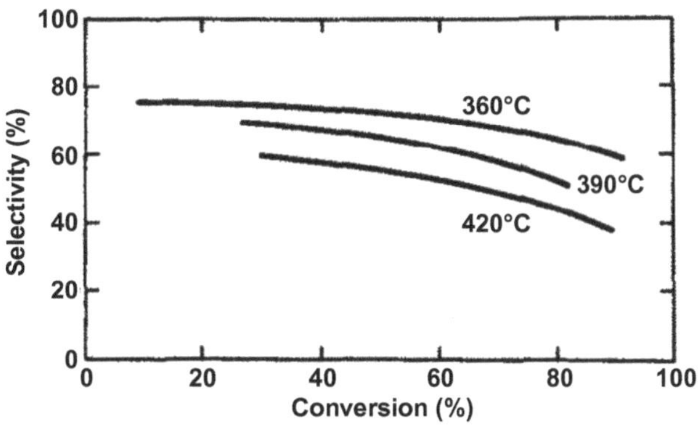

Figure 14.18

Selectivity of *n*-butane to maleic anhydride in a recirculating solids reactor at various conversions and temperatures. *Reproduced from reference of footnote 15, with kind permission of Elsevier.*

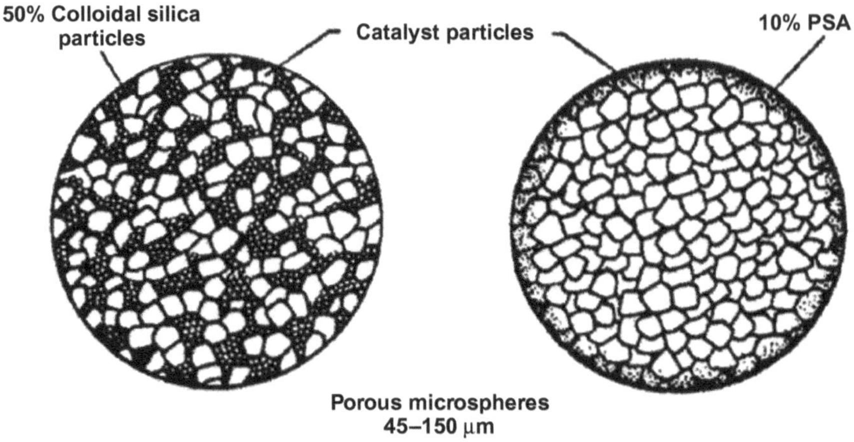

Figure 14.19
Schematic diagram of two types of spray-dried catalyst particles: material prepared with 50% of colloidal silica; and material prepared with 10% silica derived from polysilicic acid. *Reproduced from reference of footnote 15, with kind permission of Elsevier.*

Task 14.7 Literature on Maleic Anhydride Production

Carry out a literature study related to the catalysts for the selective oxidation of butane to MA. Examine also the various industrial processes used for the reaction.

For information on the catalyst structure, use the following as your source reference: C.J. Kiely, A. Burrows, S. Sajip, G.J. Hutchings, M.T. Sananes, A. Tuel, J.-C. Volta, Characterisation of variations in vanadium phosphate catalyst microstructure with preparation route, J. Catal. 162 (1996) 31–47.

The following reference gives information on the processes that were in commercial use in the 1980s: R.M. Contractor, A.W. Sleight, Maleic anhydride from C-4 feedstocks using fluidized bed reactors, Catal. Today 1 (1987) 587–607. Use it to source other papers dealing with this topic.

A useful recent article on the whole subject is: E. Bordes-Richards, A. Shekari, G.S. Patience, Vanadium-phosphorus oxide catalysts for n-butane selective oxidation: from catalyst synthesis to the industrial process, in: D. Duprez, F. Cavani (Eds.), Handbook of Advanced Methods and Processes in Oxidation Catalysis: From Laboratory to Industry, Imperial College Press, 2014, pp. 549–585.

14.3.5 Miscellaneous Oxidation Processes

The whole field of catalytic oxidation is enormous and no attempt is made here to cover every aspect of the subject. However, there are still a number of reactions and processes that deserve mention.

Production of phenol. The first of these reactions is called the AlphOx process. Developed in collaboration between the Boreskov Institute of Catalysis in Novosibirsk (Russia) and Monsanto, this is based on work by Panov and his colleagues who had discovered that it is possible to oxidize benzene to phenol using N_2O as oxidant over an iron-containing zeolite. The existing technology for phenol production is the Cumene process that involves a Friedel–Crafts alkylation of benzene with propene to give cumene (isopropylbenzene), followed by oxidation of this molecule to give cumyl hydroperoxide, and then the acidic decomposition of the latter to give phenol and acetone:

$$C_6H_6 + C_3H_6 \rightarrow C_6H_5 - CH(CH_3)_2 \rightarrow C_6H_5 - C(CH_3)_2OOH \rightarrow C_6H_5 - OH + (CH_3)_2C = O \tag{14.14}$$

The economics of the cumene process depends on the value of the acetone produced and this can vary significantly. This variation does not apply to the production route using N_2O in the AlphOx process. However, the viability of that route depends on the availability of N_2O. The sole commercial source of N_2O is currently the process used for adipic acid production, according to the following equation:

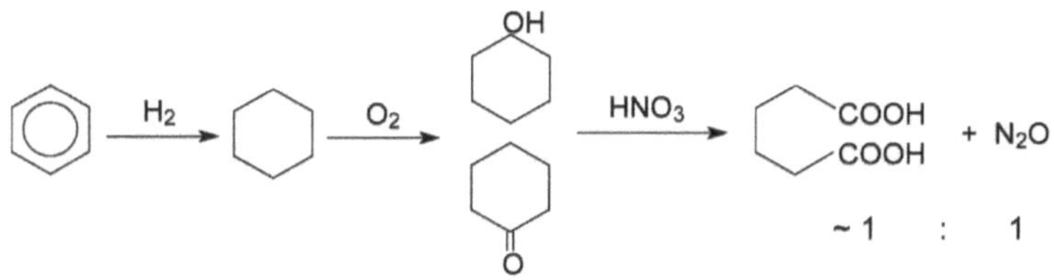

The supply of N_2O is therefore restricted to sites where adipic acid is manufactured and this is a severe limitation. Work is in progress to find other routes for the synthesis of N_2O (e.g., by the selective oxidation of ammonia) but this approach not so far been successful.

Another approach to the production of phenol is based on the discovery by Eni of the zeolite TS-1, a zeolite structure in which titanium species are incorporated in a silicalite structure as isolated terahedrally coordinated T species (see Task 1.14). This remarkable material has been found to catalyze a number of useful reactions, among them the oxidation of benzene to phenol using H_2O_2 as oxidant. Other reactions catalyzed by TS-1 include propene epoxidation, phenol hydroxylation, ammoxidation of cyclohexanone to ε-caprolactam (used for nylon-6 production), and hydrogen peroxide production.[14]

[14] For a detailed description of some of the processes developed by Eni, refer to the article: R. Buzzoni, M. Ricco, S. Rossini, C. Parego, Selective oxidations at Eni," in: D. Duprez, F. Cavani (Eds.), Handbook of Advanced Methods and Processes in Oxidation Catalysis: From Laboratory to Industry, Imperial College Press, 2014, pp. 353–381; C. Parego, A. Carati, P. Ingallina, M.A. Mantegazza, G. Bellussi, Production of titanium containing molecular sieves and their application in catalysis, Appl. Catal. A (General) 221 (2001) 63–72.

Oxidation processes catalyzed by gold. We have already considered the use of gold as a catalyst for hydrogenation processes. It has also been found that it is a very selective catalyst for the epoxidation of propylene to give propene oxide when the reaction is carried out in the presence of molecular hydrogen:

$$\underset{\text{propylene}}{CH_3CH{=}CH_2} + O_2 + H_2 \rightarrow \underset{\text{propylene oxide}}{CH_3\underset{\diagdown O \diagup}{CH{-}CH_2}} + H_2O$$

(Propylene oxide is an intermediate in the formation of polyurethane and polyols.) Haruta and his coworkers have reported that the support used for the gold as a catalyst for this reaction is extremely important. Although they found that TiO_2 as support gave reasonable results (Fig. 14.20), they have reported that much better results were obtained using TS-1 and other Ti-containing zeolites. The results for the TiO_2 support show quite clearly that preparation method is quite critical in obtaining the correct surface geometry. Haruta and his colleagues have speculated that peroxo species are formed on the catalyst surface, as shown in Fig. 14.21, and that these are critical for the formation of the propylene oxide.

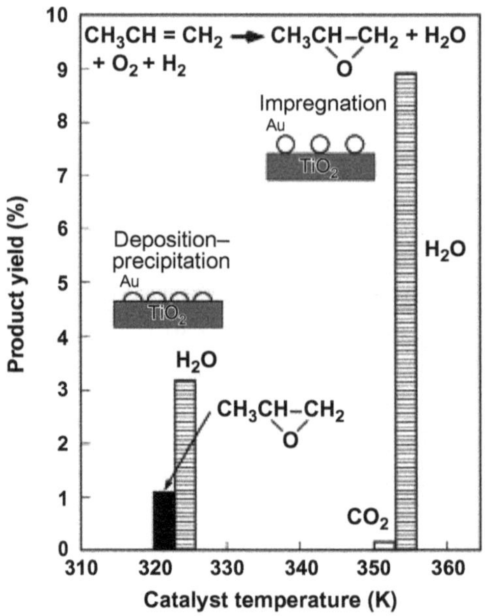

Figure 14.20

Epoxidation of propylene over two different types of Au/TiO_2 catalysts, showing that the preparation method is critical. *Reproduced from M. Haruta, M. Daté, Advances in the catalysis of Au nanoparticles, Appl. Catal. A (General) 222 (2001) 427–437, with kind permission of Elsevier.*

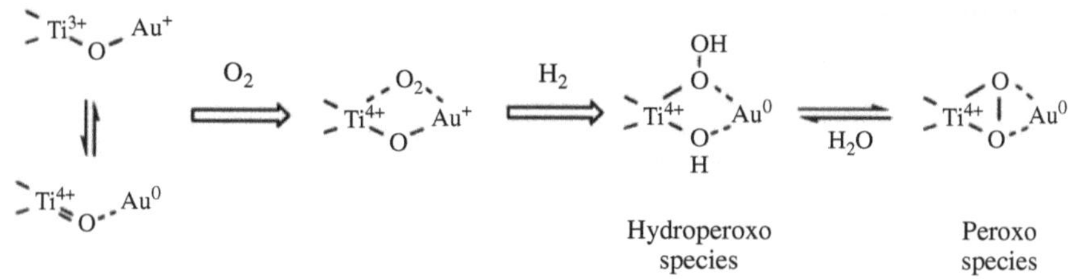

Figure 14.21
Speculative scheme for the formation of peroxo species on the surface of a Au–TiO_2 catalyst. *Reproduced from T. Hayashi, K. Tanaka, M. Haruta, Selective vapor-phase epoxidation of propylene over Au/TiO2 catalysts in the presence of oxygen and hydrogen, J. Catal. 178 (1998) 566–75, with kind permission of Elsevier.*

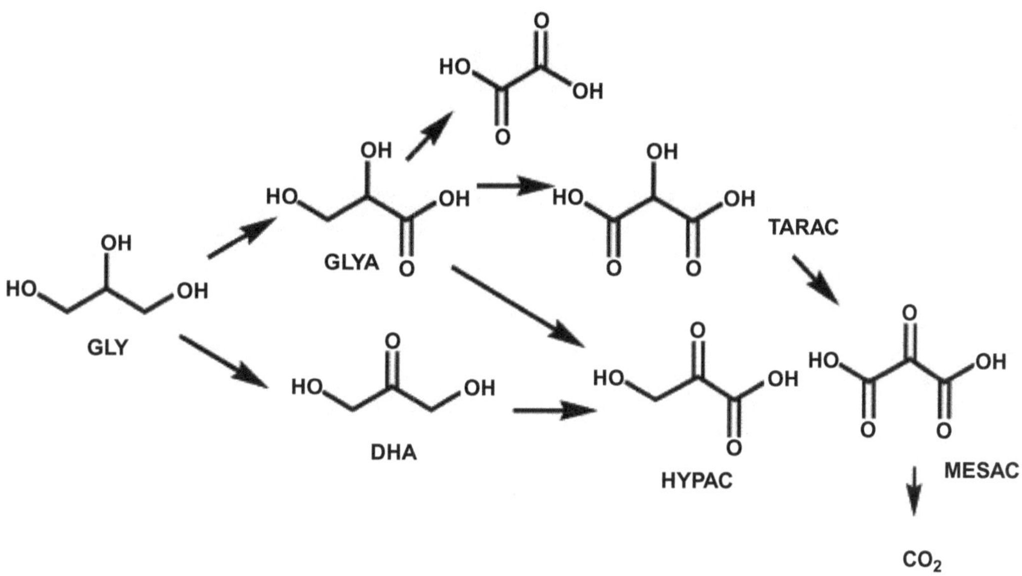

Figure 14.22
Reaction scheme for the oxidation of glycerol. *Reproduced from the reference of footnote 17, with kind permission of Elsevier.*

Other important oxidation reactions that can be carried out by gold catalysts are the oxidation of alcohols (including diols) and polyols (typically glucose and sorbitol).
Fig. 14.22 shows the various products that can be formed from glycerol. Results reported by Hutchings[15] show that when the reaction is carried out in an autoclave with added NaOH as base and using a 1%Au/C catalyst, the reaction was 100% selective to glyceric acid. (See also Task 14.8.)

[15] G.J. Hutchings, Catalysis by gold, Catal. Today 100 (2005) 55–61.

Task 14.8 Catalysis by Gold

Using the reference of footnote 17 as starting point, find out what you can about modern developments in the use of Au-containing catalysts. Concentrate on oxidation reactions but pay some attention also to the many other applications of gold catalysts.

Other oxidation processes. There are many other oxidation processes that could have been considered in this chapter. Many of these are used in various ways in fine chemicals production. Task 14.9 gives you the opportunity to examine some of these.

Task 14.9 Oxidation Catalysis in Fine Chemicals Production

For several useful reviews of the area of oxidation, you are referred to:

R.A. Sheldon, Heterogeneous catalytic oxidation in the manufacture of fine chemicals, in: M. Guisnet et al. (Eds.), Heterogeneous Catalysis and Fine Chemicals, II, Studies in Surface Science and Catalysis, 1991, pp. 33–54.

R.A. Sheldon, J. Dakka, Heterogeneous catalytic oxidations in the manufacture of fine chemicals, Catal. Today 19 (1994) 215–245.

R.A. Sheldon, I.W.C.E. Arends, A. Dijksman, New developments in catalytic alcohol oxidation, Catal. Today 57 (2000) 157–166.

You should also carry out citation-based searches on these papers in order to identify other relevant literature.

CHAPTER 15

Catalysis in Biomass Conversion

Chapter Outline

15.1 Introduction

This chapter deals with a currently very important topic: biomass conversion. This subject has received a great deal of attention over the last few years as it is generally recognized that, with depleting oil reserves, the world has to seek other sources of energy and also of some of the chemicals currently produced from hydrocarbons. While solar, wind, hydro, and tidal resources are all feasible sources of energy that are ready to be tapped in many regions of the world, they are not available universally and the engineering problems that need to be overcome for their application in all situations are very considerable. Although nuclear fission or fusion can also be used to provide a proportion of our energy needs, there are strong political pressures against the wider use of fission energy while fusion energy is still something for the future. Hence, the use of biomass has become of great importance, at least as an interim measure until alternative energy sources become fully and universally available. As a result, there are political pressures to introduce biomass into our energy palate. For example, in Europe, there is a mandatory target of 10% of energy requirements from biomass by 2020; in the United States, the Energy Independence and Security Act of 2007 requires 36 billion gallons of renewable fuels per year by 2022.

In principal, biomass can be used for the production of both fuels and chemicals. Indeed, processes have existed for many years for the production of chemicals from biomass; a well-known example is the production of alcohol from plants such as potatoes, corn, and sugar cane, all these containing sugar or starch. Alcohol production from sugar cane has become a very important process in Brazil where a significant proportion of motor fuel is provided by ethanol produced by fermentation. Some proportion of Europe's energy supplies already comes from biomass but these are dependent on such technologies as

Contemporary Catalysis.
DOI: https://doi.org/10.1016/B978-0-444-63474-0.00015-1

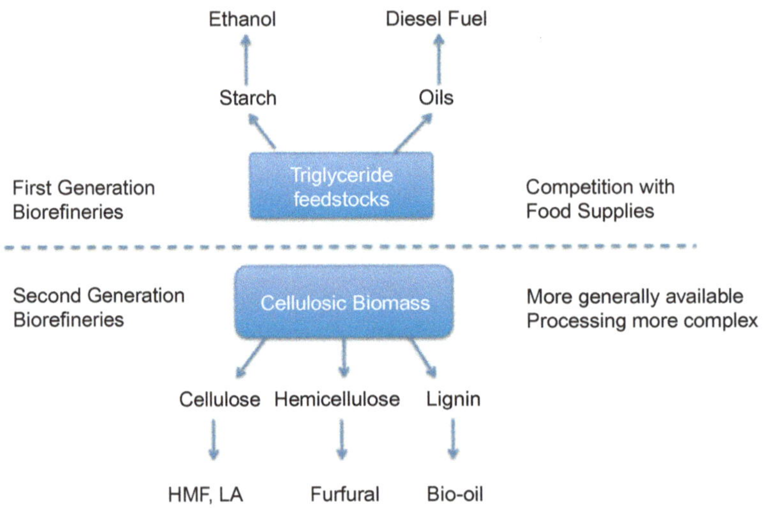

Figure 15.1
The so-called first and second generation biorefineries.

biomass combustion (i.e., burning wood and other woody biomass instead of coal and oil (see Chapter 9)) or from the use of biodiesel from so-called "first generation" biorefineries (see Fig. 15.1). As we will see in the following section (Section 15.2), these biorefineries use processes such as ethanol production or the catalytic transesterification of triglycerides contained in the raw biomass, this giving diesel fuels. The first-generation processes have therefore come under attack because of the competition they represent between the methods of supply of food and energy. Further, it has been argued that the use of these crops can have an all-over negative effect on green-house gas production as the NO_x emissions arising from the application of nitrate fertilizers to encourage plant growth will have a greater negative effect on global warming than the positive effect arising from the reduction of the use of fossil fuels.

Second-generation biorefineries (see Fig. 15.1) operate using lingo-cellulosic biomass as feedstock. This material can be derived largely from nonfood sources (although it is present in all crops) and is often waste material such as agricultural residues (e.g., straw, sugar cane bagasse), municipal waste and waste from wood processing. It may also include specially grown and easily harvested crops such as switch grass or miscanthus grass, these being plants that can be grown on set-aside land with minimal application of fertilizer. Lignocellulose is a complex mix of three separate components: cellulose, hemicellulose, and lignin. The linear polymeric sugar structure of the cellulose component that gives the structural strength to the material is surrounded by the hemicellulose and the lignin; in order to make use of it, the structure has first to be broken down into its separate components. The vegetation used for such processes generally contains only small amounts of the sucrose and the other related molecules predominating in the first generation materials.[1] The cellulosic biopolymer is made up of chains of sugar units which are so configured that there are strong hydrogen bonds between the chains as well as covalent bonding between the sugar rings (see Fig. 15.2); this hydrogen bonding is not found in the starch structure typical of the first-generation feedstock, this being made up of amylose and amylopectin units. It should be noted that the amylose structure is similar to the building units in the cellulose; however, it lacks the β-configuration in the linkage between the sugar rings necessary for hydrogen bonding between the chains that is found in cellulose and it is therefore relatively easily hydrolyzed. Because of the hydrogen bonding between the linear chains, the cellulose structure is notoriously difficult to break down (depolymerize) and so the most critical part of any second-generation plant is likely to be the preliminary cellulose treatment unit. The processes involved in second-generation plants (Fig. 15.1) will be considered in Section 15.3 and some of the subsequent processes used to produce useful chemicals such as hydroxymethylfurfural (HMF), levulinic acid (LA), furfural, and bio-oil will also be discussed. Particular attention will also be given (Section 15.4) to the production of hydrogen from formic acid, an important by-product from several second-generation processes. (See also Task 15.1.)

Task 15.1 Legislation Concerning Biomass as a Source of Energy

Carry out a web search regarding the specific requirements of different countries in relation to the supply of energy from biomass; include your own country in the search. How many of these also mention the supply of chemicals from biomass?

BOX 15.1 Reviews on Biomass Conversion

A useful review on biomass usage and the relevant processes is given by D.J. Hayes in: "An examination of biorefining processes, catalysts and challenges", Catal. Today 145 (2009) 138–151. A review by G.W. Huber et al. covers many aspects of the background to biomass conversion as well as giving an excellent overview of biomass conversion processes: G.W. Huber, S. Iborra, and A. Corma, "Synthesis of transport fuels from biomass: chemistry, catalysts and engineering", Chem. Rev. 106 (2006) 4044–4098. A further review by the same group gives additional material: A. Corma, S. Iborra, and A. Velty, "Chemical routes for the transformation of biomass into chemicals", Chem. Rev. 107 (2007) 2411–2502. These latter two reviews are available as open-access items. Further, the references given in the reviews provide useful entries by applying citation searching to the various different topics covered.

[1] Second generation feedstocks include specially grown crops such as miscanthus grass or switch grass but they may be the residues from first generation production such as sugar bagasse or straw. Some second generation processes such as Biofine (see Section 15.3) can also handle waste paper and municipal wastes.

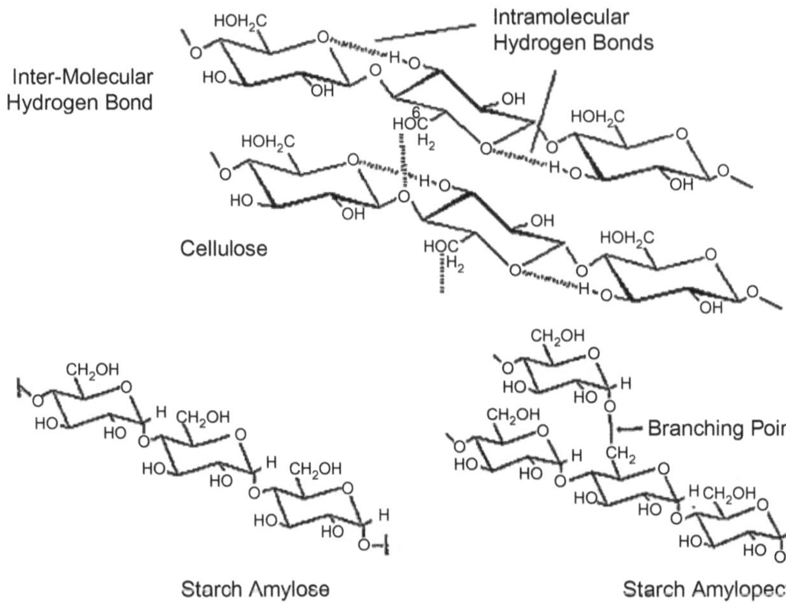

Figure 15.2
The structures of cellulose and of the components of starch, amylase, and amylopectin. *Reproduced from D.J. Hayes, Catal. Today 145 (2009) 138–151, with kind permission of Elsevier.*

15.2 First Generation Biorefineries

Fig. 15.3 shows schematically some of the processes used in a typical first-generation biorefinery. These are subdivided into chemical, thermochemical, and biochemical. We will now consider the chemical and thermochemical routes in turn; it should be recognized that some of these routes are also applicable in second-generation plants. Biochemical routes will not be dealt with here, apart from noting that digestion of the biomass gives biogas and methane, used for energy purposes, while fermentation gives ethanol and/or butanol.

The main chemical transformation route in a first generation biorefinery involves so-called transesterification (Fig. 15.4). This involves treatment of the triglyceride from the biomass in either acidic or basic solution to hydrolyze the ester linkages; this is done in the presence of an alcohol (generally methanol, ethanol, or butanol) to bring about esterification of the long-chain groups from the triglycerides. After purification, the resultant straight-chain esters are suitable for use as biodiesel. Glycerol is formed in the initial hydrolysis step and it is separated and used for the production of chemicals or fuels; in common with many of the side-products of other biomass-based processes, it can also be steam reformed to produce syngas and hence fuels or chemicals.

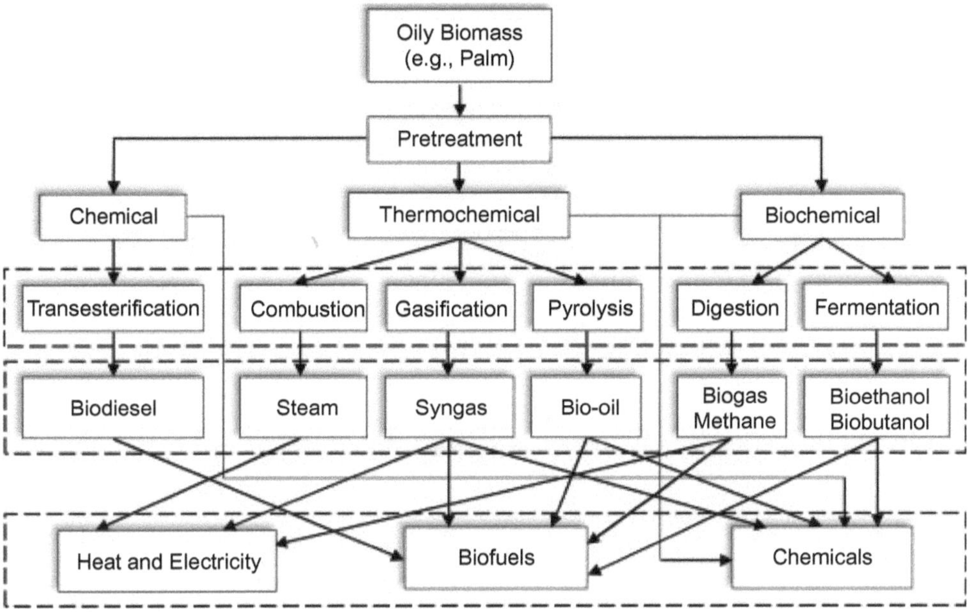

Figure 15.3

Biomass transformation pathways used for first-generation biorefineries together with the key products and their main uses. *Reproduced from C.A.C. Alzate, J.C.S. Toro, A.G. Pena, Fermentation, thermochemical and catalytic processes in the transformation of biomass through efficient biorefineries, Catal. Today 302 (2018) 61–72, with kind permission of Elsevier.*

Thermochemical routes (also usable with second-generation crops) involve combustion, gasification, or pyrolysis. Combustion is used to produce thermal energy when required and this requires no further discussion. Gasification is similar to the process used for the production of syngas for the Fischer–Tropsch and similar processes and is a well-used option. It requires processing of the biomass at high temperatures and pressures. Fig. 15.5 is a schematic representation of the options available for gasification technology; this is taken from a comprehensive review on the subject by G.J. Stiegel and R.C. Maxwell: "Gasification technologies: the path to clean, affordable energy in the 21st century", Fuel Process. Technol. 71 (2001) 79–97.[2]

While syngas production itself does not require catalysis, there are often unwanted product molecules, including methane, hydrocarbons, and tars, in the syngas ($CO + H_2$) produced and these must be removed by including a catalyst in the gasification system. This type of gas clean-up is quite specific to gasification technology. A wide variety of different catalysts have been examined for this purpose, these including dolomite, alkali-metal

[2] A number of the citations to this article that will be found in a citation search refer directly to biomass gasification.

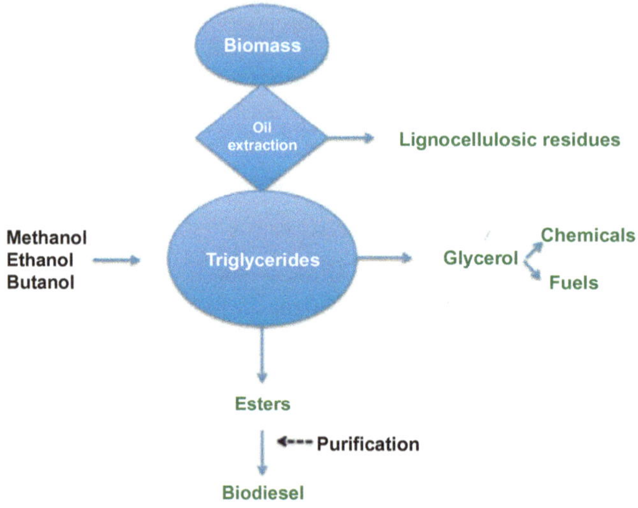

Figure 15.4
Transesterification of biomass-derived triglycerides for the production of biodiesel. *Figure modified from D.M. Alonso, J.Q. Bond, J.A. Dumesic, Catalytic conversion of biomass, Green Chem. 12 (2010) 1493–1513.*

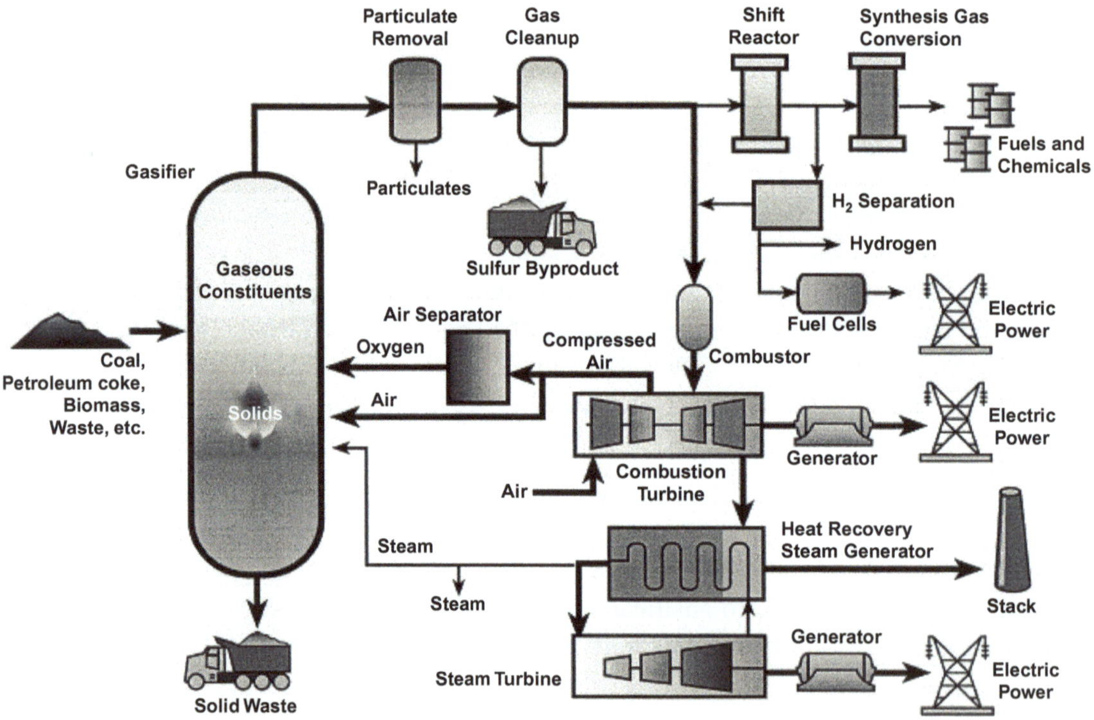

Figure 15.5
Gasification technology options. *Reproduced from G.J. Stiegel, R.C. Maxwell, Gasification technologies: The path to clean, affordable energy in the 21st century, Fuel Process. Technol. 71 (2001) 79–97, with kind permission of Elsevier.*

containing materials (such as potassium and sodium carbonate) on various supports and Ni-containing catalysts; the Ni catalysts are often of the type used for steam reforming. The literature on this subject has been reviewed by Sutton, Kelleher, and Ross; see Task 15.2. Catalysis also has a significant role to play in two of the other steps shown in Fig. 15.6: the shift reaction and synthesis gas conversion. We have already discussed the shift reaction and synthesis gas conversion (Chapter 12: Catalytic Reactions involving Syngas, Hydrogen or Carbon Monoxide for the Production of Intermediates and Chemicals), technologies that are applied elsewhere.

Task 15.2 Catalysts used for Biomass Gasification

The subject of the cleanup of the syngas produced by gasification has been reviewed by D. Sutton, B. Kelleher, and J.R.H. Ross, Review of literature on catalysts for biomass gasification, Fuel Process. Technol. 73 (2001) 155–173. Another paper that reports a reactor system for dealing with the conversion of the tars produced is: P. Ammendiola, B. Piriou, L. Lisi, G. Ruoppolo, R. Chicore, G. Russo, Dual bed reactor for the study of catalytic biomass tars conversion, Exp. Therm. Fluid Sci. 34 (2010) 264–274. Use these papers as sources from which to examine the earlier literature on this subject and, using a suitable citation tool such as Scopus, the relevant literature citing them.

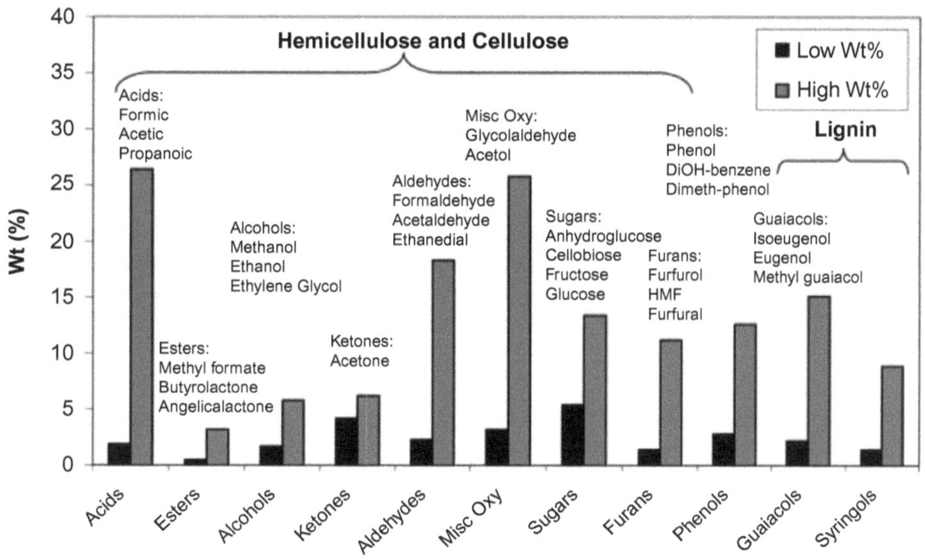

Figure 15.6

The chemical composition of a typical bio-oil produced by pyrolysis. *Figure taken from G.W. Huber, S. Iborra, A. Corma, Synthesis of transportation fuels from biomass: Chemistry, catalysts and engineering, Chem. Rev. 106 (2006) 4044–4098, with kind permission of the American Chemical Society. [Originally from T.A. Milne, F. Agblevor, M. Davis, S. Deutch, D. Johnson, in "Developments in Thermal Biomass Conversion", A.V. Bridgewater, D.G.B. Bookock (Eds.), Blackie (1997)].*

Pyrolysis of biomass is also carried out at high pressures but requires lower temperatures than those used for gasification (generally between about 400 and 500°C). The process yields of up to 70% of a bio-oil containing a vast array of products including acids, aldehydes, alcohols, sugars, esters, ketones, and aromatics. The reaction requires lower temperatures (generally between 400 and 500°C) than those used for gasification and short residence times (hence the term "fast pyrolysis"); with longer pyrolysis times, coke tends to form and this is usable as a solid fuel to provide process heat when required. The pyrolysis process is a relatively inexpensive option compared with the hydrolyis option discussed later but the complex nature of the product mixture together with its acidic and hence corrosive composition means that the products are more difficult to process and use. The breakdown of the vast number of different components found in a typical bio-oil resulting from the pyrolysis of a biomass material containing hemicellulose and cellulose as well as lignin is shown in Fig. 15.6. It should be noted that all the products contain oxygen and that acids such as formic and acetic acid are present in significant quantities. Further treatment of the catalysis literature on this vast subject is beyond the scope of this text. Interested readers follow up the references given in the legend of Fig. 15.6. (See also Box 15.2 and Task 15.3.)

BOX 15.2 Fast Gasification

Biomass pyrolysis is a very useful route for the production of a wide variety of different chemicals. Depending on the conditions used (temperature, pressure, and residence time), different proportions of pyrolysis oil, gas, and char are produced. An excellent review of the effects of different conditions has been published by A.V. Bridgwater: Renewable fuels and chemicals by thermal processing of biomass, Chem. Eng. J. 91 (2003) 87–102. In this, he discusses in particular the operation of the fast pyrolysis process that produces—at several seconds residence time and at a temperature of about 500°C—a mixture of 75% liquids, 12% char, and 13% gas. The liquid is a relatively unstable microemulsion that contains about 25% water and is made up of a mixture of oxygenates (particularly acetic acid), a solid char and alkali metals originating from the ash present. In his review, Bridgwater discusses the nature of the bio-oil and some of the products that can be obtained from it, particularly by hydrotreating or by cracking over zeolites. In the context of this chapter, it is probably of greatest relevance to note that significant amounts of work have been carried out on the upgrading of acetic acid derived from bio-oil. An example of such work is included in Task 15.3.

Task 15.3 Steam Reforming of Acetic Acid from Bio-oil

A well-cited paper on steam reforming of acetic acid obtained from bio-oil is the following: T. Takanabe, K.I. Aika, K. Seshan, L. Lefferts, Sustainable hydrogen from bio-oil: Steam reforming of acetic acid as a model oxygenate, J. Catal. 227 (2004) 101–108. Use this article as an entrance point to the literature on this subject, examining the types of catalysts used for such work and the best results obtained for hydrogen selectivities and yields.

15.3 Second Generation Biorefineries

We will now proceed to discuss the hydrolysis routes involved in second-generation biorefineries. As shown schematically in Fig. 15.7, such biorefineries also involve the gasification and pyrolysis routes discussed in the previous section (Section 15.2) and so these topics will not be discussed further here. Fig. 15.8 shows the make-up of a refinery in somewhat more detail. This figure shows the pretreatment steps for both the hydrolyis (upper half of the diagram) and thermochemical processes (lower half) as well as typical products from each; Hayes gives details of a number of these pretreatment systems in his review paper listed in Box 15.1. It should also be noted that various biochemical processes have also been introduced, these producing some of the chemicals shown in the upper half of Fig. 15.9. We will now concentrate on the hydrolysis processes applied in these refineries, paying particular attention to the Biofine Process that is used to produce levulinic acid (LA, 4-oxopentanoic acid, $CH_3COCH_2CH_2COOH$) and other valuable products, the "Platform Molecules" shown in Fig. 15.9.

As shown in the upper half of Fig. 15.8, the cellulose fraction can on hydrolysis give free mono-saccharides and these can then be converted to ethanol or butanol. However, more complete hydrolysis (as in the Biofine Process[3]) gives equimolar proportions of LA and formic acid. We will discuss some of the possible products from LA later. During the hydrolysis process, the lignin component of the lignocellulose, together with unconverted pentoses, form a "char" material. Gasification of this char (see the lower half of Fig. 8.25

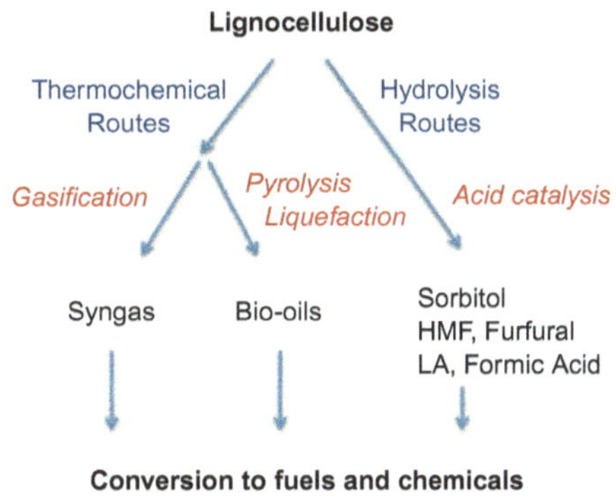

Figure 15.7
The conversion of lingo-cellulose to fuels and chemicals in second-generation biorefineries.

[3] J.J. Bozell, L. Moens, D.C. Elliott, Y. Wang, G.G. Neuenscwander, S.W. Fitzpatrick, R.J. Bilski, J.L. Jarnefeld, Production of levulinic acid and use as a platform chemical for derived products, Resour., Conserv. Recycl. 28 (2000) 227–239.

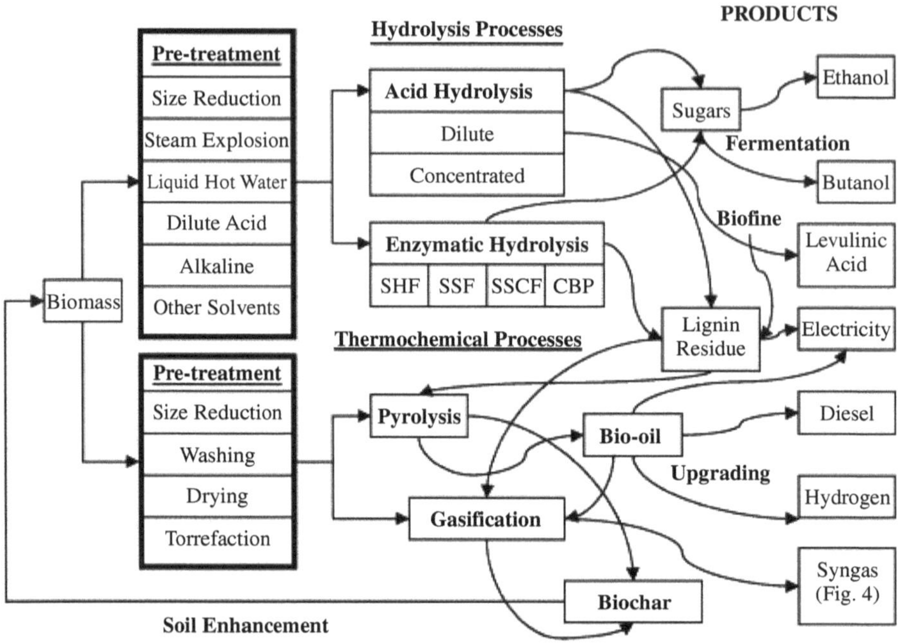

Figure 15.8

Lignocellulosic biomass conversion to useful products. *Reproduced from D.J. Hayes, Catal. Today 145 (2009) 138–151, with kind permission of Elsevier.*

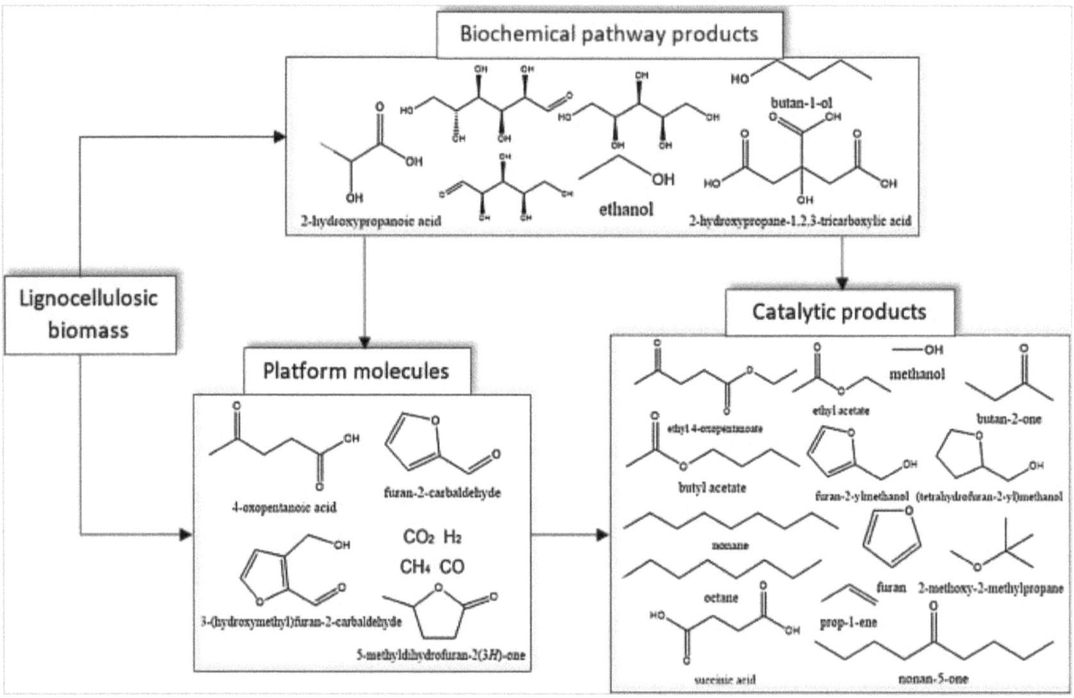

Figure 15.9

Conversion of lignocellulosic biomass to various products by catalytic routes. *Reproduced from C.A.C. Alzate, J.C.S. Toro, A.G. Pena, Catal. Today 302 (2018) 61–72, with kind permission of Elsevier*

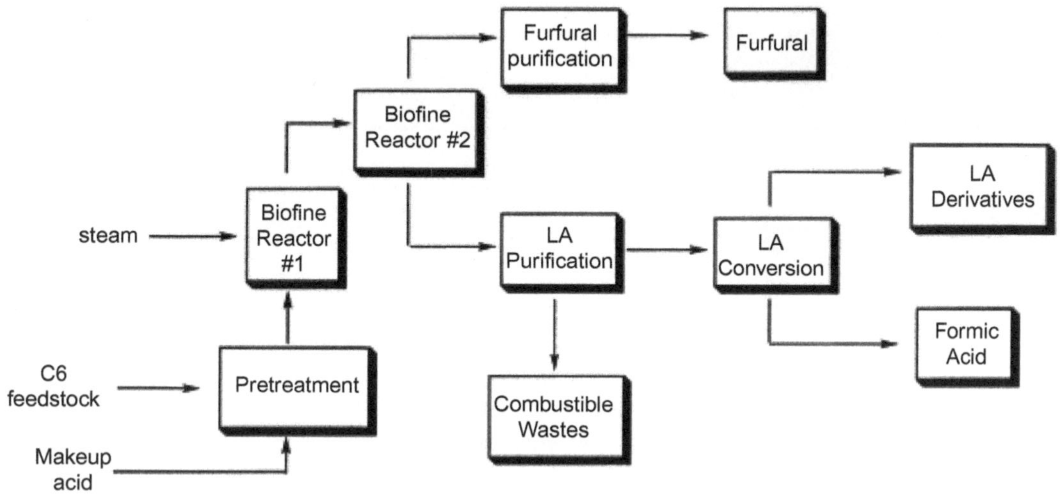

Figure 15.10
The arrangement of the most important steps in the Biofine Process.[3] *From the reference of footnote 3 and reproduced by kind permission of Elsevier.*

and later) is used to produce syngas which provides the energy needed to drive the process; if the scale is sufficiently high, the syngas can also be used for the production of chemicals.

The set-up of the Biofine Process is shown very schematically in Fig. 15.10. In essence, the plant is based on two separate reactors in sequence. The lingo-cellulose material is fed to the first reactor where it is hydrolyzed at 210–230°C for 13–25 s in the presence of a dilute mineral acid (1%–5%). The first step produces HMF that is then supplied to a second reactor operating at a slightly lower temperature (195–215°C) with a longer residence time (15–30 min). The LA, produced in this second reactor, with a yield of more than 60%, is then removed and, following a series of purification steps, can be used to produce a variety of important products, as shown in Fig. 15.11[4].

Fig. 15.12 shows various parts of the LA plant that was constructed by Biofine in Caserta in Italy and started operation in about 2005. This plant was later acquired by GF Biochemicals (www.gfbiochemicals.com). The latter company, after investing significant amounts in further research on the process at the University of Pisa and the Polytechnic University of Milan, have expanded significantly by acquiring Segetis, a producer of LA derivatives. The GF Biochemicals web-site (see Task 15.5) gives details of some of the molecules that can be produced from LA.

[4] This figure is taken from a comprehensive review of the Biofine Process by D.J. Hayes, J.R.H. Ross, M.H.B. Hayes, and S.W. Fitzpatrick, "The Biofine Process: Production of Levulinic Acid, Furfural and Formic Acid from Lignocellulosic Feedstocks", in Biorefineries Industrial Processes and Products, B. Kamm, P.R. Gruber, M. Kamm (Eds.), Wiley-VCH, 2006.

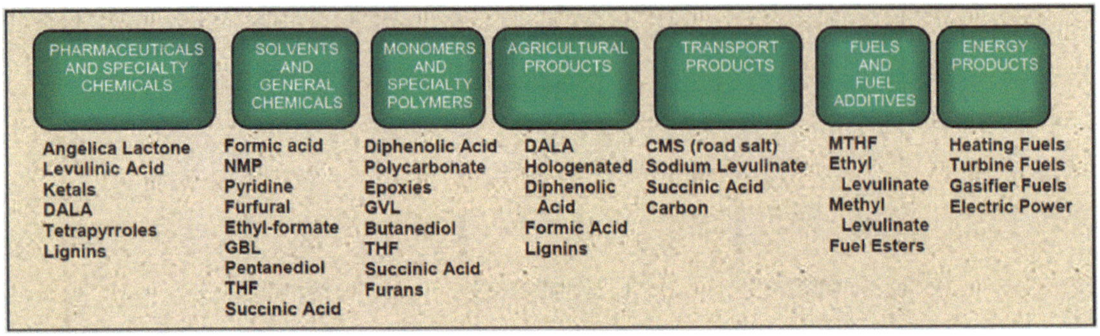

Figure 15.11

Possible markets and saleable products from the Biofine Process. *From the reference of footnote 4, reproduced by kind permission of Wiley-VCH.*

Figure 15.12

Some details of the Biofine plant at Caserta, near Naples, Italy: (A) recovery vessels; (B) mixing tank; (C) second reactor; (D) aerial view of the plant. *From the reference of footnote 4, reproduced with kind permission of Wiley-VCH.*

Fig. 15.13 shows one of the routes for the conversion of LA: the production of methyltetrahydrofuran, a molecule that can be used as a petroleum additive at concentrations up to 30%. A number of the other possible products are also discussed in the paper by Bozell et al. Shell has shown that it is possible to make so-called "valeric biofuels" from LA.[5] (See also Tasks 15.4 and 15.5.)

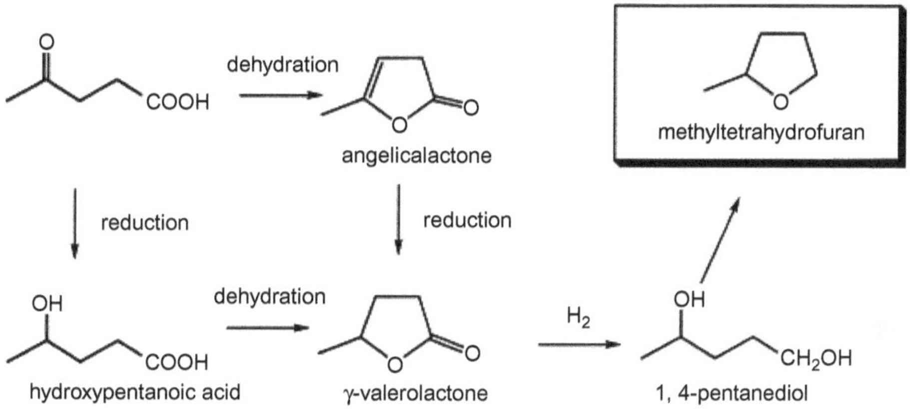

Figure 15.13
Catalytic conversion of LA produced in the Biofine Process to give various products. *Reproduced from reference of Fig. 15.10 with kind permission of Elsevier.*

Task 15.4 Catalysis in the Conversion of Levulinic Acid

Use the papers by Bozell et al. (see footnote 3) and Hayes (see Fig. 15.2) as sources for citation searches on the conversion of LA to valuable products. Include in your material the results shown in the paper of footnote 5. Include in your search papers by L.E. Manzer, formerly of DuPont, who has made major contributions to the field of biomass valorization.

Task 15.5 GF Biochemicals

Carry out a web-based search on GF Biochemicals and find out more about the history and current activities of this firm. In particular, note the entrepreneurial activities of its founders. Examine any patents or other publications that you can find regarding the improvements in the Biofine Process that this company has introduced.

[5] J.-P. Lange, R. Price, P.M. Ayoub, J. Louis, L. Petrus, L. Clarke and H. Gosselink, Valeric biofuels: A platform of cellulosic biofuels, Angew. Chem., Int. Ed. 49 (2010) 4479–4483.

15.4 Uses of Formic Acid Derived From Biomass

We have already in Chapter 12, discussed the production of formic acid by chemical routes and we have also included a brief consideration of the kinetics of the reaction in Chapter 7. As we have seen in the preceding sections of this chapter (Sections 15.2 and 15.3), formic acid is likely to be available in large quantities from the operation of biorefineries, not only ones such as the Biofine Process but also from those using pyrolysis procedures for biomass conversion. The discussion of the uses of formic acid that follows has therefore been included in this chapter rather than in Chapter 12 since much of the recent work on the use of formic acid has been stimulated by the increased interest in biomass conversion.

The main interest in formic acid conversion has focused on formic acid decomposition (dehydrogenation) according to the reaction:

$$HCOOH \rightarrow H_2 + CO_2 \tag{15.1}$$

We have already considered this reaction in the earlier discussion of the history of catalysis in the introductory Chapter 1, (Box 1.5), where it was shown that formic acid decomposition has been used as a tool in establishing activity patterns for a range of metal catalysts. The dehydration reaction:

$$HCOOH \rightarrow H_2O + CO \tag{15.2}$$

is in many practical situations undesirable since CO is a poison for many catalysts that might be used for subsequent reactions using the hydrogen produced in reaction (15.1), particularly in fuel cell applications. Any catalyst to be used for the production of hydrogen from formic acid must therefore either produce it under conditions in which no CO will be allowed thermodynamically or the catalyst must bring about reaction (15.1) selectively and must therefore not catalyze the reverse water–gas shift reaction:

$$CO_2 + H_2 \leftrightarrows CO + H_2O \tag{15.3}$$

A calculation carried out using HSC Chemistry (Fig. 15.14) shows the proportions of H_2, CO_2, H_2O, and CO that would be in equilibrium at temperatures in the range 0–300°C. The calculation was performed in relation to the decomposition of HCOOH and the results show that the concentration of the acid (shown by the red curve, coinciding with the base-line) is very close to zero over the whole temperature range.[6] More importantly from the point of view of this discussion, the amount of CO that can be formed below

[6] This last observation allows one to conclude that the synthesis of formic acid from CO_2 and H_2 does not occur to any significant extent in this temperature range (see Chapter 12: Catalytic Reactions involving Syngas, Hydrogen or Carbon Monoxide for the Production of Intermediates and Chemicals).

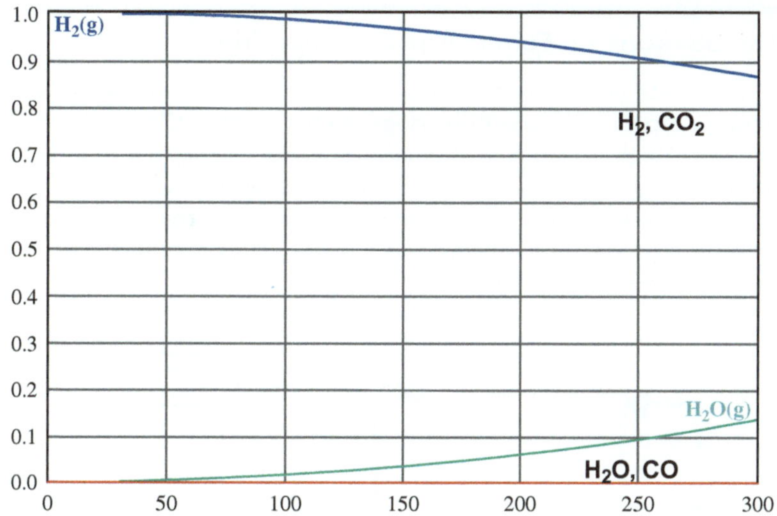

Figure 15.14
Thermodynamic equilibrium data calculated for the decomposition of formic acid to give products, allowing the reverse water–gas shift reaction [Eq. (15.3)] to occur.

about 50°C is very low and so kinetic selectivity is unlikely to be necessary at these temperatures.[7]

It is not the intention to review here all the work that has been carried out recently on formic acid decomposition. The text will instead be focused largely on results from the author's laboratory that illustrate some of the types of measurements that have been made. (See also Task 15.6.)

Fig. 15.15 shows some results for the formic acid decomposition reaction occurring over two Pd/C catalysts with different Pd loadings and two gold-containing samples, one supported on C and the other on TiO_2.[8] Fig. 15.15A shows that a catalyst containing 1% Pd was the most active of those examined, giving complete conversion above about 400K (c. 130°C). At these temperatures, the selectivity is still of some concern and the selectivities to H_2 and CO_2 (equimolar proportions) as a function of temperature for the same catalysts are shown in Fig. 15.15B. Also shown in the latter figure is the equilibrium amount of these products expected if Eq. (15.3) is at equilibrium (again calculated using HSC Chemistry). It can be seen that the selectivities for all but the 1% Au/TiO_2 sample (for which the selectivity is very poor) are somewhat below 100% but are greater than those calculated for the equilibrium situation.

[7] This is a factor that has seemingly not been considered by many researchers working on formic acid decomposition and many still claim that they have found new selective catalysts for the reaction even though their experiments were carried out at temperatures when no CO would be expected!

[8] D.A. Bulushev, S. Beloshapkin, Julian R.H. Ross, Hydrogen from formic acid decomposition over Pd and Au catalysts, Catal. Today 154 (2010) 7–12.

Task 15.6 The Literature on Formic Acid Decomposition

There have been a number of detailed reviews on formic acid decomposition although most of these concentrate more on the use of homogeneous catalysts. One useful relatively recent review is that by A.K. Singh, S. Singh, A. Kumar, Catal. Sci. Technol. 6 (2016) 12–40, entitled "Hydrogen Energy Future with Formic Acid: A Renewable Chemical Hydrogen Storage System". This review, already well cited, not only covers the uses of formic acid as a fuel for fuel cells but also that of formic acid as a hydrogen storage material, as shown schematically in the following diagram:

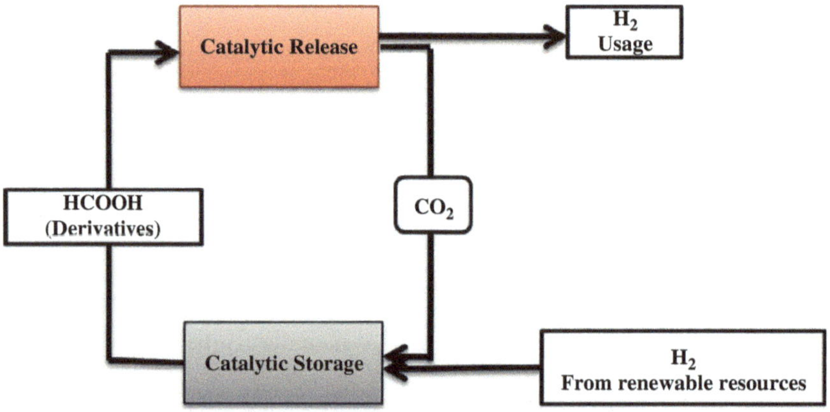

Figure reproduced with kind permission of the Royal Society of Chemistry.

In this system, the formic acid is either formed electrocatalytically or by using a derivative such as an amine, and the hydrogen is then released when required using catalytic dehydrogenation as discussed in the main text.

Use the references listed in the review (available from Scopus or another citation tool) to access other publications on the subject of formic acid decomposition and, if you can access it, also study the material of the review itself.

Hence, it can be concluded that these catalysts have some level of selectivity but that this is insufficient to allow the product gas to be used immediately in a fuel-cell application (requiring $<0.1\%$ CO). The use of the TiO_2 support for Au seems to encourage the decomposition reaction to proceed predominantly along the dehydration route [Eq. (15.2)].

When the catalyst containing 1 wt% Pd was doped with various alkali metal ions (as carbonate), there was a very significant improvement of the activities of the resultant materials by a factor of 1–2 orders of magnitude.[9] Fig. 15.16A shows the steady-state

[9] L. Jia, D.A. Bulushev, Beloschapkin, J.R.H. Ross, Hydrogen production from formic acid vapor over a Pd/C catalyst doped by potassium salts: evidence for participation of buffer-like solution in the pores of the catalyst, Appl. Catal., B 160 (2014) 35–43.

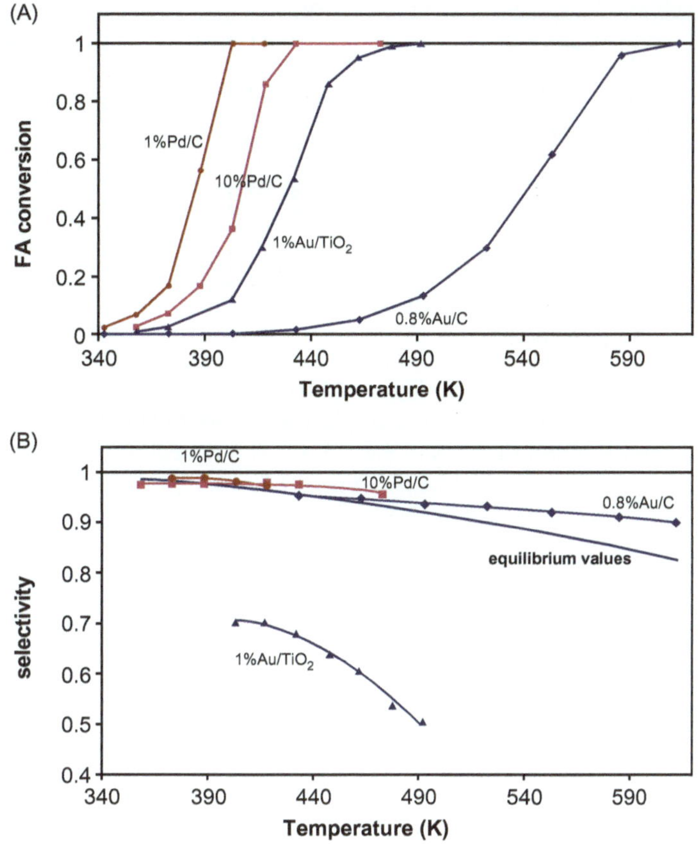

Figure 15.15
Formic acid decomposition over Pd/C, Au/C, and Au/TiO_2 catalysts: (A) formic acid conversion and (B) selectivity to CO_2 as a function of reaction temperature. Reaction conditions: 2.4% HCOOH in He, catalyst weight chosen to give constant metal molar content. *From reference of footnote 8; reproduced with kind permission of Elsevier.*

conversions as a function of temperature for the samples containing 10 wt% of K, Cs, Na, and Li compared with the undoped sample. The most effective dopant was potassium. When the effect of K content was examined (Fig. 15.16B), the most effective sample was found to be the 10 wt% sample. It was also found that the activation energy for the decomposition reaction was significantly changed by doping, from 65 kJ mol^{-1} for the undoped material to 97 kJ mol^{-1} for all of the doped materials; further, doping caused a change in the order of reaction with respect to formic acid concentration from first order to zero order. The selectivities of all the doped samples were very close to 100%. It was therefore quite clear that a very significant change had taken place following the doping with K^+ species. The cause of this change was therefore examined by carefully studying the changes occurring when commencing the reaction over fresh samples at 80°C. It was

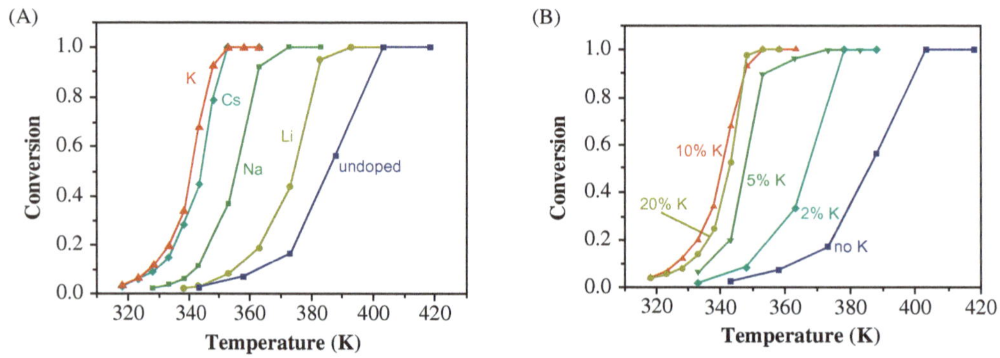

Figure 15.16
The effect on the conversion of formic acid of (A) the nature of the added cation and (B) the concentration of added K; conditions as in Fig. 15.13. *A is reproduced from reference of footnote 9; with kind permission of Elsevier.*

found that during the first period of reaction there was a significantly greater evolution of CO_2 than after the steady state had been achieved, this evolution corresponding to a higher conversion of formic acid. Careful analysis of the results showed that the extra CO_2 corresponded to what would be evolved if the following reaction were to be occurring:

$$KHCO_3 + HCOOH \rightarrow KHCOO + H_2O + CO_2 \quad (15.4)$$

It was therefore concluded that the bicarbonate ions initially present in the pores of the Pd/C catalyst were being replaced by formate ions and that the steady state behavior was attributable to the presence in the pores of the catalyst of a buffer-like solution of formic acid and K formate. The rate-determining step for the reaction for the promoted samples is now the rate of decomposition of formate ions, these immediately being replaced by further dissociation of formic acid. As the formate ion concentration remains constant and is dictated by the K^+ concentration, the rate is zero order. Additional work showed that this behavior was not restricted to carbon supports as similar promotional effects were found with Pd supported on alumina and silica.[10] Based on infrared measurements (diffuse reflectance infrared Fourier transform spectroscopy, DRIFTS) made with a silica-supported sample, the model shown in Fig. 15.17 was proposed. (See also Task 15.7.)

As already discussed in Chapter 7 (The Kinetics and Mechanisms of Catalytic Reactions), a very large body of work has been carried out on the decomposition of formic acid using

[10] I. Jia, D.A. Bulushev, J.R.H. Ross, Formic acid decomposition over palladium based catalysts doped by potassium carbonate, Catal. Today 259 (2016) 453–459.

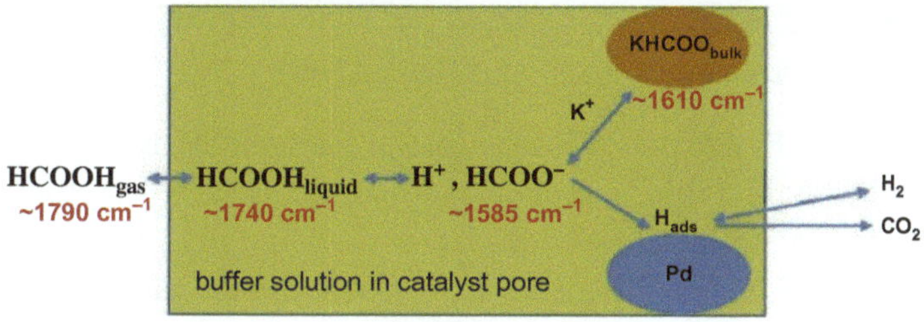

Figure 15.17

Model for the action of the formic acid/formate buffer solution in the pores of the Pd/SiO_2 catalyst. *Reproduced from reference of footnote 10 (graphical abstract); with kind permission of Elsevier.*

Task 15.7 Formic Acid as a Source of Hydrogen for Other Reactions

Formic acid can be used as a source of hydrogen for use in other reactions. For example, it has been used to carry out the hydrogenation of olefins [D.A. Bulushev, J.R.H. Ross, Catal. Today 163 (2011) 42–46] or for the hydrogenation of glycerol to propanediols [A. Martin, U. Ambruster, I. Gandarias, P.L. Arias, Eur. J. Lipid Sci. Technol. 115 (2013) 9–27] without the need for the liberation of molecular hydrogen. In a different sort of application, it can also used as a catalyst in the conversion of hemicellulose sugars [K. Dussan, B. Girisuta, M. Lopez, J.J. Leahy, M.H.B. Hayes, ChemSusChem 8 (2015) 1411–28]. Use these references to find other literature that covers these different uses of formic acid.

various catalysts (mostly containing nanocrystals of Pd on different supports) suspended in aqueous solutions.[11] The vast majority of such work has been carried out using stirred reactors of the type described in Box 6.3 and so the reaction is not studied continuously; instead, if any aspect of the ageing of the catalyst under operating conditions is studied, a series of consecutive experiments are reported. Further, much of the work appears to have been directed at developing more and more active catalysts that can perform many cycles but little work has been reported of attempts to carry out the reaction continuously. Of greatest importance in relation to the work discussed earlier is that although the reaction of aqueous formic acid alone has been reported, many of the papers report results obtained with mixtures of formic acid and a formate, usually sodium formate.[12] It is also relevant to note that the majority of these papers show that negligible quantities of CO are formed at

[11] A significant amount of work has also been carried out using homogeneous catalysts but this will not be discussed here. Interested readers are referred to a detailed review by K. Sordakis, C. Tang, L.K. Vogt, H. Junge, P.J. Dyson, M. Beller, G. Laurenczy, Homogeneous catalysis for sustainable hydrogen storage in formic acid and alcphols, Chem. Rev. 118 (2018) 372–433.

[12] A short description of some of these results was given in Chapter 7, The Kinetics and Mechanisms of Catalytic Reactions, in relation to the kinetics of the reaction.

the low temperatures used in the work reported but none seem to recognize that this is largely a matter of thermodynamics. (See also Task 15.8.)

It has been reported by T.K. Oh that it is possible to generate hydrogen continuously from formic acid in such a way that this hydrogen can then be used in a proton exchange membrane fuel cell.[13] Fig. 15.18 shows the system used and Fig. 5.20 shows the results obtained in a test lasting 3 hours. The catalyst used was a material consisting of Pd supported on a mesoporous graphitic carbon nitride made by an elaborate templating procedure which contained 6.7 wt% Pd. The feed was of 98% HCOOH containing 2% water. The reaction temperature was between 52.3 and 60.1°C and the product gas had a temperature of 25−25.7°C. It can be seen from the results of Fig. 15.19 that the

Task 15.8 The Decomposition of Formic Acid Solutions at Ambient Temperatures

An important paper in the development of catalysts for the decomposition of formic acid at room temperature was one by Tsang and his colleagues [K. Tedsree et al., Hydrogen production from formic acid at room temperature using an Ag-Pd core−shell nanocaatalyst Nat. Nanotechnol. 6 (2011) 302−307], who described the preparation of a novel catalyst structure involving a core of silver atoms surrounded by a shell of between 1 and 10 layers of Pd. This work has since inspired a great deal more work on related catalysts. Use this reference as a starting point, carry out a citation-based search for other work on such complex catalysts for formic acid decomposition at ambient temperatures.

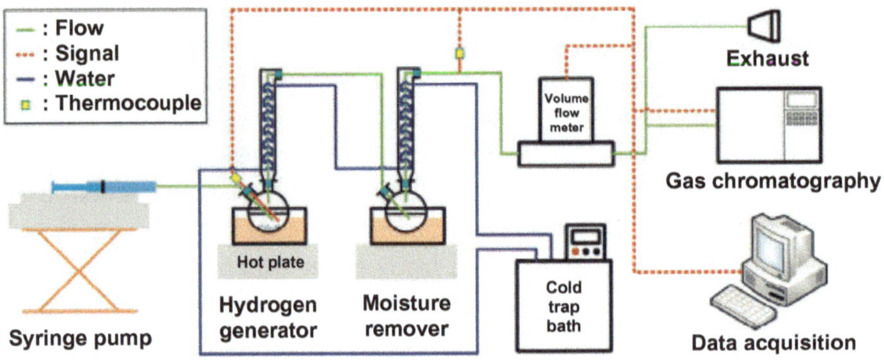

Figure 15.18
Experimental setup used for the generation of hydrogen for use in a 10 W PEMFC system.
Figure reproduced from the reference given in footnote 13, with kind permission of Elsevier.

[13] T.H. Oh, A formic acid hydrogen generator using Pd/C3N4 catalyst for mobile proton exchange membrane fuel cell systems, Energy 112 (2016) 679−685.

temperature dropped over the first few minutes of reaction and then stayed relatively steady, giving a constant conversion rate of about 320 mL/min, above the 280 mL/min required for operation of the proton-exchange membrane fuel cell (PEMFC). No CO could be detected in the product gas. Oh compares these results with those that might be expected with two other "hydrogen carriers" (sodium borohydride and liquid hydrogen) and concludes that the use of formic acid is a preferable option. Oh did not test the inclusion of alkaline formates in the reacting solution, as discussed in Section 15.5, but undoubtedly that would improve the performance. (See also Task 15.9.)

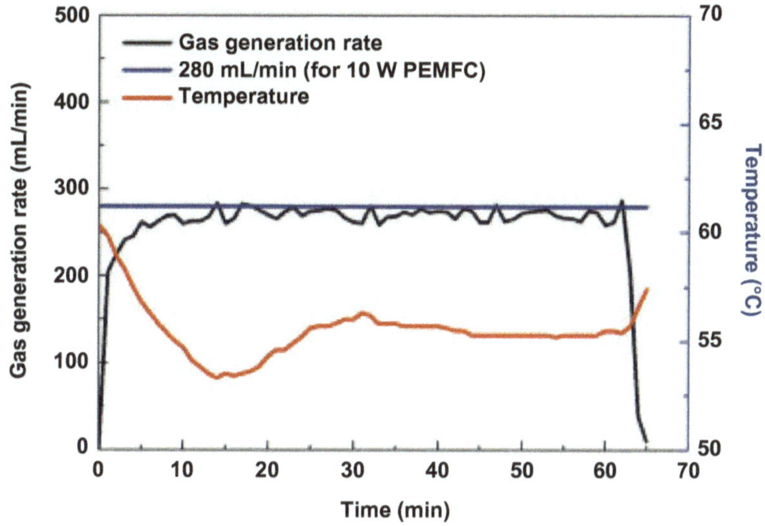

Figure 15.19

Results obtained using a Pd/C3N4 catalyst for the conversion of a 98% solution of formic acid to give hydrogen; the generation rate required to power a 10 W PEMFC is shown by the horizontal line (280 mL/min). *Figure reproduced from the reference given in footnote 13, with kind permission of Elsevier.*

Task 15.9 Continuous Generation of H_2 using Aqueous Formic Acid Solutions

There have been several papers describing the use of homogeneous catalyst systems for the continuous generation of hydrogen from formic acid. For example, A. Majewski, D.M. Morris, K. Kendall, M. Wills [ChemSusChem 3 (2010) 431–434] have described some experiments in which hydrogen was formed continuously for several hours at a time over a period of several days using an Ru complex as catalyst. More recently, M. Graseman and G. Laurenczy [Energy Environ. Sci. 5 (2012) 8171–8181] have described experiments using a homogeneous $RuCl_3$ catalyst in a continuous system that gave the equivalent of a power output of 1 kW for over a year without any need to exchange or replenish the catalyst system. Use these references as starting points in a citation study in order to find if there have been other attempts to use continuous systems for hydrogen generation.

15.5 Conclusions

It has been shown in this chapter that biomass can be used as a source of valuable chemicals, including replacements for fossil fuels, raw materials for polymer synthesis and hydrogen production. No attempt has been made to cover all aspects of the application of catalysis in biomass conversion but it is hoped the contents of this chapter will inspire at least some of its readers to carry out new research in this important field.

APPENDIX

The Use of the Catalysis Literature

Chapter Outline

A.1 Historic Development of the Literature on Catalysis

When a scientist carries out a piece of fundamental research,[1] only he/she is aware of the new knowledge gained until it is transmitted to his/her peers. In the early days of scientific discovery, the transmission of knowledge was generally by personal letter or by presentation to learned societies such as the Royal Society (founded in 1660) or the French Academy of Sciences (founded in 1666). The earliest journals were records of the proceedings of such societies (*Proc. R. Soc., Comptes Rendus*). These "proceedings" were later supplemented by journals published by professional associations such as the Chemical Society (later, the *Royal Society of Chemistry*) and the American Chemical Society. It has only been in the last 50–60 years that an explosion of the chemical literature has occurred with the launch of many new specialized journals by commercial scientific publishing houses such as Pergamon Press, Academic Press, and Elsevier Science Publishers. Box A.1 lists a number of catalysis journals. This list was obtained from the Web of Science using InCites by entering the search term "catalysis" and so it does not include journals publishing catalysis papers but which do not include catalysis in their titles. (See also Task A.1.)

[1] Applied research has often been communicated in a completely different way, through patents (see also Chapter 1: An Introduction to Heterogeneous Catalysis—Chemistry in Two Dimensions). A patent is a description of a discovery written in such a way that anyone reading it should be able to reproduce the results described; it is a claim by the discoverer that he or she has found something new and, in turn, for describing the discovery in detail, the inventor is granted rights to exploit that discovery. Patents are very important to those working in industry since they not only describe in detail the current state of the art in relation to any branch of knowledge but they also define clearly the rights of the patentee, e.g., to "the opposition." Hence, patents are written very carefully in order to provide legal protection for a new discovery but in such a way as not to give away too clear a picture of what is being done in actual practice. In order to be able to obtain its own patents in a given field, the opposition then has to "discover" nonobvious modifications of the methodology (or sometimes, materials) described in the original patent. Industrial catalysis experts frequently argue that academics in the field pay too little attention to patents to which the response is often that patents give useful background information on the state of the art but seldom say what is actually being done.

BOX A.1 Some Catalysis Journals

Journal Title	Publisher	Journal Impact Factor[a]
ACS Catalysis	American Chemical Society	10.61
Advanced Synthesis and Catalysis	Wiley-VCH	5.65
Advances in Catalysis	Academic Press/Elsevier	6.80
Applied Catalysis A—General	Elsevier	4.34
Applied Catalysis B—Environmental	Elsevier	9.45
Biocatalysis and Biotransformation	Taylor and Francis	0.84
Catalysis Communications	Elsevier	3.30
Catalysis Letters	Springer	2.80
Catalysis Reviews—Science and Engineering	Taylor and Francis	6.14
Catalysis Science and Technology	Royal Society of Chemistry	5.77
Catalysis Surveys from Asia	Springer/Plenum	3.22
Catalysis Surveys from Japan	Springer/Plenum	1.63
Catalysis Today	Elsevier	4.64
Chinese Journal of Catalysis	Science Press	2.81
Electrocatalysis	Springer	2.40
Journal of Catalysis	Academic Press/Elsevier	6.84
Journal of Molecular Catalysis A—Chemical	Elsevier	4.21
Journal of Molecular Catalysis B—Enzymatic	Elsevier	2.67
Kinetics and Catalysis	Maik Nauka/Interperiodica/ Springer	0.91
Reaction Kinetics and Catalysis Letters	Springer	0.93
Reaction Kinetics Mechanisms and Catalysis	Springer	1.26
Studies in Surface Science and Catalysis	Elsevier	0.31
Topics in Catalysis	Springer/Plenum	2.49

[a]The Journal Impact Factors are taken from the 2016 Journal Citation Reports Science Edition (Clarivate Analytics, 2018); the Impact Factor is calculated from the sum of number of citations to all the articles published in the journal in question in 2016–17 divided by the total number of articles published in that journal in the same period.

Data from Clarivate.

Task A.1 Impact Factors

Determine the impact factors for Nature, Science and some of the journals of the professional bodies discussed in this section (e.g. J. Amer. Chem. Soc., J. Chem. Soc., Faraday Trans. I and II) that do not contain catalysis in their titles.

Since the subject of catalysis was first recognized, the results of research on the topic have been published using all of the means discussed earlier. Some of the early research work described in the historical section of Chapter 1, An Introduction to Heterogeneous Catalysis—Chemistry in Two Dimensions, was communicated to fellow scientists by personal letter or was presented at learned gatherings. Later, during the period when the science of catalysis was becoming fully established and the current wide range of industrial processes was being developed, many of the most important papers appeared in the journals of the learned societies. For example, some of the keynote papers tracing some of the early developments of catalysis appeared in the 1933, 1950, and 1966 volumes of the Discussions of the Faraday Society (the latter now a "Division" of the Royal Society of Chemistry but then a separate organization).

As a scientist starting his career in the 1960s, my only way of keeping up with current literature was to go each week to the library in order to scan the contents pages of general chemical (or chemical engineering) journals in which important papers on catalysis might be found. As outlined in Chapter 1, the first "commercial" journal in the field was the *Journal of Catalysis.* At this time, Russian catalysis papers were published in *Kinetika i Kataliz* (translated some 12 months later into an English version with the title *Kinetics and Catalysis*) and Japanese catalysis papers appeared in *Shokobai*, these containing an English abstract. *Journal of Molecular Catalysis* was launched by Elsevier in 1975 to take papers dealing with molecular aspects of catalysis, particularly enzymatic and homogeneous catalysis but also molecular aspects of heterogeneous catalysis. It was only in 1981 that *Applied Catalysis* first appeared and, as discussed in more detail in Chapter 1, this was followed by the launch of a range of other journals devoted to the results of catalysis research.

A consequence of the launch of these new journals has been that the vast proportion of the publications on catalysis are now concentrated in a relatively small number of specialized journals. Hence, a scientist working on catalysis has potentially an easier task than did his or her equivalent 50 year ago. However, in parallel to the developments in publishing, there has been a very significant increase in the total number of catalysis-related publications each year, this being coupled by a large increase in the number of meetings, workshops, and conferences devoted to the subject. As a result, the current-awareness task of the catalytic scientist has become more rather than less complex. Indeed, the task of keeping up with the literature would be well-nigh impossible if it were not for the availability of modern electronic bibliometric tools. These will be discussed further in the following text.

A.2 Abstracting and Databases

In the 1960s, when I commenced my research career, a chemist was trained to make use of Chemical Abstracts, published by the American Chemical Society, to find papers relevant to his or her research interests. Chemical Abstracts (Chem. Abs.), a vast paper-based resource, enabled one to look up a specific paper (e.g., a paper encountered as a reference in a publication of interest) and read an abstract of this[2] in order to decide whether or not one wished to obtain a copy of the full document from an appropriate lending library. While it was also possible to use the most current issues of Chem. Abs. as a current-awareness tool, this was not so frequently done, partly because of the time delay involved in the production of the abstracts and partly because of the inconvenience of browsing through extensive abstract entries; as a result, Chem. Abs. was used mainly as a means of *looking back* at older literature. When Chem. Abs. was computerized in the 1970s, it became in principle possible to do current-awareness searches much more effectively. However, at least in the early days, this required the use of large main-frame institutional computers and complicated data handling methods.[3]

Literature searching methods took a large step forward with the foundation (in 1960) by Eugene Garfield of the Institute of Scientific Information. ISI produced two invaluable tools. The first of these was Current Contents; this provided copies of the contents pages (with contact addresses for the authors) of the most important journals in each field. The second was the Science Citation Index (SCI); this, as discussed further in the following text, gave information on newly published papers that cited older work. Both these tools later (in the 1980s) became available in electronic form and were therefore more generally used and accepted; the SCI was later renamed "The Web of Science" and is now published by Thompson Reuters. A more recently launched alternative to the SCI is "Scopus"; produced by Elsevier; this now claims to be the largest database of its kind currently available and, as mentioned earlier, has the advantage over other databases that it is directly linked to Science Direct, this giving access to the full text of all Elsevier's published papers. Yet another citation-based approach is Google Scholar, a source that is particularly valuable when dealing with items not fully covered by Scopus or Web of Science such as patents or books and book chapters. However, Google Scholar coverage can be somewhat excessive as it also includes items such as book reviews and, in my own case short News Brief entries!

---

2 Abstracts are generally available about a year after the original publication.

3 My own institution at the time received every two weeks a series of "tapes" that included the results of a profile that had been set up, in collaboration with a specialist librarian, to cover my research interests. To access and then store the results of the search and to weed out irrelevant citations required me to visit another building, log in to a computer terminal and then to read and edit the output. Thereafter, if I wished to consult the data again, I had to follow the same procedure. Needless to say, the system was not fully utilized!

A.3 The Use of Citation Indexing in Literature Searching

As we will use Scopus (or alternatively Web of Science or Google Scholar) as an integral part of this text, particularly in relation to the various tasks, it is appropriate now to say a little more about these citation-based methods as a means of researching the scientific literature. When a newly published paper is indexed, the papers that it cites are linked to the new paper and full details of the authors and the abstract of the paper are also stored. Links are also provided to full-text versions of the papers, access to these being limited by the subscription under which the search is being carried out. Because of the citation listing, it is possible to work forward in the literature from a source article. Suppose that one is aware of an important source article (perhaps a comprehensive review of a particular subject) published some 10 years ago and one wishes to find out what has been added to the literature since then, one looks up that article and gets a list of the papers that have cited it. Depending on the number of citing papers, one can then either look up each of those papers (or at least read the abstracts) or refine the list before going further in order to give a number of "key references" by excluding titles that are peripheral to the subject of interest. One then works further through the literature by looking up those papers citing these key references. One can also ensure that one has not missed any earlier important references by checking the details of papers cited by the key references. (Current awareness of a specific topic can also be achieved by storing a number of the latter key references on that subject and requesting electronic alerts to any papers citing the key references.[4]) An example of such a search, based on a review article by the present author, is given in Box A.2.

The student using this book is asked to use Scopus (or another search engine) in various literature studies which are included as an integral part of this text. In each chapter, a number of topics are suggested for further study and details of authors publishing in that area or of several important papers or reviews in that area of research are given. The student is then encouraged to follow up on that topic using a citation index such as Scopus (and the links available from it) to select the most recent and significant relevant

[4] It should be noted that citation analysis has been adopted as a method of assessing scientific output as it is very easy to find the number of times that a particular author has been cited and to determine the so-called Hirsch Index (h-index), the number of papers, h, which have been cited h or more times. It should be remembered, but often is not, that the value of h is very dependent on the field in which a particular scientist is working, on the popularity of the specific topic covered, and on factors such as whether or not the author has written significant reviews or methodological papers. (Someone working in a relatively obscure field and writing one or two very significant papers a year is much less likely to be cited than someone working on a "hot" topic and publishing prolifically.) The methodology is therefore wide open to abuse unless used very carefully.

BOX A.2 A Typical Literature Search Carried Out Using Scopus

We will now follow through an example of a Scopus search on the subject of methane conversion to syngas and hydrogen. This is a subject covered in Chapter 11 (Production of Hydrogen and Syngas From Methane and Some Other Reactions of Methane) and so the student is at this stage encouraged to concentrate on the methodology used rather than on the contents of the papers encountered.

The example given is based on a review paper by the current author published in *Catalysis Today* in 2005. (It should be noted that the results shown in the following screenshots are taken from Scopus at a specific time; the reader will almost certainly encounter different information if repeating the search at a later date.) Start by searching in Scopus for papers by Julian R.H. Ross and then identify the review in question. Clicking on the title of that paper results in the following screen, giving the abstract of the review. Also given in this screenshot are the first two entries in the list of the publications from the reference list (all are available in the original but these arc not shown here) and, on the right-hand side, a list of related documents and the first two of a list of documents citing the review (top right-hand corner).

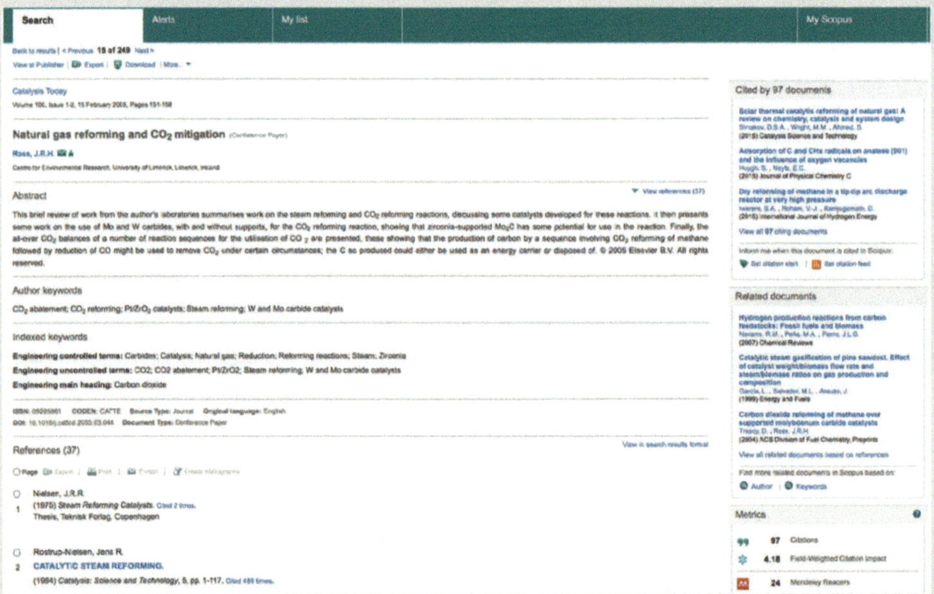

To find the most important documents citing that paper, one clicks on "View all 97 citing documents" and then clicks on "cited by" to obtain the following list in descending order of citation:

(*Continued*)

BOX A.2 A Typical Literature Search Carried Out Using Scopus (Continued)

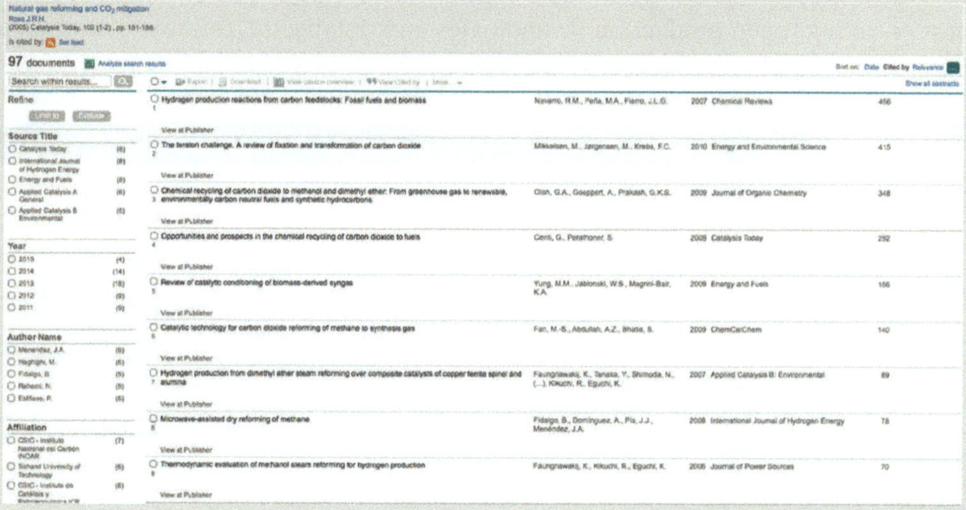

The title of the first paper on this list, a review by Novara and his colleagues, looks particularly relevant to the topic of hydrogen production. To make sure that it should be followed up, one obtains the abstract and other details (see the following screenshot) by clicking on the title.

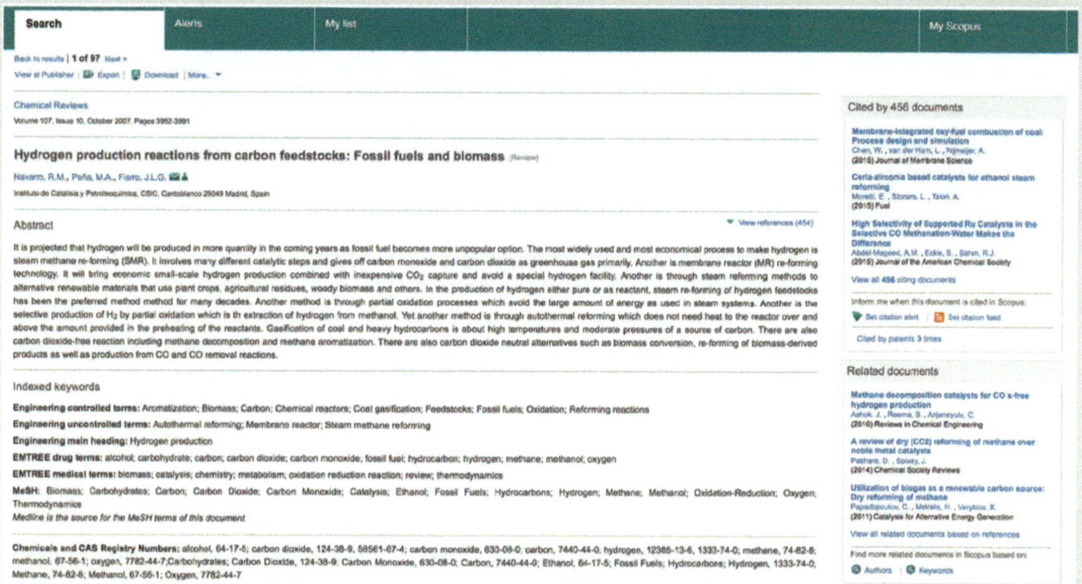

Finally, having clicked on "View all 456 citing papers" and having ordered them according to numbers of citations, one locates an interesting paper by Holladay and his

(*Continued*)

BOX A.2 A Typical Literature Search Carried Out Using Scopus (Continued)

colleagues on hydrogen production technologies and obtains the following abstract and links:

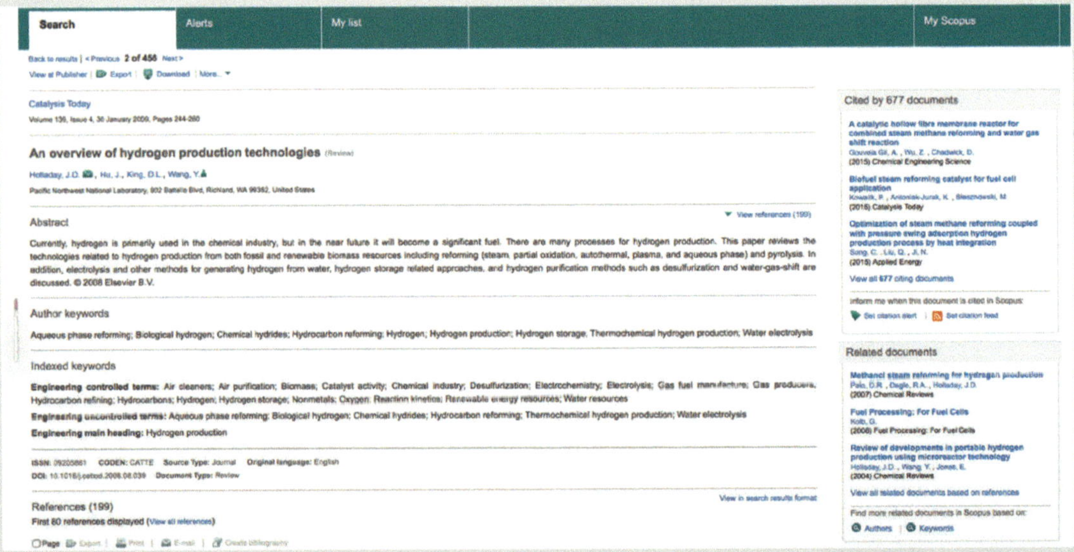

Clearly, the illustration given earlier shows only one of many routes that could have been followed and many of the other papers in the list of citations on the second screen could have been used as a starting point. By several iterations of the type illustrated, one can quickly get a very good picture of some of the most extensively cited papers in the field. As reviews tend to be cited more frequently than normal papers, this approach of looking at the most cited papers will frequently lead to reviews published soon after the cited paper. The list of citing articles shows the most recent further down the list but it is worth checking to see if any very recent reviews have appeared.

publications. These may be either reviews or, less frequently, full scientific articles; in most cases, such searches should generally be limited to review articles as they tend to give all the necessary background information but the introductions to the scientific articles can also give very useful details. Each student will follow up the subject in a different way, depending on his or her own interests and specific requirements: one will be more interested in catalyst materials, another in process aspects of a reaction, another in economic assessment of new processes, etc. When the text is used in conjunction with a course of lectures, this approach also allows the instructor to have an input in the choice of topics or to suggest new ones relevant to the research interests of the department in which the course is being given. The instructor can also, if required, use these studies as the basis for assessment rather than setting formal examinations; in the view of this author, that is the preferable approach since the methodology suggested allows the student to exhibit his or her own initiative and understanding of a subject much better than would be possible under examination conditions[5]. (See also Tasks A.2 and A.3.)

Task A.2 Search for Papers Related to Review by Centi and Parthoner

To give some practice in such searches, select the review by Centi and Parathoner shown in the second screenshot of Box A.2 and follow up some of the references given in it, identifying other more recent significant reviews.

Task A.3 Search for Journals Without "Catalysis" in Their Titles

Using some of the searches associated with Box A.2, identify the titles of journals dealing with catalytic topics but not containing the word "catalysis" in their titles; cf. Box A.1. Examine the scopes of these journals (this may require you to look up the appropriate web pages) and determine their impact factors. Compare these with the values determined in Task A.1.

[5] Every student has her or his own preferred way of studying a given subject. Traditionally, teaching at all levels has involved much of the "chalk and talk" approach: the lecturer stands and talks about a particular subject and gives examples on the blackboard and the student takes it all down religiously. With the advent of modern technologies, lecturers have used a variety of new approaches, each adopting her or his or her own favored method, the most common being the use of handouts in conjunction with overhead transparencies; in the early days, these were drawn by hand but now are most frequently generated by computer (most commonly using PowerPoint). Over a career spanning 40 years, the author has used such a range of techniques, progressing from handwritten "Banda" handouts to PowerPoint presentations. The problem has always been to guage the effectiveness of the approach: what proportion of the students are benefitting from the modern approaches and what proportion would have benefitted better from the chalk and talk approach? Much has depended on the backgrounds of the students and on whether they have become accustomed to one or another approach. No matter which approach has been used, there have always appeared to be students who were turned on by the subject matter and methodology used and others who did not appear to appreciate it!

Index

Note: Page numbers followed by "*f*", "*t*" and "*b*" refer to figures, tables, and boxes, respectively.

B

C

D

E

F

Kolibri 360 GmbH
Ferdinand-Jühlke-Straße 7
99095 Erfurt, Deutschland
produktsicherheit@kolibri360.de